Land Use and Land Cover Mapping in a Changing World

Land Use and Land Cover Mapping in a Changing World

Editor

Giuseppe Pulighe

MDPI • Basel • Beijing • Wuhan • Barcelona • Belgrade • Manchester • Tokyo • Cluj • Tianjin

Editor
Giuseppe Pulighe
Research Centre for
Agricultural Policies and
Bioeconomy
CREA
Rome
Italy

Editorial Office
MDPI
St. Alban-Anlage 66
4052 Basel, Switzerland

This is a reprint of articles from the Special Issue published online in the open access journal *Land* (ISSN 2073-445X) (available at: www.mdpi.com/journal/land/special_issues/land_use_land_cover_change).

For citation purposes, cite each article independently as indicated on the article page online and as indicated below:

LastName, A.A.; LastName, B.B.; LastName, C.C. Article Title. *Journal Name* **Year**, *Volume Number*, Page Range.

ISBN 978-3-0365-6108-0 (Hbk)
ISBN 978-3-0365-6107-3 (PDF)

Contents

About the Editor

Giuseppe Pulighe

Giuseppe Pulighe currently works at CREA, Research Center for Agricultural Policies and Bioeconomy, Council for Agricultural Research and Agricultural Economy Analysis, Rome, Italy. He achieved an M.Sc. in Agricultural Science and Technology. His research activities mainly include ecohydrological modelling, agronomy, bioenergy, climate change, land use change impacts, irrigation and water management, remote sensing, and GIS. He has authored or co-authored over 50 journal publications and conference contributions. He regularly serves as a reviewer for various international scientific journals, and is an instructor of hydrological modelling with SWAT+ and GIS.

land

Editorial

Perspectives and Advancements on "Land Use and Land Cover Mapping in a Changing World"

Giuseppe Pulighe

CREA-Research Centre for Agricultural Policies and Bioeconomy, Via Barberini 36, 00187 Rome, Italy; giuseppe.pulighe@crea.gov.it

1. Introduction

It is increasingly recognized that land use and land cover changes driven by anthropogenic pressures are increasingly impacting terrestrial and aquatic ecosystems and their services, human society, and human livelihoods and well-being. Mapping and monitoring land use and land cover change dynamics are essential for preserving the environment and natural capital and ensuring the sustainability of ecosystem services.

In recent years, the rise of new geospatial technologies around computational techniques and the Internet brought forth a revolution in mapping creation, visualization, and dissemination, bringing new prospects for land mapping and monitoring and enabling near-real-time and cost-effective analysis at multiple scales.

Among others, salient mapping approaches at multiple scales were applied to investigate land use and land cover changes seeking to provide answers in the spheres of human-dominated landscapes and land-related issues (i.e., to explore, manage, organize, or predict land changes). Examples include automated cropland mapping [1], glacier inventory [2], flood inundation [3], mapping forest harvesting [4], and mapping urban agriculture at high-resolution [5].

This Special Issue contains 12 original papers covering various issues related to land use and land use change in several parts of the world (see references), with the purpose to provide a forum to exchange ideas and progress in related areas. Research topics include land use targets, dynamic modelling and mapping using satellite images, pressures from energy production, deforestation, impacts on ecosystem services, aboveground biomass evaluation, as well as investigations on libraries of legends and classification systems.

Citation: Pulighe, G. Perspectives and Advancements on "Land Use and Land Cover Mapping in a Changing World". *Land* **2022**, *11*, 2108. https://doi.org/10.3390/land11122108

Received: 16 November 2022
Accepted: 17 November 2022
Published: 23 November 2022

2. Key Findings and Insights

Measuring and mapping aboveground biomass is a critical component for carbon stock inventories and quantification (Appendix A). In the first paper of this Special Issue, Amara et al., assessed aboveground biomass distribution in a multi-use savannah landscape in southeastern Kenya using airborne laser scanning data, field surveys, and Sentinel-2 satellite images in the Google Earth Engine. Their study evidenced that fences and conservation areas can lead to reduced biomass stocks, which is a vital role of savannahs. The paper by Žoncová et al. used CORINE land cover data for mapping extent and character of land cover changes in the Low Tatras National Park in Slovakia over the last 30 years (1990–2018). This approach allowed them to exploit the potentials of CORINE data to evaluate the long-term landscape changes in protected areas.

Similarly, the study by Gu et al., analyzed land use and land cover dynamics and their impacts on ecosystem services in central Himalaya using the Google Earth Engine between 2000 and 2005. This study highlighted that the Google Earth Engine is a valuable source of data to evaluate the effects of land use and land cover changes on ecosystem service values.

Monitoring the intensity of land use and urban expansion is of great importance for environmental policies. Kim et al. determined changes in land coverage for 31 satellite cities surrounding Seoul using land cover maps from 1988 and 2018 and employing morphological spatial pattern analysis and cluster analysis. The authors suggested that their results can serve for establishing differentiated environmental policies at the local level. The paper by Nedd et al., performed a literature review to scrutinize and evaluate land and land cover definitions and classification systems at the national, regional, and global scales, highlighting the most important challenges, discrepancies, and knowledge gaps. The methodology proposed by the authors will aid the researcher in analyzing the information required in land us and land cover studies.

Deforestation and forest degradation is one of the main environmental problems in Africa. The paper by Kabuanga et al. evaluated deforestation in the Ituri-Epulu-Aru Landscape (Democratic Republic of the Congo) analyzing historical changes and future trajectories through the diachronic analysis of satellite images (2003–2010–2014–2016) and using the DINAMICA EGO platform. The study shows that observed deforestation rates remain relatively low compared to other regions, but forests are shrinking as a result of the unsustainable land use pattern. In their Perspective article, Jand and Woo reaffirm the importance of native trees and their potential for carbon sequestration and mitigation of greenhouse gas emissions. The study highlighted the importance of native trees for providing vital ecosystem services.

Renewable energies can play an important rule toward carbon neutrality. Nevertheless, they can also impact on landscape integrity. Cole et al. assessed landscape dynamics in the United Kingdom driven by pressures from energy production and forests, analyzing change patterns and land cover transitions using CORINE data (years 2006–2018). The authors reported that there has been an increase in the rate of change attributed to renewable energy infrastructure.

Remote sensing images can be efficiently used for multitemporal analysis of changes in forest ecosystems. De Oliveira et al., used high-resolution Landsat images to carry out a multitemporal analysis of changes in land use and land cover in the municipality of Floresta in Pernambuco State in Brazil. The authors analyzed impact of changes in the study area, showing a reduction in the forest and agricultural classes and an increase for exposed-soil class. In another study, Nicolau and Condessa assessed net land take in Portugal between 2007 and 201 by using the Land and Ecosystem Accounting (LEAC) system developed by the European Environment Agency. The study shows that the land use rate amounted to 7.2 ha/day.

The paper by Mushtaq et al., developed an International online catalogue for land cover legend, named Land Cover Legend Registry. This is an international platform that can contribute to development of harmonized land cover legends and datasets at various levels globally. In the final paper, Allan et al. performed a review on the drivers of land use and land cover change in urban areas (2012 to 2022). The study shows that transportation availability was the most frequent factor impacting land use and land cover change processes.

3. Conclusions

A growing body of literature has shown that land use and land cover change can impact the global ecosystem, shaping the future sustainability of natural resources. Research findings, challenges, and key insights that emerged in the cutting-edge studies in this Special Issue contribute to the literature by exploiting the full potential of land mapping in understanding the complex nexus of dynamics among land ecosystems, use of resources, and anthropogenic interaction with the land. We hope that the readers of the *Land* journal find these articles of interest and that they may help in the development of further applications of land use and land cover mapping.

Funding: This research received no external funding.

Acknowledgments: The author wish to thank *Land* for hosting this Special Issue on "Land Use and land Cover Mapping in a Changing World"; those that submitted papers to this issue; and all the anonymous reviewers for their thoughtful insights and comments.

Conflicts of Interest: The authors declare no conflict of interest.

Appendix A

1. Amara, E.; Adhikari, H.; Heiskanen, J.; Siljander, M.; Munyao, M.; Omondi, P.; Pellikka, P. Aboveground Biomass Distribution in a Multi-Use Savannah Landscape in Southeastern Kenya: Impact of Land Use and Fences. *Land* **2020**, *9*, 381. https://doi.org/10.3390/land9100381.
2. Žoncová, M.; Hronček, P.; Gregorová, B. Mapping of the Land Cover Changes in High Mountains of Western Carpathians between 1990–2018: Case Study of the Low Tatras National Park (Slovakia). *Land* **2020**, *9*, 483. https://doi.org/10.3390/land9120483.
3. Gu, C.; Zhang, Y.; Liu, L.; Li, L.; Li, S.; Zhang, B.; Cui, B.; Rai, M.K. Qualifying Land Use and Land Cover Dynamics and Their Impacts on Ecosystem Service in Central Himalaya Transboundary Landscape Based on Google Earth Engine. *Land* **2021**, *10*, 173. https://doi.org/10.3390/land10020173.
4. Kim, J.-H.; Kwon, O.-S.; Ra, J.-H. Urban Type Classification and Characteristic Analysis through Time-Series Environmental Changes for Land Use Management for 31 Satellite Cities around Seoul, South Korea. *Land* **2021**, *10*, 799. https://doi.org/10.3390/land10080799.
5. Nedd, R.; Light, K.; Owens, M.; James, N.; Johnson, E.; Anandhi, A. A Synthesis of Land Use/Land Cover Studies: Definitions, Classification Systems, Meta-Studies, Challenges and Knowledge Gaps on a Global Landscape. *Land* **2021**, *10*, 994. https://doi.org/10.3390/land10090994.
6. Kabuanga, J.M.; Kankonda, O.M.; Saqalli, M.; Maestripieri, N.; Bilintoh, T.M.; Mweru, J.-P.M.; Liama, A.B.; Nishuli, R.; Mané, L. Historical Changes and Future Trajectories of Deforestation in the Ituri-Epulu-Aru Landscape (Democratic Republic of the Congo). *Land* **2021**, *10*, 1042. https://doi.org/10.3390/land10101042.
7. Jang, J.; Woo, S.-Y. Native Trees as a Provider of Vital Urban Ecosystem Services in Urbanizing New Zealand: Status Quo, Challenges and Prospects. *Land* **2022**, *11*, 92. https://doi.org/10.3390/land11010092.
8. Pulighe, G.; Altobelli, F.; Bonati, G.; Lupia, F. Challenges and Opportunities for Growing Bioenergy Crops in the EU: Linking Support Schemes With Sustainability Issues Towards Carbon Neutrality. In *Comprehensive Renewable Energy*; Elsevier: Amsterdam, The Netherlands, 2022; pp. 22–33.
9. Cole, B.; Smith, G.; de la Barreda-Bautista, B.; Hamer, A.; Payne, M.; Codd, T.; Johnson, S.C.M.; Chan, L.Y.; Balzter, H. Dynamic Landscapes in the UK Driven by Pressures from Energy Production and Forestry—Results of the CORINE Land Cover Map 2018. *Land* **2022**, *11*, 192. https://doi.org/10.3390/land11020192.
10. de Oliveira, C.P.; de Lima, R.B.; Alves Junior, F.T.; de Lima Pessoa, M.M.; da Silva, A.F.; dos Santos, N.A.T.; Lopes, I.J.C.; de Melo, C.L.S.-M.S.; Silva, E.A.; da Silva, J.A.A.; et al. Dynamic Modeling of Land Use and Coverage Changes in the Dryland Pernambuco, Brazil. *Land* **2022**, *11*, 998. https://doi.org/10.3390/land11070998.
11. Nicolau, R.; Condessa, B. Monitoring Net Land Take: Is Mainland Portugal on Track to Meet the 2050 Target? *Land* **2022**, *11*, 1005. https://doi.org/10.3390/land11071005.
12. Mushtaq, F.; Henry, M.; O'Brien, C.D.; Di Gregorio, A.; Jalal, R.; Latham, J.; Muchoney, D.; Hill, C.T.; Mosca, N.; Tefera, M.G.; et al. An International Library for Land Cover Legends: The Land Cover Legend Registry. *Land* **2022**, *11*, 1083. https://doi.org/10.3390/land11071083.
13. Allan, A.; Soltani, A.; Abdi, M.H.; Zarei, M. Driving Forces behind Land Use and Land Cover Change: A Systematic and Bibliometric Review. *Land* **2022**, *11*, 1222. https://doi.org/10.3390/land11081222.

References

1. Xiong, J.; Thenkabail, P.S.; Gumma, M.K.; Teluguntla, P.; Poehnelt, J.; Congalton, R.G.; Yadav, K.; Thau, D. Automated Cropland Mapping of Continental Africa Using Google Earth Engine Cloud Computing. *ISPRS J. Photogramm. Remote Sens.* **2017**, *126*, 225–244. [CrossRef]
2. Mölg, N.; Bolch, T.; Rastner, P.; Strozzi, T.; Paul, F. A Consistent Glacier Inventory for Karakoram and Pamir Derived from Landsat Data: Distribution of Debris Cover and Mapping Challenges. *Earth Syst. Sci. Data* **2018**, *10*, 1807–1827. [CrossRef]
3. Jiang, X.; Liang, S.; He, X.; Ziegler, A.D.; Lin, P.; Pan, M.; Wang, D.; Zou, J.; Hao, D.; Mao, G.; et al. Rapid and Large-Scale Mapping of Flood Inundation via Integrating Spaceborne Synthetic Aperture Radar Imagery with Unsupervised Deep Learning. *ISPRS J. Photogramm. Remote Sens.* **2021**, *178*, 36–50. [CrossRef]
4. Zhao, F.; Sun, R.; Zhong, L.; Meng, R.; Huang, C.; Zeng, X.; Wang, M.; Li, Y.; Wang, Z. Monthly Mapping of Forest Harvesting Using Dense Time Series Sentinel-1 SAR Imagery and Deep Learning. *Remote Sens. Environ.* **2022**, *269*, 112822. [CrossRef]
5. Pulighe, G.; Lupia, F. Mapping Spatial Patterns of Urban Agriculture in Rome (Italy) Using Google Earth and Web-Mapping Services. *Land Use Policy* **2016**, *59*, 49–58. [CrossRef]

land

MDPI

Review

Driving Forces behind Land Use and Land Cover Change: A Systematic and Bibliometric Review

Andrew Allan [1], Ali Soltani [2,3,*], Mohammad Hamed Abdi [4] and Melika Zarei [3]

1 UniSA Creative, University of South Australia, Adelaide 5001, Australia; andrew.allan@unisa.edu.au
2 UniSA Business, University of South Australia, Adelaide 5001, Australia
3 Faculty of Art and Architecture, Shiraz University, Shiraz 73, Iran; zareimelika28@gmail.com
4 School of Architecture, Universidad Politécnica de Madrid, 28040 Madrid, Spain; mh.abdi@alumnos.upm.es
* Correspondence: ali.soltani@unisa.edu.au

Abstract: This paper is based on reviewing the literature in the past 10 years on the drivers of land use and land cover change (LULCC) in urban areas. It combines quantitative and qualitative keyword analysis of papers drawn out from the Scopus database. The analysis is primarily based on the number of mentions of keywords in the titles and abstracts of the papers, in addition to the number of keywords appearing in the papers. On the basis of content analysis, a three-level structural categorization of the driving factors was developed. These are presented in a schematic diagram, where the contextual factors are shown as influencing economic and financial factors and policy and regulation, which in turn influences transportation investments and availability, and industrial and residential location choices. Transportation availability was seen as the most frequent factor identified in the literature. This research contends that LULCC is mostly determined by interactions among these four themes in a three-level structure, and on this basis, a model is presented that illustrates LULCC drivers based on local circumstances across the globe.

Keywords: urban growth; land use change; land cover change; driving forces

Citation: Allan, A.; Soltani, A.; Abdi, M.H.; Zarei, M. Driving Forces behind Land Use and Land Cover Change: A Systematic and Bibliometric Review. *Land* **2022**, *11*, 1222. https://doi.org/10.3390/land11081222

Academic Editor: Giusppe Pulighe

Received: 4 July 2022
Accepted: 25 July 2022
Published: 2 August 2022

1. Introduction

Land Use and Land Cover Change (LULCC) is the most prevalent and dynamic landscape phenomena on the surface of the planet, and it plays a key role in reflecting regional and global environmental changes. Urban regions, in particular, have seen the most extreme alterations and transitions between urban vegetation, built land, water bodies, and other forms of land [1]. Hence, urbanized places reflect the most dramatic changes in LULCC [2]. When the aim is to optimize land use patterns for urban development, it is critical to properly understand the factors that drive urban expansion. Because urban expansion is a complex spatiotemporal activity, it is influenced by a variety of factors including society, economy, geography, and policy [3]. Some researchers have considered demographic factors such as population increase [4–7], population density [8–10] and migration from rural to urban areas to be key drivers in LULCC [11–13].

Other researchers have identified economic factors as critically important in the expansion of urban areas such as increase of income [6,14–16], gross domestic product per capita [10,17–19] and foreign direct investment [20–23]. Literature has also focused on geographical factors such as slope [24–27], elevation [1,10,28,29], and distance from water bodies [16,18,30–34] as key drivers. In this regard, the impacts of geomorphological landscape [35], environmental and natural risks such as volcanoes [36], flood, subsidence, unstable soils and rockfalls [37–39] were considered. On the policy side, many scholars have placed emphasis on the fact that institutional factors such as local government policy [10,21,40,41], rules and regulations [7,22,42,43] and land ownership change [11,44–48] have impacts on urban growth processes.

Although many empirical studies show that urban growth is evolving under the influence of varied and diverse factors [1,49–52] less research has been conducted on the systematic classification and explanation of motivating factors affecting LULCC of urban areas [16,53]. Hence, related work of albeit of secondary interest in related journals, scholars' specialties (including their theoretical, methodological and temporal dimension) have tended to be overlooked.

The goal of this article is to offer the foundation for a comprehensive examination and systematic analysis of chosen studies in order to determine the drivers of LUCC. To do this, the primary issue is, what are the driving factors influencing land use change and land cover during the urban development process? In this context, notable publications published over the past decade (i.e., from 2012 to 2022) were investigated. The selected publications evaluated in this study were both quantitative and qualitative. The study focuses on three key indicators at the quantitative level: study timeline, primary concepts and methods/tools, and journal characteristics. It is subsequently followed by two qualitative analyses: the identification and classification of methodological structure, as well as the identification and classification of factors affecting LUCC.

2. Materials and Methods

This article is a bibliometric and systematic review, with the aim of identifying the drivers LULCC from 2012 to 2022. The systematic review process was conducted in four steps: collecting, assessing, extracting, and explaining the data (thematic synthesis).

In the first step (collecting the data), attention was paid to academic papers published in English from 2012 to 2022 selected from the prominent scientific Scopus database which contain a significant number of contributions in the fields of urban development, urbanization, urban growth, land use and land cover change. In order to ensure homogeneity and consistency, conference papers, book chapters and dissertations and grey literature were excluded from this process. To address the major research question and find peer-reviewed articles published in Scopus, several keywords were then queried using the following components of search formula in the title, abstract or keywords sections (Table 1).

Table 1. Components of search formula.

Item	Sub-Item	Details
Keywords	Main keywords	Land Use Change, Land Cover Change, Land Use and Land Cover Change, Land Use/Land Cover Change, Land Use/Land Cover, Land Use, Land Cover
	Supplemented Key-Words	Urban Growth, Urbanization, Urban Development, Urban Area, Urban Planning, Urban Sprawl, Urban Expansion, Expansion, Land Use Planning, Planning
Operators		"OR", "AND"
Time period		2012–2022
Language		English
Document type		Journal paper

Following the collection of papers, the second phase (document assessment) was followed by five steps (Figure 1). The initial collection of 1541 studies based on the searched database was reduced to 1,121 after duplications were removed. By eliminating ambiguous or irrelevant titles, the data set was reduced to 883 records. Subsequently, 432 records were excluded through abstract screening yielding 451 pre-final records. These records were centered on LULCC, providing the basis for an additional bibliometric study. The principal eligibility criterion (encompassing the driving reasons for LULCC) was used to generate the final data set list of research encompassing 110 articles for a full-text content analysis in order to develop the study's synthesizing themes and conceptual model. The data was last updated on 20 June 2022.

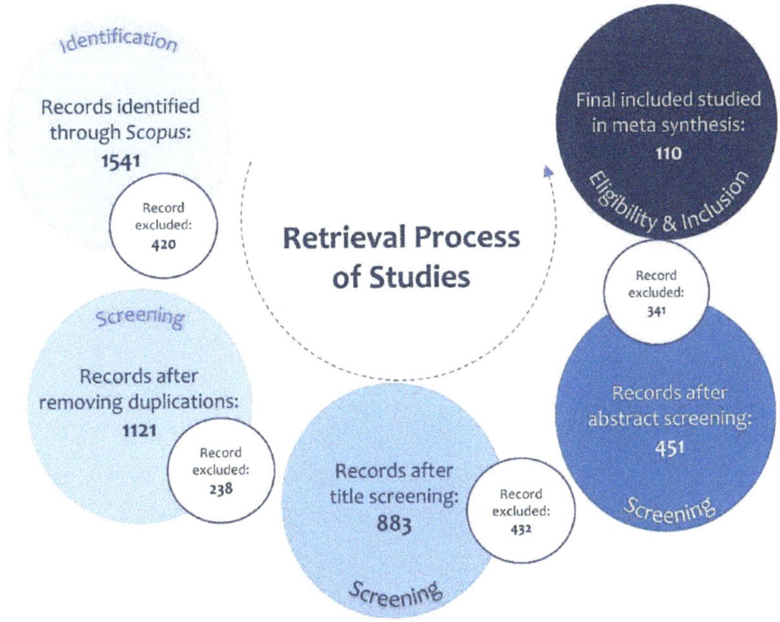

Figure 1. Flowchart for retrieval of studies.

To review all of the selected publications, both quantitative and qualitative methodologies were used. In the case of the former, the following analyses were carried out using the VOSviewer (version 1.6.15), developed by Leiden University, The Netherlands, 2022:

1. Study timeline: number of papers
2. The co-occurrence of fundamental concepts and methodological rules.
3. Journal specialisation and distribution: publications, citations, average citation/publication

In terms of content analysis, the full-texts contained were fed into MAXQDA (version 12.3.3), by VERBI GmbH, Berlin, Germany 2022. Using this method, the codes were taken from the text of the studies (first-order coding) and then re-coded, resulting in the formulation of the ideas (second-order coding). Finally, during the third-order coding procedure, the concepts were synthesised and categories (i.e., theme and sub-theme) were formed. As a result, the evaluation includes the following two key analyses:

1. Methodological approach: Type of methods, data collection, data analysis, and analytical software.
2. Theme of studies: Thematic framework, dimensions and frequency.

3. Results

Several approaches, such as citation analysis and publication count by authors, institutions, universities, or nations, are commonly employed to do this [54]. In this study, a larger sample of articles (n = 451) was assessed using VOSviewer for the number of papers published each year, occurrences of main codes (concepts), methodological codes, and source journals.

3.1. Timeline of Studies

The number of papers published annually varied from 2012 to 2022, but it witnessed a rise as of 2016 with 48 articles, and reached a peak in 2019 with 60 published articles. Figure 2 depicts the annual trends in publications on this topic based on a sample of 451 articles gathered on 25 June 2022.

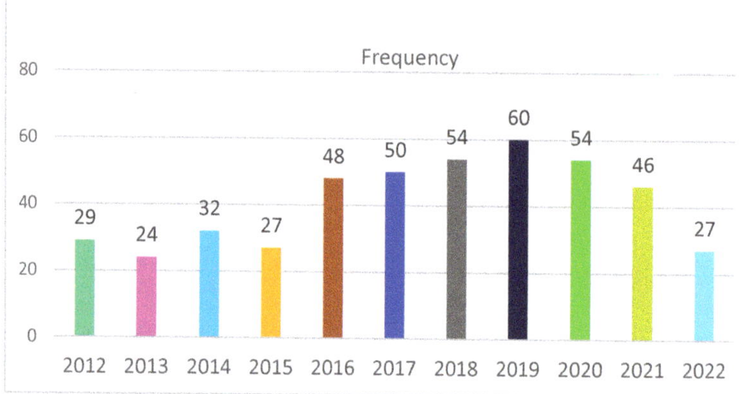

Figure 2. Publication by year (2012–2022).

3.2. Primary Concept and Methodological Codes

The studies selected by abstract screening included primary codes, as illustrated in the Figure 3 below. According to this, "urban growth", "urbanization", "urban expansions", "management", "region", "land", "environment" were among major codes, in other words, primary driving factors behind LULCC. They were thematically synthesized in the next stage, qualitative meta synthesis, resulting in the study themes and sub-themes.

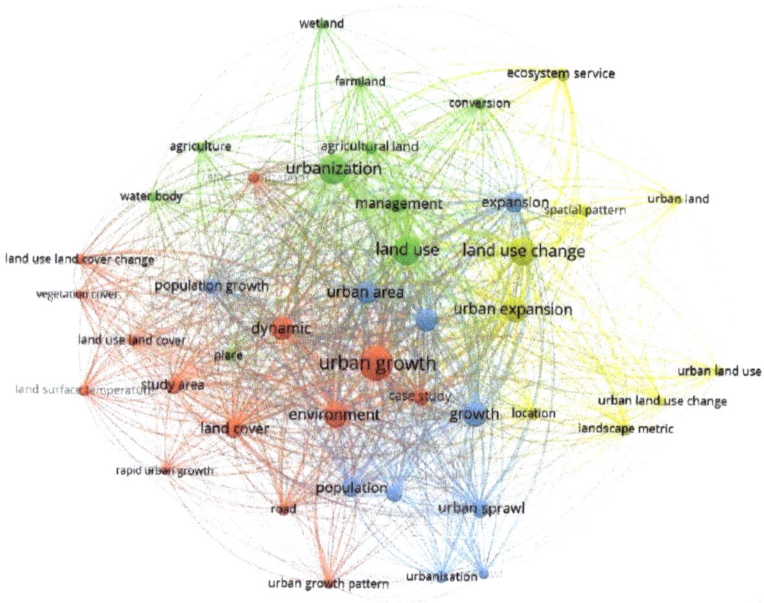

Figure 3. Primary codes (factors driving LULCC) found in the 451 selected records by abstract.

Finding the methodological codes given in the titles and abstracts of the papers was another source of analysis. Figure 4 depicts this, indicating that modelling, scenario building, modeling, mapping, and so on are among the most important methodologies and tools. They cannot, however, represent the methodological approach and instruments utilized in the focused research on variables causing LULCC, which were subsequently produced in the first part of the meta synthesis section.

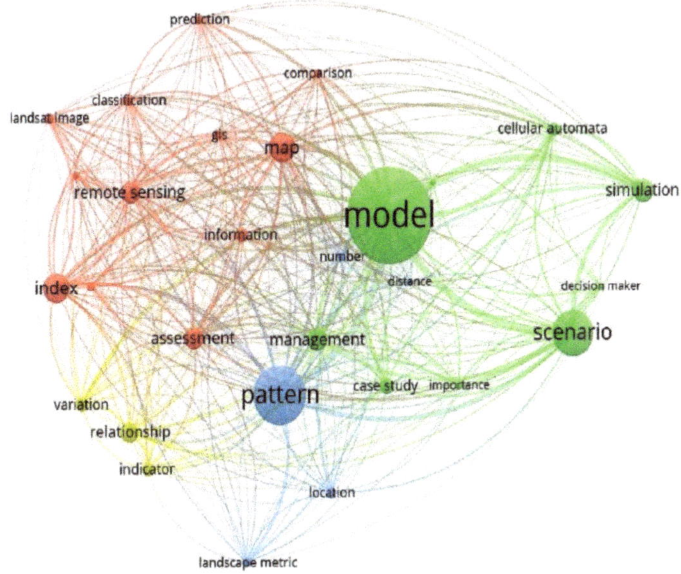

Figure 4. Major methodological codes found in the LULCC studies.

3.3. *Leading Journals*

According to Dzikowski [55], a journal will have more impact if a greater number of papers are published in it and the more the number of citations it possesses. On this, the number of publications and citations as well as average citation per publication of all journals were used to analyze the source journals. The results of top-ranked journals portrayed that the journals of *Computers, Environment and Urban Systems*, *Ecological Indicators*, *Environmental Monitoring and Assessment* and *Land Use Policy* were among the top-three journal with the highest record of publications in the field of study (Table 2).

Table 2. Top-eight source journals, their number of publications and citations.

Journal Title	Number of Papers	Number of Citations	Citation per Paper
Computers, Environment and Urban Systems	34	381	11.21
Ecological Indicators	22	458	20.82
Environmental Monitoring and Assessment	22	227	10.32
Land Use Policy	22	621	28.23
Landscape and Urban Planning	19	745	39.21
Remote Sensing	17	241	14.18
Science of the Total Environment	14	992	70.86
Sustainability	12	593	49.42

3.4. *Methodological Approach*

Another source of analysis was locating the methodological codes listed in the titles and abstracts of the studies. Figure 4 demonstrates this, revealing that among the most essential approaches and tools are modelling, scenario building, modelling, mapping, and so on. They cannot, however, represent the methodological approach and instruments utilized in the focused research on factors that cause LULCC, which was created later in the first part of the meta synthesis section.

According to the findings, 68 studies (62 percent) of the total number of selected papers were done quantitatively, 7 studies (6 percent) qualitatively, and 35 studies (32 percent) utilising the combined method. In relation to data collection, the majority of research (80 studies, 73 percent of total chosen papers) utilized primary data, 29 studies (26 percent) relied on secondary data sources, and just one study applied mixed data collecting. In terms of data analysis, their approach was based on an analytical technique consistent with the study techniques used. The majority of the time, statistical analysis, geographical analysis, descriptive analysis, and qualitative content analysis were used. The qualitative methods mostly include: focus group; interview; policy review; case study research and content analysis. Table 3 outlines the analytical tools used in the chosen LULCC-centered papers.

Table 3. Methodological Analysis of Selected Articles in LULCC.

Research Approach	Number and Percentage of Papers	Data Collection Method	Data Analysis Method	Analytical Tools/Software (Some Examples)
Quantitative	68 (62%)	Primary data	Spatial analysis; Spatiotemporal analysis & Simulation	-Cellular automata, SLUETH: [34] -Imperviousness Change Analysis Tool (I-CAT) and MCE: [28] -Satellite Image analysis: [40,56,57]
			Statistical analysis	-Descriptive comparison: [13,58] - Regression analysis: [20,46,59,60]
			Mixed (Spatial analysis, Statistical analysis & Descriptive analysis)	-Spatial auto-correlation: [61] -Scenario-building: [62] -IDRISI image analyser: [63] -MCE (ANP): [64] -Cellular automata [65,66]
Qualitative	7 (6%)	Primary or	Descriptive analysis	-Narrative: [67]
		Secondary data		-Case study research: [48]
		Mixed (Primary and secondary data)	Qualitative content analysis	-Focus group discussion (FGD), Questionnaire survey and Interview: [47,68] -Systematic review: [16]
Mixed	35 (32%)	Primary or secondary data	Spatial analysis and Statistical analysis	-Satellite image analysis and Regression analysis: Policy review [10,69]
		Mixed (Primary and secondary data)	Spatial analysis, Statistical analysis & Descriptive analysis	-Policy-analysis: [70] -Mixed method [71]
			Spatial analysis	-Principal component analysis (PCA): Policy review [5]
			Mixed (Spatial analysis, Statistical analysis & Descriptive analysis)	-Cross-tabulation: [72] -Spatial Statistics and Logit Regression: Policy review [73]

3.5. *The Study Themes: Driving Factors of LULCC*

Table 4 displays the core result of the systematic review including the factors driving LULCC, categorized into themes, sub-themes, codes (factors), and the share of repeating the codes within the papers investigated. A total of 64 final factors, 11 sub-themes and four main themes titled *Urban growth Factors, Policy and Regulation Factors, Economic and Financial Factors*, and *Contextual Factors* were acquired hierarchically (Figure 5).

Table 4. Factors driving land use and land cover change process.

Theme	Sub-Theme	Code (Factor)	Sample Studies	Frequency	Share
Urban growth factors	Transport infrastructure	Airport	Kamh et al. (2012); Banzhaf et al. (2013); Nassar et al. (2014); Chen et al. (2018); Essien & Cyrus (2019)	5	4.55
		Bridge	Geymen (2013); Cao et al. (2021); Chu et al. (2021); Jawarneh et al. (2015); Han & Jia (2017)	5	4.55
		High-speed rail	Zhang et al. (2020)	1	0.91
		Highway	Feng & Wang (2021); Hanlon et al. (2012); Sandhya Kiran & Joshi (2013); Geymen (2013); Nassar et al. (2014); Jawarneh et al. (2015); Kong et al. (2017); Chen et al. (2018); Wang & Zhou (2018); Colsaet et al. (2018); Nadafianshahamabadi et al. (2021); Meyer & Früh-Müller (2020); Chu et al. (2021); Pratama et al. (2022); Schumacher et al. (2019); Inouye et al. (2015); Wu et al. (2021)	17	15.47
		Light rail transit	Hurst, et al. (2014); Wang, et al. (2020); Wu, et al. (2021); Moghadam et al. (2018)	3	3.64
		Railway	Feng & Wang (2021); Chu et al. (2021); Wang & Zhou, (2018); Chen et al. (2018); Kong et al. (2017); Jawarneh et al. (2015); Zhang & Xu (2015); Li et al. (2014); Zhao & Shen (2019)	9	8.19
		Road network	Nassar et al. (2014); Chen et al. (2018); Colsaet et al. (2018); McGarigal et al. (2018); Nadafianshahamabadi et al. (2021); Islam et al. (2021); Liu et al. (2020); Kasraian et al. (2020); Tavares et al. (2019); Sunde et al. (2014); Kontgis et al. (2014) Li et al. (2014); Fitawok et al. (2020); Bajracharya et al. (2020); Shafizadeh Moghadam & Helbich (2013); Xu et al. (2013); Jawarneh et al. (2015); Gallardo & Martinezvega (2016); de la Luz Hernández-Flores et al. (2017); Fen (2017); Kong et al. (2017); Essien & Cyrus (2019); Schumacher et al. (2019); Daunt et al. (2021); Deslatte et al. (2022); Lal et al. (2017); Inouye et al. (2015); Gerten et al. (2019); Ma, (2020); Leyk et al. (2020)	30	27.30
		Subway and Subway station	Nassar et al. (2014); Feng & wang, (2021); Ahmad et al. (2016); Wu et al. (2021)	4	3.64
		Traffic service	Wenner & Thierstein (2021); Wu et al. (2021); Wang & Zhou (2018); Deng & Srinivasan (2016)	4	3.64
		Wharf	Cao et al. (2021); Nassar et al. (2014); Daunt et al. (2021); Inouye et al. (2015)	4	3.64
		Train	Meyer & Früh-Müller (2020); Wu et al. (2021)	2	1.82
	Industry	Technological progress and industrial transformation	Din & Mak (2021); Cao et al. (2021); Dong et al. (2020); Liu et al. (2019); Dai et al. (2018); Li et al. (2017); Kontgis et al. (2014); Nassar et al. (2014), Xu et al. (2013); Hasan et al. (2019); Wang et al. (2018); Li et al. (2014); Jawarneh et al. (2015); Leyk et al. (2020); Chu et al. (2021); Dou & Han (2021); Feng & Wang (2021); Tavares et al. (2019); Sandhya Kiran & Joshi (2013); Kleemann et al. (2017); de la Luz Hernández-Flores et al. (2017); Inouye et al. (2015)	22	20.02
		Industrial parks/sites	Cheng (2021), Kang et al. (2019), Zambon et al. (2019), Shin & Chae (2018), Han & Jia (2017), Zhang & Xu (2015)	6	5.45
		Factories	Wu et al. (2021); Shin & Chae (2018); Ul Din & Mak (2021)	3	2.73
	Accessibility	Proximity to the city/county/megacity centre	Han & Jia (2017); Deng & Srinivasan (2016); Li et al. (2014); Lal et al. (2017); Wang & Zhou (2018); Fitawok et al. (2020); Nguyen et al. (2018)	7	6.36
		Commercial /leisure centre/park	Gallardo & Martinezvega (2016); Chen et al. (2018); Kong et al. (2017); Bajracharya et al. (2020); Han & Jia (2017); Wu et al. (2021)	6	5.45
		Education and research	Wu et al. (2021); Liu et al. (2020); Cao et al. (2021); Li et al. (2015); de la Luz Hernández-Flores et al. (2017); Zhang & Xu, (2015)	6	5.45
		Hotel	Chen et al. (2018); Essien & Cyrus, (2019); Wu et al. (2021)	3	2.73
		Neighbouring effect	Luo et al. (2018)	1	0.91
		Distance from built-up areas	Shafizadeh Moghadam & Helbich (2013); Xu et al. (2013)	2	1.82
		Medical care	de la Luz Hernández-Flores et al. (2017);	1	0/91
		Accessibility to public facilities	Han & Jia (2017); Kong et al. (2017)	2	1.82
	Residence	Constructing residential settlements	Meyer & Früh-Müller, (2020); Ponstingel (2020), Bajracharya et al. (2020); Sandhya Kiran & Joshi (2013)	1	3.64
Policy and regulation factors	Urban/land use policies	Administrative division adjustment	Feng & Wang (2021); Feng & Wang (2022)	2	1.82
		Urban administrative hierarchy	Dong et al. (2020); Li et al. (2015)	2	1.82
		Local government policy	Xu et al. (2013); Nassar et al. (2014); Kontgis et al. (2014); Xu et al. (2015); Luo et al. (2018); Cheng, (2021); Meyer & Früh-Müller (2020); Ponstingel (2020); Wang et al. (2018); Wadduwage (2018); Deslatte et al. (2022); Dou & Han (2021); Dai et al. (2018); Yue et al. (2014); Cao et al. (2021); Kuang, (2020); Essien & Cyrus (2019); Gerten et al. (2019); Chen et al. (2018); Kong et al. (2012); Kleemann et al. (2017); Li et al. (2015)	22	20.02
		Private enterprise	Hamnett (2020); Soria et al. (2020)	2	1.82
		User (property owner, developers, real estate companies)	Deslatte et al. (2022); Fitawok et al. (2020); Soria et al. (2020); Colsaet et al. (2018); Zhang et al. (2015); Nassar et al. (2014)	6	5.45
		Changing land ownership	Kleemann et al. (2017); Schumacher et al. (2019); Whiteside (2020); De Tong et al. (2018); Adam (2019); Zhang et al. (2015)	6	5.45
		Zoning	Colsaet et al. (2018); McGarigal et al. (2018)	2	1.82
		Land use policies	Daunt et al. (2021); Deslatte et al. (2022)	2	1.82
		Developable land	Deslatte et al. (2022); Deng & Srinivasan (2016)	2	1.82
	Regulations	Impact property tax	Bimonte & Stabile (2015); Deslatte et al. (2022); Colsaet et al. (2018); Kontgis et al. (2014)	4	3.64
		Municipalities regulation	Deslatte et al. (2022)	1	0.91
		Urban planning regulation	Feng & Wang (2022); Fitawok et al. (2020); Dai et al. (2018); Yue et al. (2014); Banzhaf et al. (2013); Kong et al. (2012); Kong et al. (2017)	7	6.36
		Regulation of residential Land use	Tiitu (2018); Daunt et al. (2021); Colsaet et al. (2018)	3	2.73

Table 4. *Cont.*

Theme	Sub-Theme	Code (Factor)	Sample Studies	Frequency	Share
Economic and Financial factors	Investment	Foreign direct investment	Li et al. (2015); Kontgis et al. (2014); Dai et al. (2018); Asabere et al. (2020); Dou & Han (2021)	5	4.55
		Investment attraction	Dou & Han (2021); Deslatte et al. (2022); Kuang (2020); Chen et al. (2018); Admaus (2015)	5	4.55
	Urban Economy	Market power/incentives	Hamnett (2020); Chen et al. (2018);	2	1.82
		Land market	Simwanda et al. (2020); Yue et al. (2014)	2	1.82
		Land price	Magliocca et al. (2015); Hasan et al. (2019)	2	1.82
		Land price distribution	Hu et al. (2012); Hanlon et al. (2012)	2	1.82
		Housing price	Magliocca et al. (2015); Daunt et al. (2021)	2	1.82
		Tourism development	Kamh et al. (2012); Sang et al. (2019); Colsaet et al. (2018); Nassar et al. (2014); Chu et al. (2021); Daunt et al. (2021)	6	5.46
		Economic opportunities (trade, industrial)	Simwanda et al. (2020); Tavares et al. (2019); Sandhya Kiran & Joshi (2013); Nguyen et al. (2018)	4	3.64
Contextual factors	Demographic	Rural population migration	Kleemann et al. (2017); Ul Din & Mak (2021); Cao et al. (2021); Islam et al. (2021); Asabere et al. (2020); Xu et al. (2020); Gerten et al. (2019); Essien & Cyrus (2019); Simwanda et al. (2020); Fitawok et al. (2020)	10	9.1
		Labor migration	Shin & Chae, (2018); Essien & Cyrus (2019); Kleemann et al. (2017); Dai et al. (2018); Simwanda et al. (2020); Nassar et al. (2014); Sang et al. (2019); Azhdari et al. (2019)	7	6.36
		Internal migration	Colsaet et al. (2018); Kang et al. (2019); Liu et al. (2019); Jawarneh. et al. (2015); Skog & Steinnes (2016); Kamh et al. (2012), Abulibdeh et al. (2019).	6	5.45
		Increase in urban population	Li et al. (2022); Dou & Han (2022); Daunt et al. (2021); Din & Mak (2021); Cao et al. (2021); Leyk et al. (2020); Xu et al. (2020); Bajracharya et al. (2020); Fitawok et al. (2020); Gerten et al. (2019); Tavares et al. (2019); Luo et al. (2018); Kleemann et al. (2017); Skog & Steinnes (2016); Sandhya Kiran & Joshi (2013); Banzhaf et al. (2013); Li et al. (2014); Nassar et al. (2014); Sunde et al. (2014); Zhang & Xu (2015); Lal et al. (2017); de la Luz Hernández-Flores et al. (2017); Essien & Cyrus (2019); Essien & Cyrus (2018); Tiitu (2018); Jawarneh et al. (2015); Kamh et al. (2012); Colsaet et al. (2018)	28	25.48
		Population density	Banzhaf et al. (2013); Lal et al. (2017); de la Luz Hernández-Flores et al. (2017); Xu et al. (2013); Liu et al. (2020); Meyer & Früh-Müller (2020)	6	5.45
	Socio-economic features	Lifestyle	Kleemann et al. (2017)	1	0.91
		Gross Domestic Production (GDP)	Xu et al. (2013); Jiang et al. (2013); Li et al. (2014); Gong et al. (2014); Luo et al. (2018); Colsaet et al. (2018); Liu et al. (2019); Hasan et al. (2019); Dong et al. (2020); Kuang, (2020); Cao et al. (2021); Liu et al. (2020); Chu et al. (2021); Dou & Han (2021); Ul Din & Mak (2021)	15	13.65
		Increased income	Hasan et al. (2019); Ponstingel (2020); Colsaet et al. (2018)	3	2.73
		Economic down-turn/unemployment rate	Meyer & Früh-Müller (2020); Tomao et al. (2021); Salvati (2019); Kang et al. (2019)	4	3.64
	Environment and natural resources	Slope	Kamh et al. (2012); Shafizadeh Moghadam & Helbich (2013); Xu et al. (2013); Sunde et al. (2014); Han & Jia (2017); Kong et al. (2017); Wadduwage (2018); Wang & Zhou (2018); Colsaet et al. (2018); Liu et al. (2020); Fitawok et al. (2020); de la Luz Hernández-Flores et al. (2017); Wu et al. (2021); Gerten et al. (2019); Jawarneh et al. (2015)	15	13.65
		Elevation	Xu et al. (2013); Sunde et al. (2014); Han & Jia (2017); Wang & Zhou (2018); Liu et al. (2020); Wu et al. (2021); Gerten et al. (2019); Jawarneh et al. (2015)	8	7.27
		Climate	Yan et al. (2013); Colsaet et al. (2018); Wang et al. (2018); Admaus (2015)	4	3.64
		Geographical location	Hasan et al. (2019); Dai et al. (2018); Ul Din & Mak (2021); Kamh et al. (2012); Nguyen et al (2018)	5	4.55
		Flood prone areas	Jawarneh et al. (2015)	2	0.91
		Sea shoreline	Kamh et al. (2012); Leyk et al. (2020)	2	1.82
		Distance from water	Han & Jia (2017); Feng (2017); Li et al. (2014); Shafizadeh Moghadam & Helbich (2013); Kong et al. (2017); Sunde et al. (2014); Colsaet et al. (2018); Deslatte et al. (2022); Leyk et al. (2020); Bajracharya et al. (2020)	10	9.09
		Resource	Ma, (2020)	1	0.91
		Oil resource	Li et al. (2014); Nassar et al. (2014); Daunt et al. (2021); Essien & Cyrus, (2019)	4	3.64
		Mine	Lal et al. (2017); de la Luz Hernández-Flores et al. (2017); Wu et al. (2021)	3	2.73
		Ecosystem services	Pan et al. (2021); Peng et al. (2021)	2	1.82

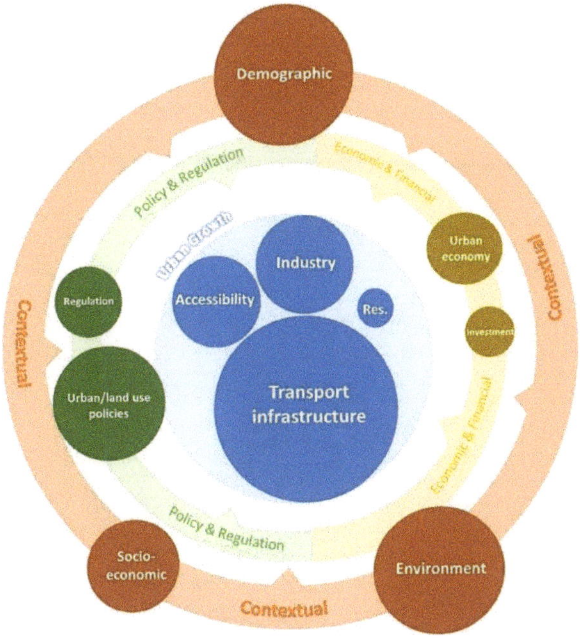

Figure 5. Driving forces causing LULCC, their frequency and interrelations.

4. Discussion

4.1. The Interacting Model

Apart from the driving factors identified above, the consequences determined the frequency of factors among the selected studies. In total, they referred to different terms 373 times. Accordingly, *urban growth factors-* with about 40% of the total references—account for more than double the number of references to *policy and regulation factors* and a little more than *contextual factors*. With regard to the sub-theme level, the most frequently cited items are *transport infrastructure* (an urban growth factors), by a considerable margin, and then *demographic* (a *contextual factor*) with about 23% and 15% of the total references, respectively. *Accessibility* and *industry* subthemes in the urban growth theme are similar with *socio-economic* (a *contextual factor*), in terms of the number of references. This is also the case for *environment* subtheme (a *contextual factor*) and *urban/land use policies*, as the most frequent cited subtheme in *policy and regulation factors*. Figure 5 schematically portrays the extent to which themes and sub-themes are frequent by proportionally sized squares.

Beyond theme synthesis and frequency computation, the results expanded on the relationships between driving elements. This helps in understanding inter-factor processes and side effects, which are highly interdependent. Using placement, level grouping, and arrows, the picture above reveals complicated links between analytical categories (i.e., themes and sub-themes). They may be studied in a three-level interaction on this basis. In the center, direct, place-based urban development initiatives (i.e., building transportation infrastructure, industries, housing, and services) create LULCC in urban areas. The second tier drives urban processes through the creation of policies, regulations, and financing of urban development projects, which is facilitated via various agents, entities and operational processes. Finally, the outer tier, *contextual*, is perceived as a set of effective factors (i.e., demographic, socio-economic, environment) through which the process of LULCC of an urban area is developed. In other words, these factors drive urban growth through decisions on urban policies and other operations (i.e., the second level or immediate inner circle in Figure 5). The theme and sub-themes, and factors (codes) are shown on Figure 6A (top), and B (bottom) respectively.

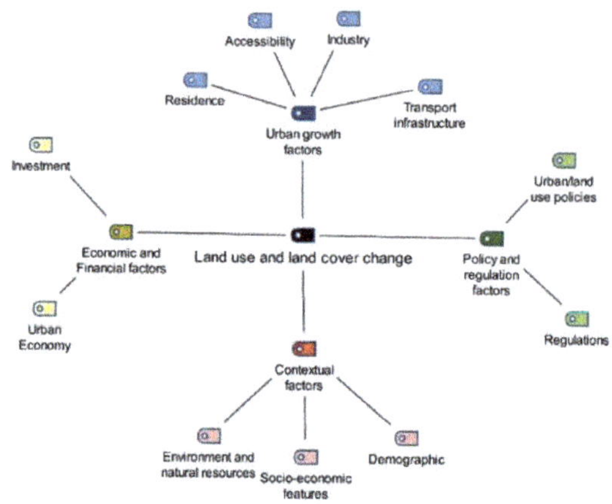

(A)

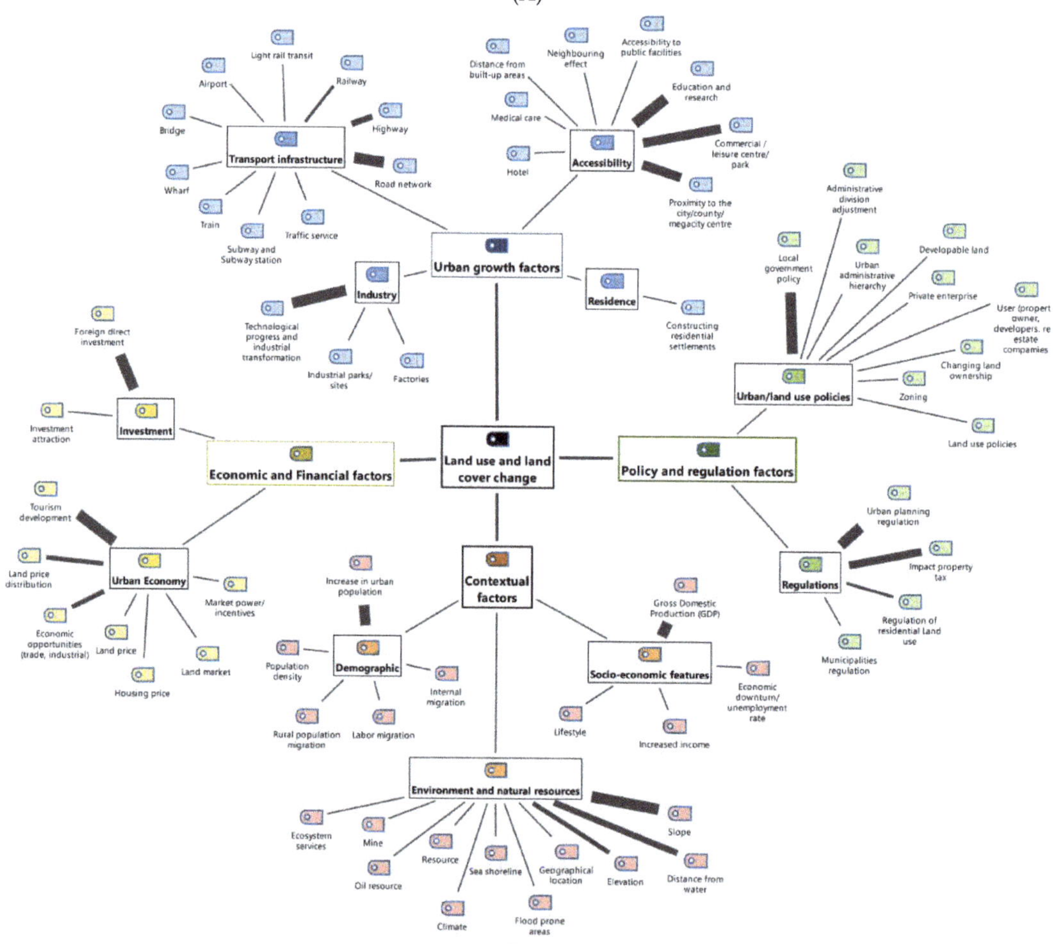

(B)

Figure 6. Components of the driving forces system causing LULCC: (**A**) themes and sub-themes (top); (**B**) theme, sub-themes, and factors (codes) (bottom).

4.2. Urban Growth Factors

These factors explore the driving forces of urban growth that contribute to changing the spatial structure and LULC of urban areas. This theme includes physical factors and growth of transportation infrastructure, industry, accessibility to services, and residential development.

4.2.1. Transport Infrastructure

Transport infrastructure is the most frequently cited factor in LULCC, which refers to the large effect of transportation development on a city spatial structure. In this way, transportation networks such as subways [1,3,40,74], can provide a new access model for the city and upset existing spatial equilibrium. Besides development potentials associated with the operation of a subway system [74], the potential for development in the areas around stations are affected by the presence of developable/vacant lands, plot size, urban fabric and pedestrian access.

Another factor is the development potential of rapid bus transit (BRT), light-rail transit [1,75–77], highspeed rail and stations, in value capturing and added value to adjacent properties and spaces. This is related to the dual functions of transit stations, facilitating accessibility to mass transit and multi-modal connections (i.e., as a transit node) [72], but also characterized by mixed-use development, a diversity of architecture and planned open spaces (i.e., transit place). These functional characteristics of transit stations are the key reasons that they are able to be catalysts for increased urban development potential within a larger urban system resulting in higher development intensity and providing structure to urban form [73,78]. Accordingly, a railway station is not an ordinary station, rather, it is a place where various activities take place [79,80] and can completely affect the surrounding space and change the type and composition of established uses. Such modifications can have a significant influence; for example, urban planning regulations and codes allow some activities to take place in residential settings, transferring these activities to these locations, freeing residential areas from everyday traffic disruptions. In general, transportation networks not only facilitate the flow of commodities and passengers, but they also have an impact on urban growth at different scales.

The review confirms that the quality of transit systems such as fast and low-cost rail transportation networks can also play a role in driving urban growth [1,3,18,27,32,56,79,81,82] which can change the growth of the city from a nuclear, centralized form to a multi-centre city through with multiple (employment) centres. Improving the quality, type and speed of access to various urban areas in a metropolitan characterized by distance between areas, is a major driver determining the rate of urban growth over time.

The effect of access networks on urban/regional development is markedly different for road networks, and ring roads when compared to mass transit networks [5,9,10,12, 13,16,18,21,24,26–29,31–33,44,57,63,66,83–92], or highway [1,3,16,24,27,32,40,44,57,66,69,79–81,84,93–95]. Road networks are catalysts for residential, office, and commercial development, by facilitating development opportunities through ubiquitous transport connections and accessibility, being particularly suited to Road based transport modes such as motor vehicles, cycling and walking.

Additional factors were identified in the transport infrastructure sub-theme, which were also linked to physical-spatial change in urban areas. Wharfs, ferries, harbors, and ports characterized with special functions and coordinates can increase the speed of urban expansion in coastal cities [40,66,71,88]. Similarly, airports in convenient location contribute to the growth of urban and complementary transport infrastructure, and occasionally, when located near the core of a city can encourage substantial urban growth, that subsequently affects urban form and structure across a metropolitan area [1,8,13,24,25,40,57]. In the case of large-scale transport infrastructure projects, this can lead to the expansion of socio-economic factors such as GDP, industries, increasing investments in real estate, and the development of other complementary transport assets [27,30,71,81,94].

4.2.2. Accessibility

This factor originally refers to the index of distance from other regions/destinations, which has an impact on the development of urban areas. Proximity to the city centre [18,30,73,79,88,90,96] and distance from built-up areas [10,94], accessibility to public facilities such as public transport stations [30,32], access to education and research centres (such as colleges, universities, school, etc.) [1,9,20,56,85], commercial/leisure centre/park [1,30,56,68,96–98], hotel [1,13,57], neighboring effects [16,99], medical care e.g., hospital [1,9] are all considered to be crucial in driving urban growth. This factor refers not only to the physical distance of one region/destination from another, but also to the functional distance or distance to access a region/destination. Indeed, it relates to the tendency and potential of a population to live, work, recreate and invest, which are determinants in attracting development to a particular location. As in Burgess's model of a centralized nuclear city, lower-income households move from the centre to the suburbs as their financial capacity increases and they seek larger dwellings. Apart from the "location" factor, new transportation networks and systems affect the distribution of residential development by providing access to potential job opportunities. However, as the city grows, transportation costs increase, either due to the expansion of the city, the increasing complexity of new transport technologies, demands for increasing transport sophistication or due to the costs of congestion. The role of transport in shaping urban form in the future is however uncertain as the relevance of current forms of transport modes and infrastructure are challenged with increasing uptake of digital technologies incorporating innovative mobility solutions such as shared mobility, micro mobility, electric motor vehicles and autonomous vehicles (including land based and aerial drones).

4.2.3. Industrial Development

The second most frequent factor in urban growth factors is industry. Accordingly, industrial parks or sites [1,30,56,68,97,99], technological progress and industrial transformation [2,3,5,9–11,14,17–19,21,22,24,27,33,40,41,66,71,81,93,99–102], and factories [1,68,100], were cited as influencing factors on changing the spatial structure and LULC of urban areas.

Indeed, this component has played a critical role in the development of under developed areas, because the factors of production in the industrial sector, as opposed to agriculture, have higher potential for change with regard to environmental, regional, and national circumstances. As a result, development centres are industry-based, particularly in the global south and in places with limited agricultural development potential. Thus, urban growth is a direct outcome of the Industrial Revolution and the establishment of the capitalist economy, which occurred first in the developed world and later in the developing world. Many new industrial cities in nineteenth-century England, for example, such as Manchester and Birmingham, grew from a hamlet or a small town into a major metropolis. Similarly, with industrialization, French cities increased rapidly in the second half of the nineteenth century, a phenomenon mirrored in German cities.

4.2.4. Residential Development

The last effective factor of LULCC, relates to developing newly developed areas on the urban periphery [15,34,80,93], subsequently resulting in a decentralized spatial structure characterized by the formation of new sub-centres outside of the main urban core. This factor relates to the functional complementarity among the various sub-centres of urban areas and the main core and sub-centres, made possible by population migration from the urban core to the outer suburbs and facilitated by investment in both road transport and mass transit infrastructures, complemented by large investments in denser, higher value urban development in these sub-centres [53].

4.3. *Policy and Regulation Factors*

These factors refer to a series of policies, rules, regulation and operational efforts on general urban issues (such as land use) and processes by which urban growth requirements are facilitated. In this way, the physical and spatial structure of cities including land use/land cover is formulated.

4.3.1. Urban/land Use Policies

On the policy side, land use policies [88,89], include a wide range of activities by which governments seek to influence land use and controlling land ownership [11,45–48], zoning [16,44,83]. The varying role of local government policies on urban growth [2,10,11,13,15, 17,20–22,29,40,41,43,57,61,62,70,71,80,89,99,103,104], is influenced by the state/provincial, national and global context. Developing countries, in particular, are increasingly dominated by government-led policies and measures, and consequently, their urbanization depends on how the government acts, predominantly within these communities. Hence, this can be regarded as one of the significant stimuli for the formation and/or change of spatial structure and LULCC. This factor also contains the availability of developable lands [73,89], private enterprise [67,105], participation and the role of property owners, developers and real estate agencies which contribute to the long-term development of the city through land supply, financing, investment, design and construction of large-scale projects and infrastructures [16,40,86,88,89,103].

Additionally, according to some other studies conducted in the context of Chinese cities, administrative division adjustments (ADA) as city country mergers [3,42] and urban administrative hierarchy-spatial system of allocating urban resources [20,101], resulted an enormous transformation in the spatial structure of cities by stimulating industrial development, infrastructure development, and accelerating urban renewal and the equitable distribution of public services.

4.3.2. Regulations

Although less significant than the previous sub-theme, the secondary dimension of regulations, includes centralized rules imposed through official plans and/or directly by governmental entities. For example, effective regulation factors in the growth of urban areas include municipal regulations [89], that impose various types of land purchase and property impact taxes [16,21,61,89], land use regulations [7,16,88] and urban planning regulations [3,8,22,32,42,43,104,106].

4.4. Economic and Financial Factors

Along with *policy and regulation factors,* these factors drive urban growth through rendering developmental projects feasible. On this basis, it is important to study the economic structure of cities as well as financial system.

4.4.1. Urban Economy

As shown in Table 4, Economic Factors investigate market power/market incentives, land market, land price, land price distribution, housing prices, tourism development and economic opportunities (trade, industrial). According to the studies selected, market power or market incentives [57,67] were identified as effective forces in the changing spatial structure of urban areas. In fact, the market plays an important role in housing development, housing density and development time. However, a recession can curb urban growth or redirect it to different locations or types of investment through imposing restrictions on housing development, in addition to increasing rents and housing prices [107]. In recent decades, the demand for urban land has increased sharply in many cities with the supply of land in order to keep up with demand, precipitating inflation of land values [14,43,64,108] and housing prices [88,108]. Land and housing prices are subject to different factors and conditions, so that it varies at different times and places [93,109]. Moreover, this inflation of development costs reduces the ease of access of government and public institutions, as well as low- and even middle-income people to the land market over time, undermining the viability of marginal businesses, which reinforces the importance of the land market in urban growth processes [43,64]. It can also promote the ability to influence other strategic axes, highlighting land management as amongst the most effective urban management tool. Despite these controls in setting the price of urban land, the price of land can be volatile in responses to speculative behaviors in markets.

Another cited factor was development of the tourism industry [16,25,40,81,88,110] as one of the effective factors in the development of relations between regions and/or nations, which is associated with creating job opportunities in the economic sector [5,64,93] and in improving socio-cultural interactions.

4.4.2. Investment

Although this factor has been less referenced in the selected papers, the role of financing and investment is crucial. This factor includes two main components: investment attraction and foreign direct investment. As the factors of urban expansion in the development of service infrastructure and urban projects [2,57,89,111] these have direct impacts on the location of the settlements and activities.

4.5. Contextual Factors

Finally, how does urban context affect LULCC; what are core contextual dimensions influencing physical-spatial structure of cities? These factors point to several external driving forces through which policies and process are directly, and urban growth are indirectly shaped.

4.5.1. Demographic

Increasing urban population is the major demographic factor that many articles take into account as the effective factors in the formation and changes of land use and land cover [2,4,5,7,9,11,13,16–18,25,27–29,33,34,56,59,61,71,88,90,93,99–101,111–113] and population density [8–10,16,19,80,85,90]. Demographic changes are the result of the improvement in the state of health and well-being of families and individuals, housing affordability, and the growth of communication technology in many regions. As a result, there has been an increasing trend of an intensification of population density in some cities and the emergence of mega cities (*i.e., 10 million or more people*) in recent years.

Another frequently-cited issue related to LUCLCC is the migration of rural populations to the city as the consequence of agricultural land transformation [11–13,23,24,29,59,64,71,88,100]. Other migration concepts such as internal migration within metropolitan regions [16, 19,25,27,97,112]; and labor migration [11,13,22,40,64,68,110], were also attributed to the changes in built environment and consequently the change in spatial distribution of job oppurtunities or urban amentities resulted in improving the attractiveness of an area to absorb migrants. Another reason for internal migration includes the disparity in wages and working conditions in different locations, which creates a labour market duality. The influx of immigrants, on the other hand, raises the demand for housing and the expense of living, leading to marginalization. Changes in family structure and lifestyle necessitate changes in housing demands, which encourages bigger households to relocate from core districts to the periphery, affecting the land market and affecting the motive for suburban land usage.

4.5.2. Socio-Economic Features

As the least referenced sub-theme, the socio-economic features include gross domestic product per capita (GDP) [2,10,14,16–19,61,71,80,85,99–101,114–116], and increased income [14–16] which have increased the demand for a luxurious consumer oriented lifestyle [11]. Often this implies status conscious, spacious, comfortable houses accessible to convenient high quality transportation in master-planned estates, thereby increasing the demand for urban land [106]. IN addition, another socio-economic dimension is the phenomenon of second homes and second houses on the urban periphery to provide a retreat or for investment purposes to increase personal capital. Economic downturns/unemployment rate is another factor [80,97,116,117] influencing urban growth that can lead to the loss of population from an urban core or declining suburbs and result in urban decay. Hence, economic recessions can have a powerful role in shaping urban spatial development.

4.5.3. Environment and Natural Resources

In conjunction with the previous contextual dimensions, environment and natural resources have the capability to change urban land use and land cover patterns. They include geographical location [14,22,25,27,96,101], flood prone areas [118,119], climate quality [16,41,111,120–122], sea shoreline [25,33], distance from water bodies such as rivers, lakes, wetlands, ponds [16,18,26,28,30–34,89,95], all of which are fundamentally important determinants of the extent, spatial distribution, and spatial expansion of urban lands. Furthermore, it can relate to the efficiency of terrestrial resources such as forestry and ecological resources [91], oil resources [13,18,40,88], minerals [1,9] and ecosystem services [65,123]. Slope [1,9,10,16,24–30,32,62,79,85,87,95] and elevation [1,10,27–30,79,85] also determines the location of physical developments within a city since the developers generally prioritise development in flatter areas.

5. Conclusions

With the global urban population rapidly increasing, further physical growth and associated land use and land cover changes are unavoidable. Hence, a critically important strategic priority in the urban planning agenda is in identifying, analysing and modelling the effective drivers underlying land use and land cover change. The work in this paper was a bibliometric and systematic review of LULCC, with the goal of identifying the drivers of land use and land cover change (2012 to 2022), as well as contributing to an analysis of the most significant concepts, methdological rules, and journals in LULCC research.

The main finding from this study is that the LULCC process is impacted by a variety of interconnected elements, ranging from transportation development to legislation, as well as contextual demographic, socioeconomic, and environmental aspects. Although they were arranged in groups and three levels of interactions, and their significance was only explored using the number of occurrences in the literature, it is worth noting that the factors are highly context-sensitive, so that their relationships and significance can change depending on factors such as time, geography, scale, and decision-making agents. It was found that transportation availability was the most frequent factor identified in the literature, although this can be detailed to include multiple dimensions of transport avilability such as provosion of mobility systems, fuel price and vehicle ownesrship area [124]. A caveat is that the frequency of topic mentions in the literature does not necessarily indicate that a factor is stronger in influencing urban growth, since the context of discussion can be supportive or critical of the role of a particular factor and the relative magnitude of a factor is often not easily ascertained from mapping the frequency of a term. Moreover, there may be a bias resulting from funding factors, or other factors that influenced the direction of research. Hence, various elements ambiguously examined in the existing body of literature in this field introduce a degree of uncertainty and have the potential to influence urban growth at various local, municipal, regional/state/provincial, national and globally levels. In terms of scale, for example, the spatial scale at which the studies were conducted has an impact on the results in such a way that human and artificial factors have the greatest impact at the micro level, and as the scale becomes larger (i.e., at the regional scale), the role of environment and natural factors becomes more pronounced, as is the case in the Beijing metropolitan area [125], in relation to altitude, distance from the river, and urbanisation rate.

This is also in line with the fact that the notion of urban growth is highly dynamic with a high level of complexity and uncertainty. Urban growth can be an unstable and discontinuous process that expands metropolitan boundaries and imposes drastic changes in land use that overwhelms social and environmental capacities and the capacity of existing plans and regulations to cope. As a result, governments and urban management systems are confronted with complex challenges, particularly in relation to the stresses to ecologies and human constructed environments arising from climate change.

Additional study is recommended to investigate the usefulness of the model of driving variables (Figure 5) in relation to its unique emphasis and local circumstances. This may include thoroughly examining the impact of particular components (such as transportation infrastructure) or drawing on aspects within each level (such as outer contextual factors). Furthermore, in light of the vast diversity of publishing landscapes globally, further review studies evaluating driving variables depending on country categories (such as global south) with a particular refrence to the social context [126,127] and city size (such as aggolormationa nd scale effects) would expand the scope of this work. Reviews of additional databases (e.g., Web of Science, Google Scholar) would also be beneficial in refining a model to determine LULCC that not only identified key drivers of change but which has predictive capabilities in response to key stressors in natural and human environments.

Author Contributions: Conceptualization, A.A.; methodology, A.A., A.S. and M.H.A.; software, M.Z.; validation, A.A. and A.S.; formal analysis, M.H.A. and M.Z.; investigation, A.S. and M.Z.; resources, A.A.; data curation, M.Z.; writing—original draft preparation, A.S., M.H.A. and M.Z.; writing—review and editing, A.A.; visualization, M.H.A. and M.Z.; supervision, A.A.; project administration, A.A.; funding acquisition, A.A. All authors have read and agreed to the published version of the manuscript.

Funding: This research received funding support from IVE: Australian Research Centre for Interactive and Virtual Environments, and UniSA Creative, University of South Australia, 2021.

Data Availability Statement: Data are available from the second author on request.

Acknowledgments: The authors wish to acknowledge IVE Centre of the University of South Australia and UniSA Creative for their support and resources.

Conflicts of Interest: The authors declare that there is no conflict of interest.

References

1. Wu, H.; Lin, A.; Xing, X.; Song, D.; Li, Y. Identifying core driving factors of urban land use change from global land cover products and POI data using the random forest method. *Int. J. Appl. Earth Obs. Geoinf.* **2021**, *103*, 102475. [CrossRef]
2. Dou, P.; Han, Z. Quantifying Land Use/Land Cover Change and Urban Expansion in Dongguan, China, From 1987 to 2020. *IEEE J. Sel. Top. Appl. Earth Obs. Remote Sens.* **2021**, *15*, 201–209. [CrossRef]
3. Feng, R.; Wang, K. Spatiotemporal effects of administrative division adjustment on urban expansion in China. *Land Use Policy* **2021**, *101*, 105143. [CrossRef]
4. Li, M.; Verburg, P.H.; van Vliet, J. Global trends and local variations in land take per person. *Landsc. Urban Plan.* **2022**, *218*, 104308. [CrossRef]
5. Tavares, A.O.; Monteiro, M.; Barros, J.L.; Santos, P.P. Long-term land-use changes in small/medium-sized cities. Enhancing the general trends and local characteristics. *Eur. Plan. Stud.* **2019**, *27*, 1432–1459. [CrossRef]
6. Nagy, R.; Lockaby, B.G. Urbanization in the Southeastern United States: Socioeconomic forces and ecological responses along an urban-rural gradient. *Urban Ecosyst.* **2011**, *14*, 71–86. [CrossRef]
7. Tiitu, M. Expansion of the built-up areas in Finnish city regions–The approach of travel-related urban zones. *Appl. Geogr.* **2018**, *101*, 1–13. [CrossRef]
8. Banzhaf, E.; Reyes-Paecke, S.; Müller, A.; Kindler, A. Do demographic and land-use changes contrast urban and suburban dynamics? A sophisticated reflection on Santiago de Chile. *Habitat Int.* **2013**, *39*, 179–191. [CrossRef]
9. De la Luz Hernández-Flores, M.; Otazo-Sánchez, E.M.; Galeana-Pizana, M.; Roldán-Cruz, E.I.; Razo-Zárate, R.; González-Ramírez, C.A.; Galindo-Castillo, E.; Gordillo-Martínez, A.J. Urban driving forces and megacity expansion threats. Study case in the Mexico City periphery. *Habitat Int.* **2017**, *64*, 109–122. [CrossRef]
10. Xu, Y.; McNamara, P.; Wu, Y.; Dong, Y. An econometric analysis of changes in arable land utilization using multinomial logit model in Pinggu district, Beijing, China. *J. Environ. Manag.* **2013**, *128*, 324–334. [CrossRef]
11. Kleemann, J.; Inkoom, J.N.; Thiel, M.; Shankar, S.; Lautenbach, S.; Fürst, C. Peri-urban land use pattern and its relation to land use planning in Ghana, West Africa. *Landsc. Urban Plan.* **2017**, *165*, 280–294. [CrossRef]
12. Islam, M.D.; Islam, K.S.; Ahasan, R.; Mia, M.R.; Haque, M.E. A data-driven machine learning-based approach for urban land cover change modeling: A case of Khulna City Corporation area. *Remote Sens. Appl. Soc. Environ.* **2021**, *24*, 100634. [CrossRef]
13. Essien, E.; Cyrus, S. Detection of urban development in Uyo (Nigeria) using remote sensing. *Land* **2019**, *8*, 102. [CrossRef]
14. Hasan, S.; Shi, W.; Zhu, X.; Abbas, S. Monitoring of land use/land cover and socioeconomic changes in south china over the last three decades using landsat and nighttime light data. *Remote Sens.* **2019**, *11*, 1658. [CrossRef]
15. Ponstingel, D. The impact of exurban development on forested areas in Kurgan City, Russia. *Land Use Policy* **2020**, *94*, 104485. [CrossRef]

16. Colsaet, A.; Laurans, Y.; Levrel, H. What drives land take and urban land expansion? A systematic review. *Land Use Policy* **2018**, *79*, 339–349. [CrossRef]
17. Zhang, Q.; Seto, K.C. Mapping urbanization dynamics at regional and global scales using multi-temporal DMSP/OLS nighttime light data. *Remote Sens. Environ.* **2011**, *115*, 2320–2329. [CrossRef]
18. Li, W.; Wu, C.; Zang, S. Modeling urban land use conversion of Daqing City, China: A comparative analysis of "top-down" and "bottom-up" approaches. *Stoch. Environ. Res. Risk Assess.* **2014**, *28*, 817–828. [CrossRef]
19. Liu, Y.; Song, W.; Deng, X. Understanding the spatiotemporal variation of urban land expansion in oasis cities by integrating remote sensing and multi-dimensional DPSIR-based indicators. *Ecol. Indic.* **2019**, *96*, 23–37. [CrossRef]
20. Li, H.; Wei, Y.D.; Liao, F.H.; Huang, Z. Administrative hierarchy and urban land expansion in transitional China. *Appl. Geogr.* **2015**, *56*, 177–186. [CrossRef]
21. Kontgis, C.; Schneider, A.; Fox, J.; Saksena, S.; Spencer, J.H.; Castrence, M. Monitoring peri-urbanization in the greater Ho Chi Minh City metropolitan area. *Appl. Geogr.* **2014**, *53*, 377–388. [CrossRef]
22. Dai, E.; Wu, Z.; Du, X. A gradient analysis on urban sprawl and urban landscape pattern between 1985 and 2000 in the Pearl River Delta, China. *Front. Earth Sci.* **2018**, *12*, 791–807. [CrossRef]
23. Asabere, S.B.; Acheampong, R.A.; Ashiagbor, G.; Beckers, S.C.; Keck, M.; Erasmi, S.; Schanze, J.; Sauer, D. Urbanization, land use transformation and spatio-environmental impacts: Analyses of trends and implications in major metropolitan regions of Ghana. *Land Use Policy* **2020**, *96*, 104707. [CrossRef]
24. Shahraki, S.Z.; Sauri, D.; Serra, P.; Modugno, S.; Seifolddini, F.; Pourahmad, A. Urban sprawl pattern and land-use change detection in Yazd, Iran. *Habitat Int.* **2011**, *35*, 521–528. [CrossRef]
25. Kamh, S.; Ashmawy, M.; Kilias, A.; Christaras, B. Evaluating urban land cover change in the Hurghada area, Egypt, by using GIS and remote sensing. *Int. J. Remote Sens.* **2012**, *33*, 41–68. [CrossRef]
26. Moghadam, H.S.; Helbich, M. Spatiotemporal urbanization processes in the megacity of Mumbai, India: A Markov chains-cellular automata urban growth model. *Appl. Geogr.* **2013**, *40*, 140–149. [CrossRef]
27. Jawarneh, R.N.; Julian, J.P.; Lookingbill, T.R. The influence of physiography on historical and future land development changes: A case study of central Arkansas (USA), 1857–2030. *Landsc. Urban Plan.* **2015**, *143*, 76–89. [CrossRef]
28. Sunde, M.G.; He, H.S.; Zhou, B.; Hubbart, J.A.; Spicci, A. Imperviousness Change Analysis Tool (I-CAT) for simulating pixel-level urban growth. *Landsc. Urban Plan.* **2014**, *124*, 104–108. [CrossRef]
29. Gerten, C.; Fina, S.; Rusche, K. The sprawling planet: Simplifying the measurement of global urbanization trends. *Front. Environ. Sci.* **2019**, *7*, 140. [CrossRef]
30. Han, Y.; Jia, H. Simulating the spatial dynamics of urban growth with an integrated modeling approach: A case study of Foshan, China. *Ecol. Model.* **2017**, *353*, 107–116. [CrossRef]
31. Feng, Y. Modeling dynamic urban land-use change with geographical cellular automata and generalized pattern search-optimized rules. *Int. J. Geogr. Inf. Sci.* **2017**, *31*, 1198–1219. [CrossRef]
32. Kong, L.; Tian, G.; Ma, B.; Liu, X. Embedding ecological sensitivity analysis and new satellite town construction in an agent-based model to simulate urban expansion in the beijing metropolitan region, China. *Ecol. Indic.* **2017**, *82*, 233–249. [CrossRef]
33. Leyk, S.; Uhl, J.H.; Connor, D.S.; Braswell, A.E.; Mietkiewicz, N.; Balch, J.K.; Gutmann, M. Two centuries of settlement and urban development in the United States. *Sci. Adv.* **2020**, *6*, eaba2937. [CrossRef] [PubMed]
34. Bajracharya, P.; Lippitt, C.D.; Sultana, S. Modeling urban growth and land cover change in Albuquerque using SLEUTH. *Prof. Geogr.* **2020**, *72*, 181–193. [CrossRef]
35. Quesada-Román, A.; Castro-Chacón, J.P.; Feoli Boraschi, S. Geomorphology, land use, and environmental impacts in a densely populated urban catchment of Costa Rica. *J. S. Am. Earth Sci.* **2021**, *112*, 103560. [CrossRef]
36. Quesada-Román, A.; Mata-Cambronero, E. The geomorphic landscape of the Barva volcano, Costa Rica. *Phys. Geogr.* **2021**, *42*, 265–282. [CrossRef]
37. García-Soriano, D.; Quesada-Román, A.; Zamorano-Orozco, J.J. Geomorphological hazards susceptibility in high-density urban areas: A case study of Mexico City. *J. S. Am. Earth Sci.* **2020**, *102*, 102667. [CrossRef]
38. Winsemius, H.C.; Aerts, J.C.; Van Beek, L.P.; Bierkens, M.F.; Bouwman, A.; Jongman, B.; Kwadijk, J.C.J.; Ligtvoet, W.; Lucas, P.L.; van Vuuren, D.P.; et al. Global drivers of future river flood risk. *Nat. Clim. Change* **2016**, *6*, 381–385. [CrossRef]
39. Shah, M.A.R.; Rahman, A.; Chowdhury, S.H. Challenges for achieving sustainable flood risk management. *J. Flood Risk Manag.* **2018**, *11*, S352–S358. [CrossRef]
40. Nassar, A.K.; Blackburn, G.A.; Whyatt, J.D. Developing the desert: The pace and process of urban growth in Dubai. *Comput. Environ. Urban Syst.* **2014**, *45*, 50–62. [CrossRef]
41. Wang, J.; Lin, Y.; Glendinning, A.; Xu, Y. Land-use changes and land policies evolution in China's urbanization processes. *Land Use Policy* **2018**, *75*, 375–387. [CrossRef]
42. Feng, R.; Wang, K. The direct and lag effects of administrative division adjustment on urban expansion patterns in Chinese mega-urban agglomerations. *Land Use Policy* **2022**, *112*, 105805. [CrossRef]
43. Yue, W.; Fan, P.; Wei, Y.D.; Qi, J. Economic development, urban expansion, and sustainable development in Shanghai. *Stoch. Environ. Res. Risk Assess.* **2014**, *28*, 783–799. [CrossRef]
44. Schumacher, M.; Durán-Díaz, P.; Kurjenoja, A.K.; Gutiérrez-Juárez, E.; González-Rivas, D.A. Evolution and Collapse of Ejidos in Mexico—To What Extent Is Communal Land Used for Urban Development? *Land* **2019**, *8*, 146. [CrossRef]

45. Whiteside, H. Privatizing Canadian government land and real estate: Railroads, reconciliation, and rip-offs. *Land Use Policy* **2020**, *99*, 104821. [CrossRef]
46. Tong, D.; Wang, X.; Wu, L.; Zhao, N. Land ownership and the likelihood of land development at the urban fringe: The case of Shenzhen, China. *Habitat Int.* **2018**, *73*, 43–52. [CrossRef]
47. Adam, A.G. Thinking outside the box and introducing land readjustment against the conventional urban land acquisition and delivery method in Ethiopia. *Land Use Policy* **2019**, *81*, 624–631. [CrossRef]
48. Zhang, S.; De Roo, G.; Van Dijk, T. Urban land changes as the interaction between self-organization and institutions. *Plan. Pract. Res.* **2015**, *30*, 160–178. [CrossRef]
49. Li, X.; Zhou, W.; Ouyang, Z. Forty years of urban expansion in Beijing: What is the relative importance of physical, socioeconomic, and neighborhood factors? *Appl. Geogr.* **2013**, *38*, 1–10. [CrossRef]
50. Plieninger, T.; Draux, H.; Fagerholm, N.; Bieling, C.; Bürgi, M.; Kizos, T.; Kuemmerle, T.; Primdahl, J.; Verburg, P.H. The driving forces of landscape change in Europe: A systematic review of the evidence. *Land Use Policy* **2016**, *57*, 204–214. [CrossRef]
51. Yang, G.; Chao, S.; Tsou, J.Y.; Zhang, Y. Satellite image-based methods of spatiotemporal analysis on sustainable urban land use change and the driving factors: A case study in caofeidian and the suburbs, China. *Sustainability* **2019**, *11*, 2927. [CrossRef]
52. Cai, Y.; Zhang, H.; Pan, W.; Chen, Y.; Wang, X. Urban expansion and its influencing factors in natural wetland distribution area in Fuzhou City, China. *Chin. Geogr. Sci.* **2012**, *22*, 568–577. [CrossRef]
53. Dadashpoor, H.; Malekzadeh, N. Driving factors of formation, development, and change of spatial structure in metropolitan areas: A systematic review. *J. Urban Manag.* **2020**, *9*, 286–297. [CrossRef]
54. Thelwall, M. Bibliometrics to webometrics. *J. Inf. Sci.* **2008**, *34*, 605–621. [CrossRef]
55. Dzikowski, P. A bibliometric analysis of born global firms. *J. Bus. Res.* **2018**, *85*, 281–294. [CrossRef]
56. Zhang, Y.; Xu, B. Spatiotemporal analysis of land use/cover changes in Nanchang area, China. *Int. J. Digit. Earth* **2015**, *8*, 312–333. [CrossRef]
57. Chen, T.; Lang, W.; Chan, E.; Philipp, C.H. Lhasa: Urbanising China in the frontier regions. *Cities* **2018**, *74*, 343–353. [CrossRef]
58. Li, Y.; Li, Y.; Karácsonyi, D.; Liu, Z.; Wang, Y.; Wang, J. Spatio-temporal pattern and driving forces of construction land change in a poverty-stricken county of China and implications for poverty-alleviation-oriented land use policies. *Land Use Policy* **2020**, *91*, 104267. [CrossRef]
59. Xu, F.; Wang, Z.; Chi, G.; Zhang, Z. The impacts of population and agglomeration development on land use intensity: New evidence behind urbanization in China. *Land Use Policy* **2020**, *95*, 104639. [CrossRef]
60. Bimonte, S.; Stabile, A. Local taxation and urban development. Testing for the side-effects of the Italian property tax. *Ecol. Econ.* **2015**, *120*, 100–107. [CrossRef]
61. Kuang, W. National urban land-use/cover change since the beginning of the 21st century and its policy implications in China. *Land Use Policy* **2020**, *97*, 104747. [CrossRef]
62. Wadduwage, S. Peri-urban agricultural land vulnerability due to urban sprawl–a multi-criteria spatially-explicit scenario analysis. *J. Land Use Sci.* **2018**, *13*, 358–374. [CrossRef]
63. Gallardo, M.; Martínez-Vega, J. Three decades of land-use changes in the region of Madrid and how they relate to territorial planning. *Eur. Plan. Stud.* **2016**, *24*, 1016–1033. [CrossRef]
64. Simwanda, M.; Murayama, Y.; Ranagalage, M. Modeling the drivers of urban land use changes in Lusaka, Zambia using multi-criteria evaluation: An analytic network process approach. *Land Use Policy* **2020**, *92*, 104441. [CrossRef]
65. Peng, K.; Jiang, W.; Ling, Z.; Hou, P.; Deng, Y. Evaluating the potential impacts of land use changes on ecosystem service value under multiple scenarios in support of SDG reporting: A case study of the Wuhan urban agglomeration. *J. Clean. Prod.* **2021**, *307*, 127321. [CrossRef]
66. Inouye, C.E.N.; de Sousa Jr, W.C.; De Freitas, D.M.; Simões, E. Modelling the spatial dynamics of urban growth and land use changes in the north coast of São Paulo, Brazil. *Ocean. Coast. Manag.* **2015**, *108*, 147–157. [CrossRef]
67. Hamnett, C. Is Chinese urbanization unique? *Urban Stud.* **2020**, *57*, 690–700. [CrossRef]
68. Shin, H.; Chae, S. Urbanisation and land use transition in a second-tier city: The emergence of small factories in Gimpo, South Korea. *Land Use Policy* **2018**, *77*, 534–541. [CrossRef]
69. Pratama, A.P.; Yudhistira, M.H.; Koomen, E. Highway expansion and urban sprawl in the Jakarta Metropolitan Area. *Land Use Policy* **2022**, *112*, 105856. [CrossRef]
70. Cheng, L. China's rural transformation under the Link Policy: A case study from Ezhou. *Land Use Policy* **2021**, *103*, 105319. [CrossRef]
71. Cao, W.; Zhou, Y.; Li, R.; Li, X.; Zhang, H. Monitoring long-term annual urban expansion (1986–2017) in the largest archipelago of China. *Sci. Total Environ.* **2021**, *776*, 146015. [CrossRef]
72. Wenner, F.; Thierstein, A. High speed rail as urban generator? An analysis of land use change around European stations. *Eur. Plan. Stud.* **2022**, *30*, 227–250. [CrossRef]
73. Deng, Y.; Srinivasan, S. Urban land use change and regional access: A case study in Beijing, China. *Habitat Int.* **2016**, *51*, 103–113. [CrossRef]
74. Ahmad, S.; Avtar, R.; Sethi, M.; Surjan, A. Delhi's land cover change in post transit era. *Cities* **2016**, *50*, 111–118. [CrossRef]
75. Hurst, N.B.; West, S.E. Public transit and urban redevelopment: The effect of light rail transit on land use in Minneapolis, Minnesota. *Reg. Sci. Urban Econ.* **2014**, *46*, 57–72. [CrossRef]

76. Wang, J.; Feng, Y.; Ye, Z.; Tong, X.; Wang, R.; Gao, C.; Chen, S.; Lei, Z.; Liu, S.; Jin, Y. Simulating the effect of urban light rail transit on urban development by coupling cellular automata and conjugate gradients. *Geocarto Int.* **2022**, *37*, 2346–2364. [CrossRef]
77. Nguyen, H.A.; Soltani, A.; Allan, A. Adelaide's East End tramline: Effects on modal shift and carbon reduction. *Travel Behav. Soc.* **2018**, *11*, 21–30. [CrossRef]
78. Abdi, M.H.; Soltani, A. Which Fabric/Scale Is Better for Transit-Oriented Urban Design: Case Studies in a Developing Country. *Sustainability* **2022**, *14*, 7338. [CrossRef]
79. Wang, K.; Zhou, W. Do local factors or teleconnections control urbanization? The shifting balance in a Chinese megaregion. *Landsc. Urban Plan.* **2018**, *180*, 179–186. [CrossRef]
80. Meyer, M.A.; Früh-Müller, A. Patterns and drivers of recent agricultural land-use change in Southern Germany. *Land Use Policy* **2020**, *99*, 104959. [CrossRef]
81. Chu, L.; Zou, Y.; Masiliūnas, D.; Blaschke, T.; Verbesselt, J. Assessing the impact of bridge construction on the land use/cover and socio-economic indicator time series: A case study of Hangzhou Bay Bridge. *GIScі. Remote Sens.* **2021**, *58*, 199–216. [CrossRef]
82. Zhao, L.; Shen, L. The impacts of rail transit on future urban land use development: A case study in Wuhan, China. *Transp. Policy* **2019**, *81*, 396–405. [CrossRef]
83. McGarigal, K.; Plunkett, E.B.; Willey, L.L.; Compton, B.W.; DeLuca, W.V.; Grand, J. Modeling non-stationary urban growth: The SPRAWL model and the ecological impacts of development. *Landsc. Urban Plan.* **2018**, *177*, 178–190. [CrossRef]
84. Nadafianshahamabadi, R.; Tayarani, M.; Rowangould, G. A closer look at urban development under the emergence of autonomous vehicles: Traffic, land use and air quality impacts. *J. Transp. Geogr.* **2021**, *94*, 103113. [CrossRef]
85. Liu, D.; Clarke, K.C.; Chen, N. Integrating spatial nonstationarity into SLEUTH for urban growth modeling: A case study in the Wuhan metropolitan area. *Comput. Environ. Urban Syst.* **2020**, *84*, 101545. [CrossRef]
86. Kasraian, D.; Raghav, S.; Miller, E.J. A multi-decade longitudinal analysis of transportation and land use co-evolution in the Greater Toronto-Hamilton Area. *J. Transp. Geogr.* **2020**, *84*, 102696. [CrossRef]
87. Fitawok, M.B.; Derudder, B.; Minale, A.S.; Van Passel, S.; Adgo, E.; Nyssen, J. Modeling the impact of urbanization on land-use change in Bahir Dar City, Ethiopia: An integrated cellular Automata–Markov Chain Approach. *Land* **2020**, *9*, 115. [CrossRef]
88. Daunt, A.B.P.; Inostroza, L.; Hersperger, A.M. The role of spatial planning in land change: An assessment of urban planning and nature conservation efficiency at the southeastern coast of Brazil. *Land Use Policy* **2021**, *111*, 105771. [CrossRef]
89. Deslatte, A.; Szmigiel-Rawska, K.; Tavares, A.F.; Ślawska, J.; Karsznia, I.; Łukomska, J. Land use institutions and social-ecological systems: A spatial analysis of local landscape changes in Poland. *Land Use Policy* **2022**, *114*, 105937. [CrossRef]
90. La, K.; Kumar, D.; Kumar, A. Spatio-temporal landscape modeling of urban growth patterns in Dhanbad Urban Agglomeration, India using geoinformatics techniques. *Egypt. J. Remote Sens. Space Sci.* **2017**, *20*, 91–102. [CrossRef]
91. Ma, Q. Integrating ecological correlation into cellular automata for urban growth simulation: A case study of Hangzhou, China. *Urban For. Urban Green.* **2020**, *51*, 126697. [CrossRef]
92. Hanlon, B.; Howland, M.; McGuire, M.P. Hotspots for Growth: Does Maryland's Priority Funding Area Program Reduce Sprawl? *J. Am. Plan. Assoc.* **2012**, *78*, 256–268. [CrossRef]
93. Sandhya Kiran, G.; Joshi, U.B. Estimation of variables explaining urbanization concomitant with land-use change: A spatial approach. *Int. J. Remote Sens.* **2013**, *34*, 824–847. [CrossRef]
94. Geymen, A. Impacts of Bosporus bridges on the Istanbul metropolitan settlement areas. *Land Degrad. Dev.* **2013**, *24*, 156–169. [CrossRef]
95. Jokar Arsanjani, J.; Helbich, M.; Vaz, E.D.N. Spatiotemporal simulation of urban growth patterns using agent-based modeling: The case of Tehran. *Cities* **2013**, *32*, 33–42. [CrossRef]
96. Moghadam, A.S.; Soltani, A.; Parolin, B.; Alidadi, M. Analysing the space-time dynamics of urban structure change using employment density and distribution data. *Cities.* **2018**, *81*, 203–213. [CrossRef]
97. Kang, S.; Choi, J.; Yoon, H.; Choi, W. Changes in the extent and distribution of urban land cover in the Democratic People's Republic of Korea (North Korea) between 1987 and 2010. *Land Degrad. Dev.* **2019**, *30*, 2009–2017. [CrossRef]
98. Zambon, I.; Cerdà, A.; Gambella, F.; Egidi, G.; Salvati, L. Industrial sprawl and residential housing: Exploring the interplay between local development and land-use change in the Valencian Community, Spain. *Land* **2019**, *8*, 143. [CrossRef]
99. Luo, J.; Xing, X.; Wu, Y.; Zhang, W.; Chen, R.S. Spatio-temporal analysis on built-up land expansion and population growth in the Yangtze River Delta Region, China: From a coordination perspective. *Appl. Geogr.* **2018**, *96*, 98–108. [CrossRef]
100. Ul Din, S.; Mak, H.W.L. Retrieval of Land-Use/Land Cover Change (LUCC) Maps and Urban Expansion Dynamics of Hyderabad, Pakistan via Landsat Datasets and Support Vector Machine Framework. *Remote Sens.* **2021**, *13*, 3337. [CrossRef]
101. Dong, Y.; Ren, Z.; Fu, Y.; Miao, Z.; Yang, R.; Sun, Y.; He, X. Recording urban land dynamic and its effects during 2000–2019 at 15-m resolution by cloud computing with Landsat series. *Remote Sens.* **2020**, *12*, 2451. [CrossRef]
102. Li, W.; Wang, D.; Li, H.; Liu, S. Urbanization-induced site condition changes of peri-urban cultivated land in the black soil region of northeast China. *Ecol. Indic.* **2017**, *80*, 215–223. [CrossRef]
103. Xu, Y.; Chan, E.H.; Yung, E.H. Overwhelming farmland conversion for urban development in transitional China: Case study of Shanghai. *J. Urban Plan. Dev.* **2015**, *141*, 05014013. [CrossRef]
104. Kong, F.; Yin, H.; Nakagoshi, N.; James, P. Simulating urban growth processes incorporating a potential model with spatial metrics. *Ecol. Indic.* **2012**, *20*, 82–91. [CrossRef]

105. Soria, K.Y.; Palacios, M.R.; Gomez, C.A.M. Governance and policy limitations for sustainable urban land planning. The case of Mexico. *J. Environ. Manag.* **2020**, *259*, 109575. [CrossRef] [PubMed]
106. Soltani, A.; Hosseinpour, M.; Hajizadeh, A. Urban sprawl in Iranian medium-sized cities; investigating the Role of Masterplans. *J. Sustain. Dev.* **2017**, *10*, 122–131. [CrossRef]
107. Lopez, J.C.G.; Arnott, R.J. Is higher-quality land developed earlier? *Environ. Plan. B Urban Anal. City Sci.* **2020**, *47*, 1560–1572. [CrossRef]
108. Magliocca, N.; McConnell, V.; Walls, M. Exploring sprawl: Results from an economic agent-based model of land and housing markets. *Ecol. Econ.* **2015**, *113*, 114–125. [CrossRef]
109. Hu, S.; Cheng, Q.; Wang, L.; Xie, S. Multifractal characterization of urban residential land price in space and time. *Appl. Geogr.* **2012**, *34*, 161–170. [CrossRef]
110. Sang, X.; Guo, Q.; Wu, X.; Fu, Y.; Xie, T.; He, C.; Zang, J. Intensity and stationarity analysis of land use change based on CART algorithm. *Sci. Rep.* **2019**, *9*, 12279. [CrossRef]
111. Admasu, T.G. Urban land use dynamics, the nexus between land use pattern and its challenges: The case of Hawassa city, Southern Ethiopia. *Land Use Policy* **2015**, *45*, 159–175. [CrossRef]
112. Skog, K.L.; Steinnes, M. How do centrality, population growth and urban sprawl impact farmland conversion in Norway? *Land Use Policy* **2016**, *59*, 185–196. [CrossRef]
113. Fuglsang, M.; Münier, B.; Hansen, H.S. Modelling land-use effects of future urbanization using cellular automata: An Eastern Danish case. *Environ. Model. Softw.* **2013**, *50*, 1–11. [CrossRef]
114. Jiang, L.; Deng, X.; Seto, K.C. The impact of urban expansion on agricultural land use intensity in China. *Land Use Policy* **2013**, *35*, 33–39. [CrossRef]
115. Gong, J.; Chen, W.; Liu, Y.; Wang, J. The intensity change of urban development land: Implications for the city master plan of Guangzhou, China. *Land Use Policy* **2014**, *40*, 91–100. [CrossRef]
116. Tomao, A.; Quaranta, G.; Salvia, R.; Vinci, S.; Salvati, L. Revisiting the 'southern mood'? Post-crisis Mediterranean urbanities between economic downturns and land-use change. *Land Use Policy* **2021**, *111*, 105740. [CrossRef]
117. Salvati, L. Urban dispersion and economic crisis: Empirical evidence from a Mediterranean region. *J. Environ. Plan. Manag.* **2019**, *62*, 1205–1226. [CrossRef]
118. Mustafa, A.; Bruwier, M.; Archambeau, P.; Erpicum, S.; Pirotton, M.; Dewals, B.; Jacques Teller, J. Effects of spatial planning on future flood risks in urban environments. *J. Environ. Manag.* **2018**, *225*, 193–204. [CrossRef]
119. Santos, P.P.; Pereira, S.; Zêzere, J.L.; Tavares, A.O.; Reis, E.; Garcia, R.A.C.; Oliveira, S.C. A comprehensive approach to understanding flood risk drivers at the municipal level. *J. Environ. Manag.* **2020**, *260*, 110127. [CrossRef]
120. Yan, D.; Schneider, U.A.; Schmid, E.; Huang, H.Q.; Pan, L.; Dilly, O. Interactions between land use change, regional development, and climate change in the Poyang Lake district from 1985 to 2035. *Agric. Syst.* **2013**, *119*, 10–21. [CrossRef]
121. Soltani, A.; Sharifi, E. Daily variation of urban heat island effect and its correlations to urban greenery: A case study of Adelaide. *Front. Archit. Res.* **2017**, *6*, 529–538. [CrossRef]
122. Azhdari, A.; Soltani, A.; Alidadi, M. Urban morphology and landscape structure effect on land surface temperature: Evidence from Shiraz, a semi-arid city. *Sustain. Cities Soc.* **2018**, *41*, 853–864. [CrossRef]
123. Pan, H.; Page, J.; Cong, C.; Barthel, S.; Kalantari, Z. How ecosystems services drive urban growth: Integrating nature-based solutions. *Anthropocene* **2021**, *35*, 100297. [CrossRef]
124. Soltani, A. Social and urban form determinants of vehicle ownership; evidence from a developing country, Transportation Research Part A: Policy and Practice. *Transp. Res. Part A Policy Pract.* **2017**, *96*, 90–100. [CrossRef]
125. Han, H.; Yang, C.; Song, J. Scenario Simulation and the Prediction of Land Use and Land Cover Change in Beijing, China. *Sustainability.* **2015**, *7*, 4260–4279. [CrossRef]
126. Azhdari, A.; Sasani, M.A.; Soltani, A. Exploring the relationship between spatial driving forces of urban expansion and socioeconomic segregation: The case of Shiraz, Habitat International. *Habitat Int.* **2018**, *81*, 33–44. [CrossRef]
127. Abulibdeh, A.; Al-Awadhi, T.; Al-Barwani, M. Comparative analysis of the driving forces and spatiotemporal patterns of urbanisation in Muscat, Doha, and Dubai, Development in Practice. *Dev. Pract.* **2019**, *29*, 606–618. [CrossRef]

land

MDPI

Article

An International Library for Land Cover Legends: The Land Cover Legend Registry

Fatima Mushtaq [1,*], Matieu Henry [1], C. Douglas O'Brien [2], Antonio Di Gregorio [1], Rashed Jalal [1], John Latham [3,4], Douglas Muchoney [1], Chris T. Hill [4], Nicola Mosca [5], Michael Golmame Tefera [1], Karl Morteo [1], Gianluca Franceschini [1], Amit Ghosh [1], Elisee Tchana [1] and Zhongxin Chen [1]

1 Food and Agriculture Organization of the United Nations, 00153 Rome, Italy; matieu.henry@fao.org (M.H.); antoniodigregorio12@gmail.com (A.D.G.); rashed.jalal@fao.org (R.J.); douglas.muchoney@fao.org (D.M.); michael.tefera@fao.org (M.G.T.); karl.morteo@fao.org (K.M.); gianluca.franceschini@fao.org (G.F.); amit.ghosh@fao.org (A.G.); elisee.tchana@fao.org (E.T.); zhongxin.chen@fao.org (Z.C.)

2 IDON Technologies Inc., 1430 Prince of Wales Dr #38036, Ottawa, ON K2C 3Y7, Canada; douglasobrien@idontech.ca

3 British Standards Institution, London W4 4AL, UK; latham.john58@gmail.com

4 GeoData Institute, University of Southampton, Southampton SO17 1BJ, UK; cth@geodata.soton.ac.uk

5 Institute of Industrial Technologies and Automation (STIIMA), Italian, National Research Council, 00185 Rome, Italy; nicola.mosca@stiima.cnr.it

* Correspondence: fatima.mushtaq@fao.org

Abstract: Information on land cover is vital to numerous United Nations (UN) missions, including achieving the Sustainable Development Goals (SDGs). Because land cover data are developed by a variety of organizations for a range of objectives, they are based on different classification schemes and have discrepancies. In addition, the sustainability for land cover is hampered by limited access to information and documentation. Accordingly, international standards for land cover are developed to improve interoperability between different land cover datasets. However, the use and development of land cover datasets are limited by various factors including availability of properly documented land cover legends in support of different applications including change assessment, comparison, and international reporting. The purpose of this article is to highlight the importance of land cover in achieving several goals and to introduce the first international platform for land cover legend, named Land Cover Legend Registry (LCLR). This registry is a contribution to the international land cover community and the UN in effort to promote and support data harmonization processes and interoperability from local to global level and vice versa. Users can not only use the registry for preparing consistent datasets, but also contribute to it by providing the latest data to ensure the long-term availability of both updated and existing datasets around the world. Moreover, building on the experience developing land cover legends with different nations, a brief explanation on the preparation of legends is also provided. Additionally, it is more important than ever to develop land cover registers to support the use, expansion, integration, and use uptake of land cover data, particularly for innovative remote sensing, machine learning, and information and communication technologies and techniques that build on existing and national contexts.

Keywords: interoperability; standards; geospatial; semantic ontology; harmonization; classification

Citation: Mushtaq, F.; Henry, M.; O'Brien, C.D.; Di Gregorio, A.; Jalal, R.; Latham, J.; Muchoney, D.; Hill, C.T.; Mosca, N.; Tefera, M.G.; et al. An International Library for Land Cover Legends: The Land Cover Legend Registry. *Land* **2022**, *11*, 1083. https://doi.org/10.3390/land11071083

Academic Editor: Giusppe Pulighe

Received: 31 May 2022
Accepted: 29 June 2022
Published: 14 July 2022

Publisher's Note: MDPI stays neutral with regard to jurisdictional claims in published maps and institutional affiliations.

1. Introduction

The critical need for monitoring natural resources has increased over time because of many factors, including rapid growth in population and climate change [1]. The impacts of climate change are not limited to one country and extend beyond national and political administrative boundaries. An assessment presented in the Intergovernmental Panel on Climate Change (IPCC) report suggested that climate change has had severe effects on freshwater, terrestrial, and marine ecosystems, including alterations in seasonal timing with negative socioeconomic consequences, heat extremes with extinction of local species, forest degradation, hydrological changes, and agricultural productivity stress. This condition exacerbates the global food and water security crisis and obstructs progress toward reaching the SDGs [2].

Land cover information is a critical component in monitoring natural resources and is recognized by the UN as one of the fourteen fundamental data layers [3]. This information is essential to support many mandates of the UN, including the United Nations Framework Convention on Climate Change (UNFCCC), United Nations Convention to Combat Desertification (UNCCD), and United Nations Forum on Forests (UNFF) [4,5], as well as many other national, regional, and/or international initiatives. The land cover information serves as a critical baseline information for land, water, and/or hazard monitoring, including agriculture production [6], rapid crisis impact assessment on agriculture sector [7], food security [8], nutrition, environmental conservation and management, climate science, and many others.

Different organizations create land cover datasets to meet specific user needs utilizing a variety of classification schemes, tools, approaches, and datasets at the global [9–11], regional [12,13], and national levels [14]. Because different land cover classification schemes are used, the interoperability between land cover datasets and their compatibility and comparability are often very limited. Such consistency is required to aggregate and harmonize results to allow cross-comparison and validation for understanding regional and global landscape trends and/or climate changes [15,16], as well as for the sustainability of the land cover monitoring efforts. Several international, regional, and national organizations and agencies have emphasized the vital need for improved, consistent, and harmonized land cover statistics and spatial data from local to global levels to address the environmental and climatic issues at a larger scale.

In this context, geospatial technologies are important in the paradigm shift and transition to climate change initiatives, greener economies, natural resource conservation, sustainable carbon emissions, catastrophe effect assessment and management, and sustainable agriculture production, among other things [17–22]. These technologies play a crucial role in the development of land cover maps and datasets using integrated efficient and cost-effective approaches including remote sensing and machine learning [14]. The use of standardized information is required for the proper processing of data and information to profit from recent technological advancements in geospatial fields.

To overcome the inconsistencies in generic classification systems, dozens of countries and international organizations have now created land cover datasets using the Food and Agriculture Organization of the United Nations (FAO) Land Cover Classification System (LCCS) based on Land Cover Meta Language (LCML-ISO 19144-2) [14]. In LCML, the land cover classes are defined using a hierarchical structure [16,23]. This system does not intend to alter or replace the previous classification systems. Nonetheless, it provides a consistent framework for the comparison and integration of data for any generic land cover classification [23], building upon the involvement of FAO in several countries. Therefore, these datasets can be used for integration and harmonization processes to analyze the results and changes at the global, regional, and national levels.

The role of the land cover legends is the initial step in the preparation of land cover maps and datasets [15]. For this, proper and complete documentation, as well as information on the land cover classes in a land cover legend, including clear and unique description of the class and/or photo-keys using field photographs and/or satellite imageries, is required. Therefore, to fill the gap, development of land cover registers is needed where information can be stored and be easily accessible to the user.

Development of land cover registers are needed (1) to facilitate the broader use and expansion of the land cover register, (2) to create sustainable ownership and technical capacity to use the register for land cover translation purposes, and (3) to support the migration of the register to incorporate new land cover and land use schemas. Accordingly, the first international platform, named the Land Cover Legend Registry (LCLR), was developed to support cataloguing land cover legend descriptions based on international standards. The register is meant to host a curated list of legends created at the global, regional, and national levels. Users can easily and freely access and download land cover legends, their datasets, and relevant documents.

The rest of the article is organized as follows. Section 2 reports on a survey of land cover and a few key findings that prompted the development of the LCLR. The focus then shifts to the latter, reporting on some aspects, like standardization, that have influenced its architecture, as well as its place in a growing ecosystem of tools developed by FAO in collaboration with other international institutions. This is followed by a discussion of important aspects that LCLR helps to address, before moving to the conclusions.

2. Land Cover Legend Registry Development

2.1. Land Cover User Need Assessment

The work on the development of the registry was inspired by a survey on the importance of land cover in different domains and applications, including change assessment, comparison, international reporting, and others at different levels. Various organizations contributed to the development of the survey questionnaire, including the FAO; Organization for Economic Co-operation and Development (OECD); United Nations Statistics Division (UNSD); Basque Centre for Climate Change (BC3Research); United States Geological Survey (USGS); National Research Council, Italy (CNR); and University of Southampton (Soton). The main aims of this survey were (1) to assess the user requirement on the use of land cover information in the context of SDGs; (2) to present limitations in the development of land cover; and (3) to contribute to the establishment of international standards, i.e., LCML and others.

Numerous academic, scientific, government, and public sector entities from the national, regional, and international levels participated in the survey, particularly from the environmental, agricultural, and climate sectors. The results revealed that land cover information is very important and is used for all SDGs at different levels and dimensions, particularly for climate action (SDG 13) and life on land (SDG 15) (Figure 1) [24,25]. Moreover, the results indicated that the lack of documentation and information on land cover and unavailability of existing datasets are the main issues in the development of consistent and/or harmonized land cover maps.

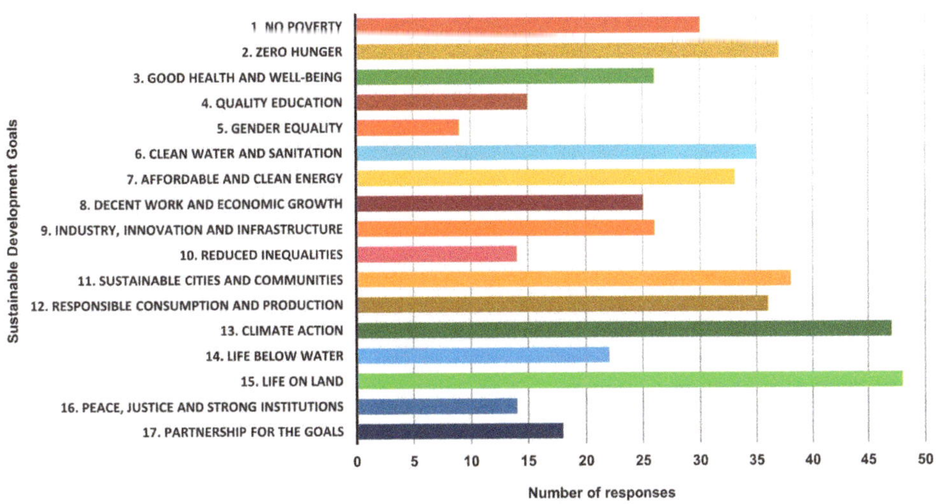

Figure 1. Land cover in the Sustainable Development Goals. Reprinted/adapted with permission from Ref. [25]. 2022, FAO and ISO/TC211 AG13.

2.2. *Land Cover Database*

A registry database was developed for land cover legends based on the registry concepts that is derived from ISO 19135-1 [26], identified in ISO 19144-1 [27], and makes use of the descriptive metalanguage described in ISO 19144-2. The LCLR was internally developed as a spreadsheet application (i.e., Microsoft EXCEL) and is a flat table linked through a primary key with different sub-registers. After a set of postprocessing, it is presented in a more compelling form. The registry consists of three main parts, i.e., (1) metadata describing the whole register (content), (2) a description of the meaning of each item in the register (content description), and (3) the registered items (categorization) (Figure 2) [28].

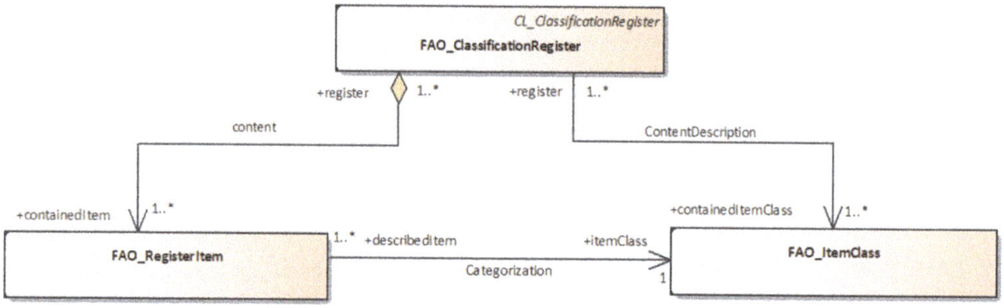

Figure 2. Structure of the land cover legend registry. Reprinted/adapted with permission from Ref. [28]. 2021, FAO, UoS and STIIMA-CNR. * Multiplicity (https://khalilstemmler.com/articles/uml-cheatsheet/, accessed on 15 June 2022): a classification register can have 1 or more register/RegisterItem/ItemClass.

The registry contains information about the land cover legend, land cover legend classes, land cover dataset, and relevant reference documentation. Several of the land cover legends in the database were prepared and translated from the original classification system into ISO 19144-2 standard using LCCS software [29]. Moreover, this database also contains land cover legends in different classification systems but marked as not yet translated. The registry is multilingual to support the adaptability of systems in local and/or national languages for data integration, comparison, and many other purposes.

2.3. *Land Cover Legends*

The land cover legends provided in land cover legend registry are at the global, regional, national, and sub-national levels. Legends are available in different formats that are used worldwide. The different legend file formats are CSV, LCCS, EAPX, HTM, and XSD, and are devised to be used under different software and platforms. Furthermore, the land cover legends in the registry were created to account for a variety of scenarios, including the following:

1. Country- or region-specific legend available and translated to LCML, e.g., non-irrigated arable land [30] translated into LCML using LCCS version three (LCCS3) software. This class was translated as "herbaceous growth forms" with the characteristics "cultivated and managed" and "rainfed". With an "optional" presence type of the LCML element "woody growth form" defined by a standard characteristic as "Orchard and other Plantation" (Figure 3);
2. Land cover legend available and translated from version 2 to version 3 using LCCS3 software, e.g., tree crop class from Himalaya land cover legend [31] (Figure 4);
3. A new legend prepared using LCCS3, e.g., "evergreen hill forest" class [32] (Figure 5).

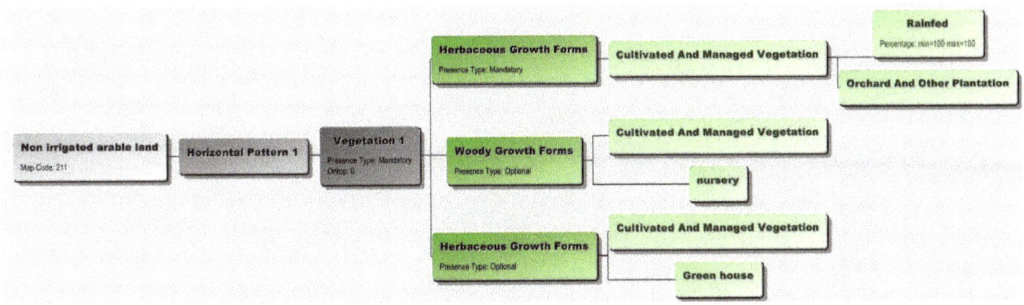

Figure 3. Non-irrigated arable land class in CORINE land cover legend using the LCCS3 tool.

Table	Structural Domains LCC Code	Mode	Level	Classifiers	User's Label	Land Cover Class Name	Map Code	User's Description
1	Cultivated and Managed Terrestrial Area(s)							
1	10001-W8	0	1	A1-W8	Tree Crop	Tree Crop(s) Crop cover Orchard(s)	1T	Tree Crop

(**A**)

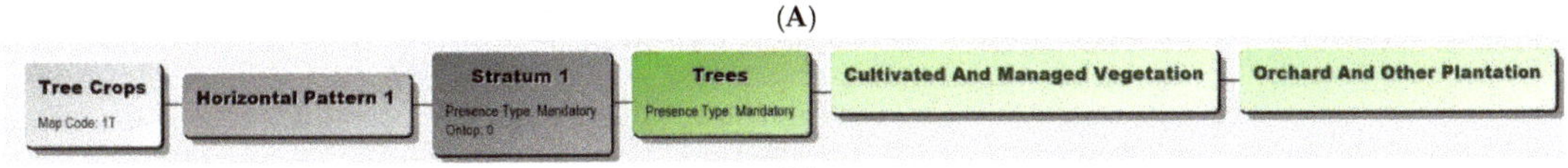

(**B**)

Figure 4. Himalaya land cover legend. (**A**) Land cover class "tree crop" using LCCS2 and (**B**) land cover class "tree crop" using LCCS3.

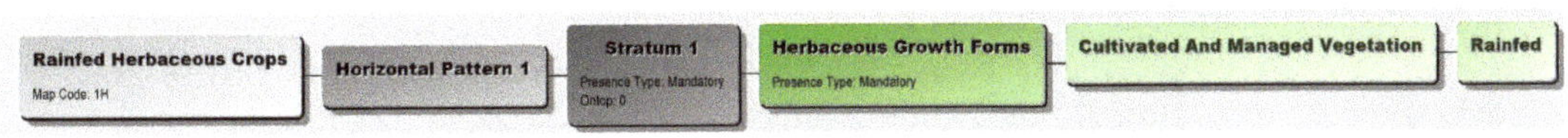

Figure 5. Rainfed herbaceous cropland class using the LCCS3 tool.

2.4. Land Cover Legend Registry Platform

The open-source LCLR platform is an online library for land cover legend and relevant products. This platform is available on the FAO Hand-In-Hand geospatial platform [33] and data can be downloaded directly in different formats. Meta data for this registry are available on the FAO CKAN [34]. The datasets on the LCLR platform are updated weekly. A user can download land cover legends in provided file formats, i.e., LCCS, CSV, EAPX, HTM, XSD, and so on. Land cover classes are prepared based on Unified Modeling Language (UML) and are available in JPEG format for download. Relevant land cover datasets are available in raster and/or vector format. All reference data for relevant land cover legends are available in PDF format. A list of land cover legends that are currently available on this platform is provided in Figure 6.

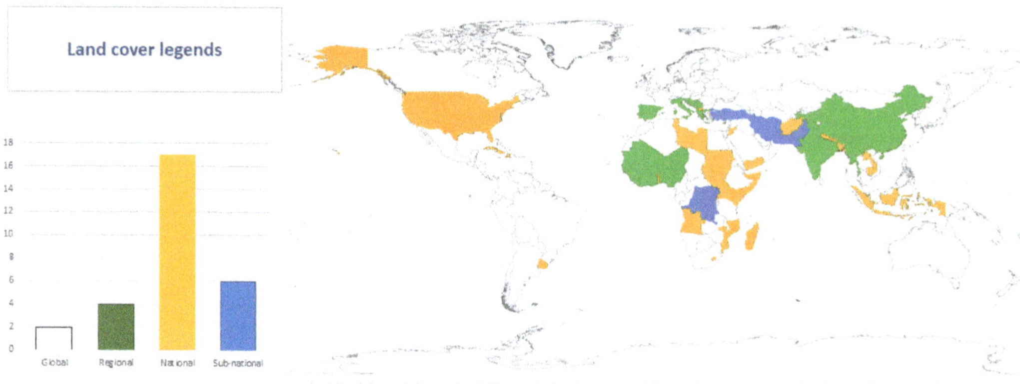

Figure 6. Status of land cover legends in the land cover legend registry platform using the LCCS3 tool.

2.5. Connectivity to Other Platforms

The Land Cover Legend Registry (LCLR) platform has the potential to link to the cloud computing platforms including System for Earth Observation Data Access, Processing, and Analysis for Land Monitoring (SEPAL) [35] and Google Earth Engine (GEE), as well as other desktop software like Enterprise Architect and many others. For example, the land cover legend in CSV format after downloading from the LCLR platform can be directly uploaded into the SEPAL platform to classify the satellite images for land cover preparation (Figure 7) using machine learning techniques including random forest (RF), support vector machine (SVM), and others.

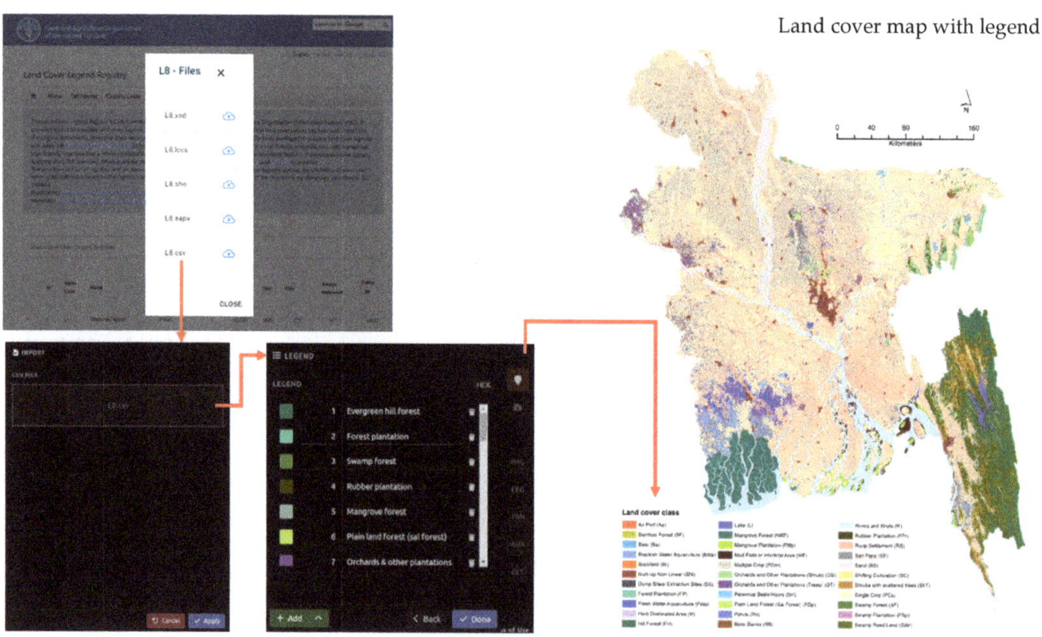

Figure 7. Land cover legend registry connectivity to SEPAL for land cover mapping.

2.6. Availability of Land Cover Datasets in the Registry

The availability of land cover datasets is as important as the development of datasets. There are several organizations and countries that have developed and are developing land cover datasets for different purposes and using different approaches and supporting documents. In most cases, land cover legends have limited documentation and the land cover legend registry provides only land cover legends that have been prepared using LCCS3. For example, the national land cover map for Bangladesh 2015 was prepared based on the LCML using the LCCS3 tool [34,36]. This interoperable system and the land cover dataset are used for a variety of applications including national forest resources assessment, estimation of REDD+ activity data, integration of biophysical and socioeconomic information, and semantic similarity assessment [14]. The land cover legend for Bangladesh is available in the registry along with a dataset and reference document. Moreover, land cover maps for Jordan [37], Afghanistan [38], and many others are also available in the registry. Therefore, a user can access existing and latest land cover datasets using this registry and can further use it for their purpose.

3. Discussion

3.1. Land Cover Data for All

Although there are several global and regional products and platforms that provide land cover products, accessibility in a user-friendly manner is always a hurdle. There is no platform that can provide updated information on existing land cover legends at the global, regional, national, or local levels. The LCLR, as an online library for land cover legend, aims at closing this gap by ensuring its wider use and accessibility by anyone around the world. This registry is based on international standards that are widely recognized by land cover experts. LCLR has potential to be used by the land cover community for different purposes and sectors including agriculture and food security monitoring, land and water resources assessment, environmental accounting, land use planning, and emergency reporting. Furthermore, there is potential for the scientific community to contribute to improve land cover at local to international level by following LCML and contributing to using or providing new legends and datasets to LCLR.

3.2. Semantic Interoperability and Comparison

Because traditional land cover legends contain symbols that are symbolic, brief, and vague, determining land cover similarity has always been difficult. The inconsistency of classification systems continues to hinder how the world is represented and managed, despite the advancements in geospatial technology that give access to new images, tools, and methods. In order to normalize geographic representation of our surroundings, nomenclatures have been developed over time [30,39], but they have not addressed the issue of sufficient representation of land cover semantic meaning.

In this context, under ISO TC211, 19,144 series comprise the set of standards that are meta languages for addressing different classification systems and approaches. For example, ISO 19144-2 (LCML) on land cover provides a common reference structure for comparison and integration of different data. It is not intended to replace any classification system. The translation of classification systems from national systems can support data integration as integration of data is required to address regional and worldwide requirements.

Classes that are coded using the LCML syntax can be easily used for land cover similarity assessment analysis. Exploiting the intrinsic modularity of the LCML standard, the similarity between land cover classes can be assessed quantitatively [15]. For instance, hosted in the Bangladesh Forest Information System (BFIS) geoportal, an object-based methodology is operationalized to make an automatic similarity assessment between LCML-derived classes present in different databases. Therefore, using LCLR, a user can use the LCML syntax-based legend information and compare the results with original datasets and/or new datasets.

3.3. Connectivity, Multi-Languages, and Multiple Formats

Land cover legend data are available in different file formats and in different languages from where a user can download data directly from the platform under the 'file section' and can be used in various platforms and tools, cloud computing, as well as desktop software, e.g., SEPAL platform, Enterprise Architect (EA), and so on. Legends in different languages make the comparability at national context more understandable. This registry has the potential to add more languages and file formats in the future. A user can use and contribute to the registry by providing legends in local language and different file formats.

3.4. Sustainability of the Registry—Future Aspect

There has been tremendous work on the development of land cover classification schemes, tools, and methods to support the land resource monitoring and to develop efficient and consistent land cover maps all around the world. The development of registers, in this context, can contribute to the international community by providing the latest existing land cover legends. For this, the first international register based on international standards has been developed [28]. In order to sustain the registry, a user can contribute by providing land cover legend using the LCCS3 tool, as described in Section 2. Moreover, a user can also provide a land cover legend in local classification schemes that can be translated using LCCS3 and will be accessible through this platform.

For the sustainability of land cover registers, existing and/or new, there is a need to develop a global register that incorporates all of the existing register to ensure the data availability to users.

4. Conclusions

Consistency in land cover datasets at national level is crucial to many applications that can be used and integrated in the analysis of local to global issues and trends. Land cover registers can fill the gaps, including limitations in data sharing and accessibility to the user. LCLR is the first international online catalogue that provides complete information on land cover legend at different levels. The registry supports multiple languages and multiple formats and is easily adaptable from local to global systems. The development of this registry is the first step towards harmonizing the different schemes using LCML and providing a platform that can facilitate end users at its best in the most common ways. It aims to support various international initiatives like ISOTC211 AG13, the FAO Hand in Hand (HIH) initiative, Global Agro Ecological Zoning (GAEZ) [40], WaPOR [41], SEPAL, and so on by providing the LCML-based land cover legend.

Using LCLR, a user can not only access a land cover legend with a proper description but can also contribute to populating the registry to ensure the sustainability of the system by engaging land cover community all around the world. They can also support the process of data transparency, consistency, and comparability through efficient ways. Moreover, this LCLR can contribute to development of harmonized land cover legends and datasets at various levels including global. Also, there is a possibility in the future to develop a registry for land cover or land use using multi-model database management system that can incorporate all existing and/or new registers with a sustainable framework and ownership.

Author Contributions: Data curation, F.M., M.H., A.D.G. and E.T.; Formal analysis, F.M., N.M. and E.T.; Funding acquisition, M.H.; Investigation, F.M., M.H., A.D.G. and E.T.; Methodology, M.H., A.D.G., C.D.O. and J.L.; Project administration, M.H.; Resources, G.F. and J.L.; Software, M.G.T., K.M., Z.C. and N.M.; Supervision, M.H., A.D.G. and C.D.O.; Validation, M.H., A.D.G. and C.D.O.; Visualization, F.M., N.M. and M.G.T.; Writing—original draft, F.M., M.H., A.D.G. and C.D.O.; Writing—review and editing, F.M., M.H., C.D.O., A.D.G., R.J., J.L., D.M., N.M., C.T.H., M.G.T., K.M., G.F., A.G., E.T. and Z.C. All authors have read and agreed to the published version of the manuscript.

Funding: This research received no external funding. The APC was funded by FAO.

Institutional Review Board Statement: Not applicable.

Informed Consent Statement: Not applicable.

Data Availability Statement: All the land cover legend can be downloaded from https://www.fao.org/hih-geospatial-platform/en/resources/land-cover-legend-registry/ (accessed 15 June 2022).

Acknowledgments: The authors would like to thank the anonymous reviewers for the comments that helped improve the manuscript.

Conflicts of Interest: The authors declare no conflict of interest.

References

1. Jhariya, M.K.; Banerjee, A.; Meena, R.S. Importance of Natural Resources Conservation: Moving toward the Sustainable World. In *Natural Resources Conservation and Advances for Sustainability*; Elsevier: Amsterdam, The Netherlands, 2022; pp. 3–27. [CrossRef]
2. The Intergovernmental Panel on Climate Change (IPCC). Climate Change 2022—Impacts, Adaptation and Vulnerability—Summary for Policymakers. 2022. Available online: https://www.ipcc.ch/report/ar6/wg2/ (accessed on 1 June 2022).
3. United Nations Committee of Experts on Global Geospatial Information Management (UN-GGIM). The Global Fundamental Geospatial Data Themes. 2019. Available online: https://ggim.un.org/documents/Fundamental%20Data%20Publication.pdf (accessed on 30 May 2022).
4. Organisation for Economic Cooperation and Development (OECD). *Towards Sustainable Land Use: Aligning Biodiversity, Climate and Food Policies*; OECD Publishing: Paris, France, 2020; p. 158. [CrossRef]
5. United Nations Framework Convention on Climate Change (UNFCCC). Adoption of the Paris Agreement. United Nations Climate Change. 2015. Available online: https://unfccc.int/resource/docs/2015/cop21/eng/l09r01.pdf (accessed on 1 May 2022).
6. De Wrachien, D. Land Use Planning: A Key to Sustainable Agriculture. In *Conservation Agriculture*; Springer: Dordrecht, The Netherlands, 2003; pp. 471–483. [CrossRef]
7. Food and Agriculture Organization of the United Nations (FAO). *Impact of the May Conflict Escalation on the Agricultural Area in the Gaza Strip*; Food and Agriculture Organization of the United Nations: Rome, Italy, 2021. [CrossRef]
8. Agidew, A.; Meta, A.; Singh, K.N. The Implications of Land Use and Land Cover Changes for Rural Household Food Insecurity in the Northeastern Highlands of Ethiopia: The Case of the Teleyayen Sub-Watershed. *Agric. Food Secur.* **2017**, *6*, 56. [CrossRef]
9. Latham, J.; Cumani, R.; Rosati, I.; Bloise, M. *Global Land Cover SHARE (GLC-SHARE) Database Beta-Release Version 1.0*; Food Agriculture Organization of the United Nations (FAO): Rome, Italy, 2014; Volume 29, pp. 1–39.
10. European Space Agency (ESA). ESA World Cover 10m 2020 V100. 2021. Available online: https://worldcover2020.esa.int/download (accessed on 4 May 2022).
11. Friedl, M.; Gray, J. *MCD12Q2 MODIS/Terra+Aqua Land Cover Dynamics Yearly L3 Global 500m SIN Grid V006*; NASA EOSDIS Land Processes DAAC: Washington, DC, USA, 2019. [CrossRef]
12. Sahara and Sahel Observatory (OSS). *Sahel and West Africa—Atlas of Land Cover Maps—Strengthening Resilience through Services Related to Innovation, Communication and Knowledge—BRICKS (Benin, Burkina Faso, Chad, Ethiopia, Ghana, Mali, Mauritania, Niger, Nigeria, Senegal, Sudan and Togo*; Sahara and Sahel Observatory (OSS): Tunis, Tunisia, 2019.
13. Comité Permanent Inter-états de Lutte contre la Sécheresse dans le Sahel (CILLS). *Landscapes of West Africa: A Window On a Chaning World. Comité Permanent Inter-états de Lutte contre la Sécheresse dans le Sahel (CSS)*; U.S. Agency for International Development (USAID): Washington, DC, USA, 2016.
14. Jalal, R.; Golam Mahboob, M.; Udita, T.S.; Aziz, T.; Masum, S.M.; Costello, L.; Saha, C.R.; Chowdhury, A.A.M.; Salam, A.; Shahrin, F.; et al. Toward Efficient Land Cover Mapping: An Overview of the National Land Representation System and Land Cover Map 2015 of Bangladesh. *IEEE J. Sel. Top. Appl. Earth Obs. Remote Sens.* **2019**, *12*, 3852–3861. [CrossRef]
15. Mosca, N.; Di Gregorio, A.; Henry, M.; Jalal, R.; Blonda, P. Object-Based Similarity Assessment Using Land Cover Meta-Language (LCML): Concept, Challenges, and Implementation. *IEEE J. Sel. Top. Appl. Earth Obs. Remote Sens.* **2020**, *13*, 3790–3805. [CrossRef]
16. Herold, M.; Latham, J.S.; Di Gregorio, A.; Schmullius, C.C. Evolving Standards in Land Cover Characterization. *J. Land Use Sci.* **2006**, *1*, 157–168. [CrossRef]
17. Avtar, R.; Aggarwal, R.; Kharrazi, A.; Kumar, P.; Kurniawan, T.A. Utilizing Geospatial Information to Implement SDGs and Monitor Their Progress. *Environ. Monit. Assess.* **2020**, *192*, 35. [CrossRef]
18. Mushtaq, F.; Ghosh, A.; Jalal, R.; Gauny, J.; Barelli, D.; Moloinyane, S. *A Rapid Geospatial Analysis of the Impact of in Madagascar, Malawi, Mozambique, Zambia, and Zimbabwe*; Food Agriculture Organization of the United Nations (FAO): Rome, Italy, 2022. [CrossRef]
19. Schulz, D.; Yin, H.; Tischbein, B.; Verleysdonk, S.; Adamou, R.; Kumar, N. Land Use Mapping Using Sentinel-1 and Sentinel-2 Time Series in a Heterogeneous Landscape in Niger, Sahel. *ISPRS J. Photogramm. Remote Sens.* **2021**, *178*, 97–111. [CrossRef]
20. Chen, B.; Xu, B.; Gong, P. Mapping Essential Urban Land Use Categories (EULUC) Using Geospatial Big Data: Progress, Challenges, and Opportunities. *Big Earth Data* **2021**, *5*, 410–441. [CrossRef]
21. Chiesi, M.; Maselli, F.; Bindi, M.; Fibbi, L.; Cherubini, P.; Arlotta, E.; Tirone, G.; Matteucci, G.; Seufert, G. Modelling Carbon Budget of Mediterranean Forests Using Ground and Remote Sensing Measurements. *Agric. For. Meteorol.* **2005**, *135*, 22–34. [CrossRef]

22. Neeff, T.; De Alencastro Graça, P.M.; Dutra, L.V.; Da Costa Freitas, C. Carbon Budget Estimation in Central Amazonia: Successional Forest Modeling from Remote Sensing Data. *Remote Sens. Environ.* **2005**, *94*, 508–522. [CrossRef]
23. Di Gregorio, A. *Land Cover Classification System—Classification Concept*; Food Agriculture Organization of the United Nations (FAO): Rome, Italy, 2016. Available online: https://www.fao.org/3/i5232e/i5232e.pdf (accessed on 31 May 2022).
24. *ISO/TC211/AG13*; Interim Report on Land Cover and Land Use Use Cases (Part 2), Report. Advisory Group 13 on Land cover and Land Use. International Organization for Standardization (ISO): Geneva, Switzerland, 2022. Available online: https://sd.iso.org/documents/open/338e880a-15cd-44da-b237-18939815016a (accessed on 31 May 2022).
25. *Interim Report on Land Cover and Land Use Use Cases (Part 1), Report. Advisory Group 13 on Land cover and Land Use*; International Organization for Standardization (ISO): Geneva, Switzerland, 2022. Available online: https://sd.iso.org/documents/open/94 4d4465-a240-4146-9a64-389afa22f1a3 (accessed on 31 May 2022).
26. ISO. *Geographic Information—Procedures for Item Registration—Part 1: Fundamentals*; International Organization for Standardization (ISO): Geneva, Switzerland, 2015. Available online: https://www.iso.org/standard/54721.html (accessed on 15 May 2022).
27. *ISO19144-1:2009-ISO/TC211*; Geographic Information—Classification Systems—Part 1: Classification System Structure. International Organization for Standardization (ISO): Geneva, Switzerland, 2009. Available online: https://www.iso.org/standard/3256 2.html (accessed on 31 May 2022).
28. O'Brien, D.C.; Di Gregorio, A.; Mushtaq, F.; Henry, M.; Tchana, E.; Mosca, N.; Blonda, P.; Hill, C.; Latham, J.; Muchoney, D. *Register Implementation for Land Cover Legends*; Food Agriculture Organization of the United Nations (FAO): Rome, Italy, 2021. [CrossRef]
29. Di Gregorio, A. *Land Cover Classification System—User Manual*; Food Agriculture Organization of the United Nations (FAO): Rome, Italy, 2016. Available online: https://www.fao.org/3/i5428e/i5428e.pdf (accessed on 31 May 2022).
30. European Environment Agency. Corine Land Cover 2018 (Vector)—Version 20. 2019. Available online: https://sdi.eea.europa.eu/catalogue/geoss/api/records/53ef1493-e7a1-4216-b043-87a7c2a5a68d (accessed on 18 May 2022).
31. Food and Agriculture Organization of the United Nations (FAO). Land Cover of Himalaya Region—FAO Map Catalog. Available online: https://data.apps.fao.org/map/catalog/srv/eng/catalog.search#/metadata/46d3c2ef-72c3-4f96-8e32-40723cd1847b (accessed on 18 May 2022).
32. GoB. *Land Cover Atlas of Bangladesh 2015*; Forest Department, Ministry of Environment, Forest and Climate Change, Government of the People's Republic of Bangladesh: Dhaka, Bangladesh, 2020. Available online: http://bfis.bforest.gov.bd/library/land-cover-atlas-of-bangladesh-2015/ (accessed on 30 May 2022).
33. Food and Agriculture Organization of the United Nations (FAO). Land Cover Legend Registry (LCLR). Available online: https://www.fao.org/hih-geospatial-platform/en/resources/land-cover-legend-registry/ (accessed on 31 May 2022).
34. Food and Agriculture Organization of the United Nations (FAO). FAO-LCLR—Datasets—"FAO Catalog". Available online: https://data.apps.fao.org/catalog/dataset/land-cover-legend-registry (accessed on 31 May 2022).
35. Food and Agriculture Organization of the United Nations (FAO). SEPAL Recipes—SEPAL Documentation. Available online: https://docs.sepal.io/en/latest/cookbook/index.html (accessed on 31 May 2022).
36. GoB. Land Representation of Bangladesh (In Support of REDD+) Forest Department, Ministry of Environment, Forest and Climate Change, Government of the People's Republic of Bangladesh. 2017. Available online: http://bfis.bforest.gov.bd/library/land-representation-system-of-bangladesh/ (accessed on 31 May 2022).
37. Food and Agriculture Organization of the United Nations (FAO). *Land Cover Atlas of Jordan*; FAO: Rome, Italy, 2019; Volume 2012, Available online: https://www.fao.org/geospatial/resources/detail/en/c/1202563/ (accessed on 31 May 2022).
38. Food and Agriculture Organization of the United Nations (FAO). *The Islamic Republic of Afghanistan—The Land Cover Atlas*; FAO: Rome, Italy, 2016. Available online: https://www.fao.org/publications/card/en/c/cc0ac143-38ed-41f0-b7c6-2342ffa7f0e6/ (accessed on 31 May 2022).
39. Comber, A.J.; Fisher, P.F.; Wadsworth, R.A. Land Cover: To Standardise or Not to Standardise? Comment on 'Evolving Standards in Land Cover Characterization' by Herold et Al. *J. Land Use Sci.* **2008**, *2*, 283–287. [CrossRef]
40. Global Agro Ecological Zones (GAEZ). *Towards Efficient and Sustainable Food and Bio-Energy Security in Thailand Global Agro Ecological Zones Key Results*; Food Agriculture Organization of the United Nations (FAO): Rome, Italy, 2022; pp. 1–2. Available online: https://www.fao.org/documents/card/en/c/CB4766EN/ (accessed on 31 May 2022).
41. Food and Agriculture Organization of the United Nations (FAO). Remote Sensing for Water Productivity | Food and Agriculture Organization of the United Nations. Available online: https://www.fao.org/in-action/remote-sensing-for-water-productivity/resources/publications/wapor-publications/en/ (accessed on 27 June 2022).

land

MDPI

Article

Monitoring Net Land Take: Is Mainland Portugal on Track to Meet the 2050 Target?

Rita Nicolau [1,*] and Beatriz Condessa [2]

1 DGT—Direção-Geral do Território, 1099-052 Lisboa, Portugal
2 IST—Center for Innovation in Territory, Urbanism and Architecture (CITUA), Department of Civil Engineering, Instituto Superior Técnico, Universidade de Lisboa, 1049-001 Lisboa, Portugal; beatriz.condessa@tecnico.ulisboa.pt
* Correspondence: rnicolau@dgterritorio.pt

Abstract: This study discusses the situation of mainland Portugal concerning the achievement of the European "no net land take" target by 2050. This target aims to curb land take by increasing the recycling of developed land and offsetting the consumption of undeveloped land by re-naturalizing an equivalent area of artificial land. Setting targets and interventions in each country to reach this goal requires monitoring land take, developed land recycling, and re-naturalization. This study assesses these processes in mainland Portugal, by NUTS III regions, for the first time, analyzing the land cover/land use changes that occurred between 2007 and 2018. In this period, the land take rate in mainland Portugal amounted to 7.2 ha/day. Re-naturalization and recycling of developed land were 1.0 and 0.2 ha/day, respectively, showing the shortage of their practices on the mainland. During said period, mainland Portugal and most of its regions experienced a reduction in population and an increase in artificial land, revealing low efficiency in urban land use. Since Portuguese legislation does not mention the European target, we believe that most decision-makers are unaware of it or have little knowledge of the practices that can contribute to its fulfillment. In this regard, the study aims to raise awareness among policymakers and public authorities about the need to limit, mitigate, and compensate for land take and to set land take targets for different levels of action. In addition, it describes how some European countries and regions are evaluating the same processes and approaching the goal under consideration.

Keywords: land consumption; urbanization; land re-naturalization; developed land recycling; urban land use efficiency

Citation: Nicolau, R.; Condessa, B. Monitoring Net Land Take: Is Mainland Portugal on Track to Meet the 2050 Target? *Land* **2022**, *11*, 1005. https://doi.org/10.3390/land11071005

Academic Editor: Giusppe Pulighe

Received: 25 May 2022
Accepted: 29 June 2022
Published: 1 July 2022

1. Introduction

In recent decades, land as an environmental resource has gained increasing importance in European policies. The goal of "no net land take" by 2050 defined in the 2011 Roadmap to a Resource Efficient Europe [1] is part of the European Union's 7th Environmental Action Program [2] and is also part of the Soil Strategy of the European Union for 2030 [3]. The strategy recommends that Member States set their own national, regional, and local targets by 2023 for net land take reduction by the year 2030 and report on their progress in order to make measurable contributions to the EU's 2050 target. The strategy also proposes the application of "land take hierarchy" principles to increase land-use efficiency through more effective land reuse—promoting reductions in land take and soil sealing.

Land take is "the loss of undeveloped land to human-developed land" [4] (p. 4). Land take can also be defined as the loss of natural and semi-natural land to urban and other artificial land developments. Accordingly, it is also known as land consumption or land artificialization [5,6]. It includes the development of areas that are sealed by construction and urban infrastructure, and non-sealed areas, such as urban green parks and some sports and leisure facilities [7]. Land take leads to the loss of natural capital and landscape fragmentation [7,8]. Soil sealing is considered the most harmful expression of land take because it is a generally irreversible process that reduces habitat space and compromises the soil's ability to provide important ecosystem services, such as biomass supply, water and nutrient cycling, and organic carbon storage [9,10]. The loss of soil's ecological functions triggers heat island effects and flooding, and can even increase soil, water, and air pollution [7,11].

The main drivers of land take are population growth and economic development [12,13]. The former is mainly related to a need for more housing, public facilities, and transportation; the latter relates to new industrial and commercial sites, the competition between municipalities to attract more investment, and the prioritization of economic development over environmental sustainability, in a broad sense. Colsaet et al. [12] carried out an extensive literature review on this topic, concluding that urban expansion "is not a mere result of market forces but is also shaped by institutions and public policies" (p. 346), including land use regulation, the legal and fiscal framework, the lack of both vertical (between administrative levels) and horizontal (between municipalities) coordination, and the strong dependence of local governments on tax revenues (such as property taxes).

The reverse process of land take (the conversion of artificial land into non-artificial land) is generally known as the re-naturalization or restoration of developed land. It is also less commonly referred to as the re-cultivation of developed land [7]. An increase in vegetation cover provided by re-naturalization translates into an increase in soil carbon stocks [14].

Net land take evaluates, for a given period, the difference between land take and re-naturalization of developed land. The "no net land take" target aims to protect soils and safeguard the services they provide through more sustainable land use, which involves reducing the consumption of undeveloped land [15]. Meeting this target by the year 2050 calls for new construction to take place on abandoned or underutilized urban land and for the non-artificial land consumed by urbanization be offset by re-naturalizing an equivalent amount of artificial land, which requires significant investment in developed land recycling [7]. Recycling of developed land aims at increasing the density of buildings (densification or infilling), building on abandoned or underused land (gray recycling), or converting developed land into green urban areas (green recycling). Both densification and gray recycling allow nations to respond to the housing needs of a growing population and diversify economic activity in areas that are already part of the urban perimeter, maximizing the use of existing infrastructure. Green recycling allows for preserving some of the natural soil functions that are vital to urban sustainability. Proper planning of green urban areas can contribute to the development of green infrastructure [7].

It should be noted that land designated herein as urban or developed refers to the Corine Land Cover (CLC) level 1 class known as artificial surfaces, and land designated as non-urban, undeveloped, or natural refers to the remaining CLC classes.

1.1. Framing the "No Net Land Take" Goal

Given the lack of a binding legal framework to achieve the goal of "no net land take" by 2050, the European Union has sought to raise awareness among Member States on the issues of land take and recycling of developed land by means of the dissemination of various studies [7,11,16,17], and more recently, through inclusion of said issues in the Soil Strategy of the European Union for 2030 [3]. Over recent decades, most European countries have adopted strategic planning guidelines to promote the sustainable development of their territories [5], following the principles of the European Spatial Development Perspective [18], the Territorial Agendas of the European Union [19–21], and the New Urban Agenda [22].

At the global level, the 2030 Agenda for Sustainable Development established 17 Sustainable Development Goals and 169 targets, which should be achieved by the year 2030 by all countries [23]. Targets 11 and 15 deal with land consumption and land degradation, respectively, which are both related to net land take. Target 11.3 aims, among other things, to improve inclusive and sustainable urbanization in all countries. The indicator chosen for monitoring this (11.3.1) is the "ratio of land consumption rate to population growth rate." As land take reflects the consumption of non-artificial land for urban development, it constitutes part of land consumption. The other part comprises the consumption of artificial land, which in this study corresponds to developed land recycling and re-naturalization. Target 15.3 aims, among other things, to combat desertification and restore degraded land. The proposed indicator for the monitoring thereof (15.3.1) measures the proportion of land that is degraded, which is also related to net land take. Land degradation can be assessed by the loss of ecosystem services provided by soil, such as a decrease in soil carbon stocks [24], and the re-cultivation of developed land usually increases these stocks [14].

1.2. *Types of Interventions to Avoid, Reduce, or Offset Land Take*

Based on the analysis of over 200 case studies from different European territorial contexts, the ESPON SUPER project (Sustainable Urbanization and Land Use Practices in European Regions) has identified the following types of interventions that can contribute to avoiding, reducing, and/or compensating for land take: densification, regeneration, urban containment, and governance and sectoral policies [25,26].

Densification aims at increasing the density of people living in built-up areas, usually by defining zones and quantitative thresholds for the closure of gaps. This may involve an increase in building volume or the reorganization of the existing urban structure [26].

Regeneration aims at reusing and improving abandoned and problematic sites, such as brownfields. Brownfields are urban areas that have been abandoned or are underused, such as former industrial sites, which may have contamination problems.

Containment initiatives aim to limit land development beyond a certain area, to reduce urban sprawl and promote more rational land use (e.g., green belts and urban growth boundaries). Normally, these interventions foster the redevelopment and densification of urban neighborhoods and improve the green spaces in the rural-urban interface.

Governance approaches to reduce land take can encompass policy goal setting, spatial planning at different levels or sectors of action, financial incentives, and environmental assessments of plans and projects aimed at urban development [27]. According to the SUPER conclusions [26], interventions that address various sectoral policies (e.g., transportation, environment, and agriculture) and their potential impacts on land use tend to promote more sustainable development.

The types of interventions mentioned can make use of more than one instrument, such as legislation (binding laws and regulations), zonal land-use regulations establishing binding principles, incentive and economic programs (policy packages aimed at a particular objective), and projects (e.g., those implemented under the URBACT III program) [25].

Given that strategies or visions are non-binding instruments, their success often depends on the existence of financial or binding instruments that make it possible to achieve targets. Accordingly, the adoption of strategies does not in itself guarantee successful interventions.

1.3. *Data Sources Used at the European Level to Estimate Land Take, Re-Naturalization, and Recycling of Developed Land*

Assessment of land cover/land use changes associated with the land take, re-naturalization, and recycling of developed land, can be achieved through a wide variety of data sources. The net land take indicators produced by the European Environment Agency (EEA) [7,11] for Europe are based on CLC maps, which are inadequate for monitoring these processes at the level of countries and their regions, as they do not make it possible to identify change areas of less than 5 ha [28]. The indicators for developed land recycling produced by the EEA are based on the Urban Atlas [29], which has a higher spatial resolution than CLC but only covers medium to large cities and their functional areas, preventing the monitoring of the phenomenon in other areas of interest. Another data source available at the European level is the Land Use/Cover Area frame Survey, known as LUCAS [30]. As the estimates for land take provided by LUCAS are based on sampling locations, they are less accurate than those resulting from full coverage datasets. The high-resolution imperviousness layers produced by Copernicus [31] capture "the spatial distribution of artificially sealed areas, including the level of sealing of the soil per area unit." Accordingly, they only make it possible to monitor the sealed land, i.e., permanent land take, but integration with other land cover or land use datasets may make it possible to estimate non-permanent land take. Given the limitations of European data sources, the evaluation in each member state of the above processes is often based on national data sources, generally more detailed than those available at the European level. The definitions of the processes and their assessment methodologies also vary substantially among countries and regions.

1.4. How Are Some Countries and Regions in Europe Approaching the "No Net Land Take" Goal?

The definition of quantitative targets aimed at achieving national objectives, such as a reduction in net land take, and the monitoring of such targets, are the responsibility of central governments. However, as land-use planning decisions are mainly taken at the regional or municipal levels, to meet a national target, translation into regional or local targets is necessary [5].

Some European countries have already set national quantitative targets for reducing land take or increasing the recycling of developed land. Among these, one should highlight Luxembourg, which limited land take to 1 ha per day by the year 2020 in its National Sustainable Development Plan, designed in 2007 [5]. As a result, the land take dropped to 0.46 ha/day between 2007 and 2018 [32]. The Luxembourg Spatial Planning Program is currently under review, and it is expected that it will reduce the land take target to 0.25 ha/day by 2035 [32]. Assessment of the land take in Luxembourg is based on orthophotos [33].

In Austria, land take is estimated based on cadastral data. In 2002, the Austrian government set a threshold of 1 ha/day by 2010 for soil sealing [16]. The report on Land Take Reduction in Austria, presented on 8 October 2019, revealed a downward trend in land take since 2010 [34]. However, the land take was still 11.5 ha/day between 2018 and 2020, and soil sealing was also above the limit set for 2010 (>4.0 ha/day) [35]. Given this situation, the Austrian government's program for 2020–2024 has set a land take target of 2.5 ha/day by the year 2030.

In Flanders (a region of Belgium), two data sources are used to monitor land take: the official federal cadastral statistics on built-up areas, available with annual updates since 1985, and the Flanders land use database [36], which is updated every 3 years but was only made available in 2013. In 2016, land take in Flanders amounted to 6 ha/day [37]. In order to reach the "no net land take" target by 2050, the Flemish Spatial Policy Plan aims to curb settlement growth so that it is zero by 2040. To meet this objective, the Flemish government intends to densify settlements, promote the multifunctional use of space, encourage management of contaminated sites, redevelop brownfield sites, and further dynamic landfill management [38].

In Wallonia (another region of Belgium), land take is also quantified on the basis of cadastral data. The target of "no net land take" by 2050 is addressed by the 2018 Walloon Spatial Development Perspective (Schéma de Développement du Territoire), which limits land take to 1.6 ha/day by 2030 in order to meet the 2050 target. In 2020, the land taken in Wallonia doubled the 2030 target. The aforementioned document also states that by 2030, 50% of new housing and 30% of economic areas should be developed in brownfields [39].

In Germany, land take was quantified based on two datasets: the "Authoritative Real Estate Cadastre Information System" and the "Digital Basic Landscape Model of the Authoritative Topographic and Cartographic Information System" [40]. In Germany, the land take evaluates the conversion of non-artificial areas into settlements and transportation networks (excluding, for example, the creation of mining sites from non-artificial areas). The first national target for curbing land take was set in 2002, to reduce it to 30 ha/day by 2020. In order to achieve this goal, the German government promoted the reuse of brownfields and the development of under-utilized urban areas [41]. In 2013, a legally binding priority for inner urban development come into force, requiring settlements to consider inner urban potential before expanding into surrounding areas. In 2016, the German Sustainable Development Strategy was relaunched, and the land take target was reset to less than 30 ha/day by 2030. The monitoring of land consumption in Germany relies on combined quantification by three indicators: the surface area occupied by settlements and transportation networks, the density of settlements, and open urban space per capita. The goal is reducing land take at the expense of increasing settlement density without decreasing urban open space per capita [42]. Meeting these targets is supported by two territorial planning instruments: the land consumption reduction action plan and the land certificate trading scheme [26]. The land certificate trading scheme was implemented in a pilot project led by the German Environment Agency, involving 87 municipalities. Each municipality received a number of certificates representing the area available for new development based on its population. Surplus or insufficient areas were tradeable between municipalities through the sale or purchase of certificates. This project showed that the trading scheme effectively reduces land take [43]. Despite the enormous importance given to land as an environmental resource by the instruments in place in Germany, undeveloped land consumed by built-up areas, open urban space, and transportation still amounted to 57.9 ha/day in 2020 [44].

Although the "no net land take" target was only introduced into French legislation in 2018, as a response to measure 1.3 of the Biodiversity Plan, which aims to limit the consumption of natural, agricultural, and forest areas [45], in 2010 the French Law on modernization of agriculture and fisheries had already set a target to halve the rate of agricultural land consumption by 2020 [5]. The French Biodiversity Plan is, however, silent on the deadline for achieving the "no net land take" target. In 2019, the French land take observatory was set up. It reports that since 2016 there has been stagnation in the land take rate in France, and that the consumption of natural, agricultural, and forest areas amounted to 54.8 ha/day between 2019 and 2020 [46]. By 2020, the French government started funding private and public brownfield redevelopment operations [15]. The data sources used in France for assessing land take are the OCS GE reference database and the cadastral tax files ("fichiers fonciers"), which enable the detection of changes in land use, in particular, the conversion of natural, agricultural, or forest areas into built-up areas. By definition, cadastral files do not cover the land in the public domain, such as the road network.

Monitoring of land transformation in Italy is carried out under Law 132/2016. Net land take is assessed annually from maps produced by means of photo-interpretation and the semiautomatic classification of remote sensing images (Sentinel-1 and Sentinel 2). In said assessment, land take covers changes from non-artificial to artificial areas (excluding the creation of green urban areas from agricultural areas) and the conversion of green urban areas into sealed areas. A clear distinction is made between permanent land take (land sealing) and reversible land take [47]. In Italy, the net land take was 14.2 ha/day between 2019 and 2020. In the same period, land re-naturalization amounted to 1.4 ha/day. The main difficulties experienced in controlling land take in Italy are the lack of a uniform policy framework at the national level and the absence of effective regulatory measures in most of the country [47,48].

Although the UK's National Planning Policy Framework makes no reference to the goal of "no net land take" [15], the re-use of previously developed land for housing has been an objective of English spatial policy since the late 1990s, intending to reduce urban sprawl and greenfield development and to densify urban areas [49]. The national target set in 1998 of having at least 60% of new housing built on brownfield land by 2008 was achieved and surpassed before 2008.

It should be noted that the above summary has sought to depict how the objective of reducing land take is being addressed in some countries and regions in Europe without looking at the effectiveness of the measures adopted, as this would require an analysis of the political and social contexts, and how the initiatives described are implemented by the legal instruments in place.

1.5. Study Aims

The main objective of this study was to quantify and characterize the non-artificial land transformed by urban development, and the artificial land that has been re-naturalized or recycled, between 2007 and 2018, in mainland Portugal. This assessment aims to contribute to:

- Deepening the knowledge on the land take, re-naturalization, and recycling of developed land, and on drivers thereof;
- Assessing the situation in mainland Portugal regarding the "no net land take" target;
- Supporting the setting of national and regional targets which, thus far, do not exist;
- Raising awareness among decision-makers and public authorities of the context of the issues addressed.

Although Portuguese legislation does not explicitly refer to the goal of "no net land take," the European guidelines on spatial planning were reflected in the Law on Public Policy on Soil Land-use Planning and Urban Planning (2014), in the National Strategy for Sustainable Urban Development 2014–2020 (2015), and finally, in the strategy and territorial model of the National Program for Spatial Planning Policy (2019). In 2015, the legal framework for spatial planning instruments and the land-use regime for municipal master plans were revised. These revisions included, among other things, the suppression of the urban expansion area category and the strengthening of the exceptional nature of the reclassification of rural land to urban land (to avoid the expansion of urban perimeters). Since then, this reclassification has required proof of economic and financial sustainability, and justification of the need for it, in the form of demographic indicators and levels of supply and demand for urban land. Portuguese Law also provides for certain tools similar to Tradeable Development Rights, which have the potential to encourage the densification and redevelopment of urban centers. However, said instruments have found little use so far. It is also worth noting the existence of different types of funding for urban regeneration. Furthermore, since the 1970s and 1980s, Portugal has had several other spatial planning instruments to protect agricultural and natural land and promote biodiversity, namely, the National Agricultural Reserve, the National Ecological Reserve, and the National Network of Protected Areas.

The monitoring of Sustainable Development Goals 2030 in Portugal is only carried out at the national level, and there are no regionally differentiated targets and criteria defined by the central government. Concerning net land take and developed land recycling, since their extent was unknown until now, this study aims to contribute to their regular follow-up, and to inform decisions regarding the establishment of national or regional quantitative objectives to meet the European target.

2. Materials and Methods

This study is based on the Land and Ecosystem Accounting (LEAC) system developed by the EEA [50–52]. This system describes land cover/land use changes through flows between land cover types, allowing us to understand how their stocks are transformed over time. Flows correspond to gains or losses due to transfers of land area between land cover/land use types [50]. When applying this accounting system to the land cover class known as artificial surfaces, a gain (or formation) is any conversion of non-artificial land to artificial land, and a loss (or consumption) is any conversion of artificial land to non-artificial land. While land take evaluates gains, re-naturalization evaluates losses. Developed land recycling evaluates the transformations between different types of artificial land.

The LEAC system provides a framework for linking changes in land use and land cover to their causes (driving forces). The processes that cause area transfers between land cover/land use types were classified into Land Cover Flows (LCFs) by the EEA [50,51]. An LCF represents a set of land cover changes grouped by a common driving force. It is essential to have a good understanding of the processes that are driving land cover changes to design policy measures that help shape future trends [50]. Analysis of LCFs by source and target land cover classes provides knowledge (quantitative and qualitative) of each land cover/land use change.

The Land Cover Flows identified by EEA [50] resulted from the analysis of the change matrix obtained from the overlay of two CLC maps. Given that the third level of the CLC nomenclature contains 44 land cover classes, the change matrix includes over 1800 changes. These were grouped into processes (LCFs) in accordance with three hierarchical levels. The first level contains nine LCFs, which at the second level are broken down into 40 [52]. Table 1 identifies the first-level LCFs causing land cover changes and the second-level processes underlying the land take, re-naturalization, and recycling of developed land. One should point out that in this table, and indeed the whole study, assumes that urban or developed land is representative of the CLC classes relating to artificial surfaces, and that non-urban or undeveloped land is representative of the remaining CLC classes.

Table 1. First-level processes (LCFs) that drive land cover changes and second-level processes linked to land take, re-naturalization, and recycling of developed land. Sources: adapted from [50,52].

Land Cover Flows (LCF)–Level 1	Land Cover Flows (LCF)–Level 2
LCF1 Urban land management	LCF11 Urban development (densification or infilling). Conversions between artificial areas: (i) Conversion of discontinuous urban fabric to continuous urban fabric, industrial, commercial and transportation units, and construction sites; (ii) Conversion of green urban areas and sport and leisure facilities to the urban fabric, industrial, commercial and transportation units, and mineral extraction, dump and construction sites. LCF12 Recycling of developed urban land (gray recycling). Conversions between artificial areas: (i) Conversion of the continuous urban fabric, industrial, commercial and transportation units, and mineral extraction, dump and construction sites to other artificial areas, with the exception of green urban areas and sport and leisure facilities; (ii) Conversion of discontinuous urban fabric to mineral extraction and dumpsites. Construction on urban Greenfields is not considered as LCF12, but LCF11. LCF13 Development of green urban areas: Conversion of urban and non-urban land in green urban areas.
LCF2 Urban residential expansion	LCF21 Urban dense residential expansion: Land uptake by means of continuous urban fabric (CLC 111) from non-urban land. LCF22 Urban diffuse residential expansion: Land uptake by means of discontinuous urban fabric (CLC 112) from non-urban land.
LCF3 Expansion of economic sites and infrastructures	LCF31 Expansion of industrial or commercial units: Non-urban land uptake by means of new industrial and commercial sites. LCF32 Expansion of transportation networks: Non-urban land uptake by means of new transportation networks. LCF33 Expansion of port areas: Development of port areas through non-urban land and sea. LCF34 Expansion of airports: Development of airports through non-urban land and sea. LCF35 Expansion of mining and quarrying areas: Non-urban land uptake by means of mines and quarries. LCF36 Expansion of dumpsites: Non-urban land uptake by means of waste dumpsites. LCF37 Expansion of construction sites: Conversion of non-urban land to construction sites. LCF38 Expansion of sport and leisure facilities: Conversion of urban and non-urban land to sport and leisure facilities.

Table 1. *Cont.*

Land Cover Flows (LCF)–Level 1	Land Cover Flows (LCF)–Level 2
LCF4 Agriculture internal conversions	–
LCF5 Conversion from other land covers to agriculture	LCF54 Conversion of developed areas to agriculture: Conversion of urban land to any type of farmland.
LCF6 Increase in forest land cover and other semi-natural areas	LCF63 Forest creation, afforestation: Forest and woodland creation from other semi-natural, wetlands, water or artificial areas.
LCF7 Forest internal land cover changes	–
LCF8 Waterbody and wetland creation and management	LCF81 Waterbody creation: Extension of water surface areas resulting from the creation of dams and reservoirs.
LCF9 Changes in land cover due to natural and multiple causes	LCF91 Semi-natural creation and rotation: Changes in natural and semi-natural land cover due to natural factors. LCF93 Coastal processes: Any process of coastal erosion or accretion. LCF99 Rare or not-applicable changes: Landcover changes that are rare, more likely improbable or not applicable due to definitions in nomenclature.

To identify the source (origin) and target (destination) classes of land cover conversions to which the LCFs refer, we subsequently aggregated the 44 CLC classes used to identify LCFs into eight land cover types, which integrate the LEAC nomenclature (see Table A1 in Appendix A). This nomenclature groups CLC classes with similar land use or environmental characteristics and simplifies the interpretation of LCF processes.

The methodology adopted to assess land take is similar to that used by the EEA for calculating the CSI 014/LSI 001 indicator [7], which assesses the amount of non-artificial surfaces (agricultural, forest, semi-natural and natural areas, wetlands, or water surfaces) converted to artificial surfaces. This methodology is explained in Figure 1 by level-2 Land Cover Flows (in green) and the changes between land cover classes (CLC-level 3), accounted for in the land take amount assessment. According to Figure 1, the processes (level-2 LCFs) driving land take (in green) are LCF21, LCF22, LCF31, LCF32, LCF33, LCF34, LCF35, LCF36, LCF37, and those parts of LCF38 and LCF13 that relate to the conversion of non-artificial areas into sport and leisure facilities and green urban areas, respectively.

The formation or creation of artificial land assesses, for a given period, the sum of the surface areas gained by each of the classes that make up the artificial surfaces (CLC-level 3). This sum comprises internal conversions between these classes and their gains in terms of surface area due to land take. According to the LEAC quantification system, the formation of artificial land is explained by four types of processes (LCF-level 1): Urban land management (LCF1); urban residential expansion (LCF2); expansion of economic facilities and infrastructures (LCF3); and land cover changes due to multiple natural causes (LCF9). As land take excludes conversions between classes of artificial surfaces, only some of the conversions between land cover classes considered by these four processes are included in land take estimation. Accordingly, the land take estimate is usually lower than the creation of artificial land that occurred in the same period.

Land re-naturalization assesses, for a given period, the surface area of artificial land converted into non-artificial land (agricultural, forest, semi-natural and natural areas, wetlands, and water bodies). The transitions between land cover classes (CLC-level 3) accounted for in the land re-naturalization quantification are shown in Table 2, which also shows the level-2 LCFs related to land re-naturalization—namely, the conversion of artificial areas to agricultural areas (LCF54), forests and woodlands (partial LCF63), dams and reservoirs (partial LCF81), semi-natural areas (partial LCF91), and intertidal flats, estuaries, or sea and ocean (partial LCF93).

Corine Land Cover classes (level 3)

Source	Target 111	112	121	122	123	124	131	132	133	141	142
111	-	LCF12	LCF12	LCF12	LCF12	LCF12	LCF12	LCF12	LCF12	LCF13	LCF38
112	LCF11	-	LCF11	LCF11	LCF11	LCF11	LCF12	LCF12	LCF11	LCF13	LCF38
121	LCF12	LCF12	-	LCF12	LCF12	LCF12	LCF12	LCF12	LCF12	LCF13	LCF38
122	LCF12	LCF12	LCF12	-	LCF12	LCF12	LCF12	LCF12	LCF12	LCF13	LCF38
123	LCF12	LCF12	LCF12	LCF12	-	LCF12	LCF12	LCF12	LCF12	LCF13	LCF38
124	LCF12	LCF12	LCF12	LCF12	LCF12	-	LCF12	LCF12	LCF12	LCF13	LCF38
131	LCF12	LCF12	LCF12	LCF12	LCF12	LCF12	-	LCF12	LCF12	LCF13	LCF38
132	LCF12	LCF12	LCF12	LCF12	LCF12	LCF12	LCF12	-	LCF12	LCF13	LCF38
133	LCF12	LCF12	LCF12	LCF12	LCF12	LCF12	LCF12	LCF12	-	LCF13	LCF38
141	LCF11	LCF11	LCF11	LCF11	LCF11	LCF11	LCF11	LCF11	LCF11	-	LCF38
142	LCF11	LCF11	LCF11	LCF11	LCF11	LCF11	LCF11	LCF11	LCF11	LCF13	-
211	LCF21	LCF22	LCF31	LCF32	LCF33	LCF34	LCF35	LCF36	LCF37	LCF13	LCF38
212	LCF21	LCF22	LCF31	LCF32	LCF33	LCF34	LCF35	LCF36	LCF37	LCF13	LCF38
213	LCF21	LCF22	LCF31	LCF32	LCF33	LCF34	LCF35	LCF36	LCF37	LCF13	LCF38
221	LCF21	LCF22	LCF31	LCF32	LCF33	LCF34	LCF35	LCF36	LCF37	LCF13	LCF38
222	LCF21	LCF22	LCF31	LCF32	LCF33	LCF34	LCF35	LCF36	LCF37	LCF13	LCF38
223	LCF21	LCF22	LCF31	LCF32	LCF33	LCF34	LCF35	LCF36	LCF37	LCF13	LCF38
231	LCF21	LCF22	LCF31	LCF32	LCF33	LCF34	LCF35	LCF36	LCF37	LCF13	LCF38
241	LCF21	LCF22	LCF31	LCF32	LCF33	LCF34	LCF35	LCF36	LCF37	LCF13	LCF38
242	LCF21	LCF22	LCF31	LCF32	LCF33	LCF34	LCF35	LCF36	LCF37	LCF13	LCF38
243	LCF21	LCF22	LCF31	LCF32	LCF33	LCF34	LCF35	LCF36	LCF37	LCF13	LCF38
244	LCF21	LCF22	LCF31	LCF32	LCF33	LCF34	LCF35	LCF36	LCF37	LCF13	LCF38
311	LCF21	LCF22	LCF31	LCF32	LCF33	LCF34	LCF35	LCF36	LCF37	LCF13	LCF38
312	LCF21	LCF22	LCF31	LCF32	LCF33	LCF34	LCF35	LCF36	LCF37	LCF13	LCF38
313	LCF21	LCF22	LCF31	LCF32	LCF33	LCF34	LCF35	LCF36	LCF37	LCF13	LCF38
321	LCF21	LCF22	LCF31	LCF32	LCF33	LCF34	LCF35	LCF36	LCF37	LCF13	LCF38
322	LCF21	LCF22	LCF31	LCF32	LCF33	LCF34	LCF35	LCF36	LCF37	LCF13	LCF38
323	LCF21	LCF22	LCF31	LCF32	LCF33	LCF34	LCF35	LCF36	LCF37	LCF13	LCF38
324	LCF21	LCF22	LCF31	LCF32	LCF33	LCF34	LCF35	LCF36	LCF37	LCF13	LCF38
331	LCF21	LCF22	LCF31	LCF32	LCF33	LCF34	LCF35	LCF36	LCF37	LCF13	LCF38
332	LCF21	LCF22	LCF31	LCF32	LCF33	LCF34	LCF35	LCF36	LCF37	LCF13	LCF38
333	LCF21	LCF22	LCF31	LCF32	LCF33	LCF34	LCF35	LCF36	LCF37	LCF13	LCF38
334	LCF21	LCF22	LCF31	LCF32	LCF33	LCF34	LCF35	LCF36	LCF37	LCF99	LCF38
335	LCF99	LCF99	LCF99	LCF99	LCF99	LCF99	LCF99	LCF99	LCF99	LCF99	LCF99
411	LCF21	LCF22	LCF31	LCF32	LCF33	LCF34	LCF35	LCF36	LCF37	LCF13	LCF38
412	LCF21	LCF22	LCF31	LCF32	LCF33	LCF34	LCF35	LCF36	LCF37	LCF13	LCF38
421	LCF21	LCF22	LCF31	LCF32	LCF33	LCF34	LCF35	LCF36	LCF37	LCF13	LCF38
422	LCF21	LCF22	LCF31	LCF32	LCF33	LCF34	LCF35	LCF36	LCF37	LCF99	LCF38
423	LCF21	LCF22	LCF31	LCF32	LCF33	LCF34	LCF35	LCF36	LCF37	LCF99	LCF38
511	LCF21	LCF22	LCF31	LCF32	LCF33	LCF34	LCF35	LCF36	LCF37	LCF13	LCF38
512	LCF21	LCF22	LCF31	LCF32	LCF33	LCF34	LCF35	LCF36	LCF37	LCF13	LCF38
521	LCF21	LCF22	LCF31	LCF32	LCF33	LCF34	LCF35	LCF36	LCF37	LCF13	LCF38
522	LCF21	LCF22	LCF31	LCF32	LCF33	LCF34	LCF35	LCF36	LCF37	LCF13	LCF38
523	LCF21	LCF22	LCF31	LCF32	LCF33	LCF34	LCF35	LCF36	LCF37	LCF13	LCF38

Figure 1. Level-2 Land Cover Flows (LCFs) and transitions between land cover classes (CLC-level 3) taken into consideration in the assessment of land take (in green) and recycling of developed land (in gray). Source: adapted from [50].

The consumption or loss of artificial land assesses, for a given period, the total area lost by each of the classes (CLC-level 3) of the artificial surfaces. This sum includes internal conversions between these classes and the surface area loss due to land re-naturalization. In accordance with the LEAC quantification system, the consumption of artificial land can be explained by four types of processes (LCF-level 1): conversion from other land covers to agricultural land (LCF5), increase in forest land cover and other semi-natural areas (LCF6), waterbody and wetland creation and management (LCF8), and land cover changes due to multiple natural causes (LCF9). As land re-naturalization excludes internal conversions between classes of artificial surfaces, only some of the conversions between land cover classes considered by these four processes are included in the land re-naturalization estimate.

Accordingly, the land re-naturalization estimate is usually lower than the consumption of artificial land that occurred in the same period.

Table 2. Transitions between land cover classes (CLC-level 3) accounted for in the land re-naturalization quantification and related LCFs processes. Source: adapted from [50,52].

Source/Origin (CLC Class Level 3)	Target/Destination (CLC Class Level 3)	Processes–LCF Level 2	Processes–LCF Level 1
1XY [1]	2XY 321	LCF54 Conversion from developed areas to agriculture	LCF5 Conversion from other land covers to agriculture
1XY [1]	31X 324	LCF63 Forest creation, afforestation	LCF6 Increase in forest land cover and other semi-natural areas
1XY [1]	322 323 33X [2]	LCF91 Semi-natural creation and rotation	LCF9 Changes in land cover due to natural and multiple causes
1XY	423 52X	LCF93 Coastal processes	
1XY	51X	LCF81 Waterbody creation	LCF8 Waterbody and wetland creation and management

Notes: X and Y represent any integers between 1 and 5, which combined, correspond to a valid CLC class-level 3; [1] excluding 141; [2] excluding 334 and 335.

As explained in the Introduction, net land take assesses the difference between land take and land re-naturalization, and the monitoring thereof allows one to evaluate how far we are from the European target [1].

From the analysis of Table 1 and Figure 1, it follows that the changes between artificial surfaces related to LCF11 and LCF12 are due to urban development (densification or infilling) and gray recycling, respectively. The changes resulting from LCF11 and LCF12, and those resulting from LCF13 (conversion of artificial and non-artificial areas into green urban areas), are integral parts of a process called urban land management (LCF1).

Our assessment of developed land recycling (in the broadest sense) covers urban development or infilling (LCF11), gray recycling (LCF12), and green recycling; and excludes the conversion of non-artificial areas into urban green areas (partial LCF13) and the conversion of construction sites into any land cover class. The latter is based on the premise that such sites represent a transitional land cover class that will evolve in the short term into a definitive one [17] (p. 42). Such sites usually include areas under construction, roads and infrastructures, and other areas under transformation, which, after completion of the construction work, may revert to land cover classes other than artificial. Figure 1 also shows (in gray) the changes between land cover classes (CLC-level 3) accounted for in the assessment of the recycling of developed land and the processes (level 2 LCFs) linked to these changes.

Our approach to assessing the recycling of developed land (in the broadest sense) differs from that used by the EEA for calculating the LSI 008 indicator [53] in that the latter takes into consideration a more comprehensive account of green recycling, which additionally includes the conversion of artificial areas into sports and leisure facilities (part of LCF38).

To assess the importance of each of the processes of developed land recycling (LCF11, LCF12, and part of LCF13), we determined their representativeness in urban land management (LCF1). Furthermore, we also evaluated the weight of developed land recycling in the creation (formation) of artificial areas, as this indicator is commonly used in international comparisons.

The processes addressed by this study are described in terms of areas per spatial unit for a given period. As spatial units have unequal sizes, we produced the following indicators to make it possible to compare a process between spatial units or to compare our results with those of analogous studies:

- Indicator reported in km^2/year—this indicator is an annual average (the area of interest, in km^2, was divided by the number of years composing the period);
- Indicator reported in m^2/year.km^2—this indicator is an annual rate (the area of interest, in m^2, was divided by the number of years in the respective period and multiplied by the spatial unit surface, in km^2).

Study Area

Portugal is a southern European state composed of three NUTS (level I) regions: mainland Portugal and the autonomous regions of Madeira and Azores. Mainland Portugal is located in the extreme southwest of the Iberian Peninsula, bordered to the north and east by Spain and to the west and south by the Atlantic Ocean.

This study assesses the land take, re-naturalization, and recycling of developed land in mainland Portugal, broken down by NUTS III, between 2007 and 2018.

This assessment was based on the 2007 and 2018 national land cover/land use maps (known as COS and hereafter referred to as COS2007 and COS 2018), which are vector maps produced by the Directorate-General for Territory. The current COS nomenclature features 83 classes organized in a four-tier hierarchical structure. Although the COS maps do not cover all the classes that make up the third level of the CLC map, it is possible to establish equivalences with CLC nomenclature. Despite the thematic similarities between COS and CLC, COS maps are based on different technical features, namely, the Minimum Mapping Unit (MMU) and the Minimum Distance Between Lines (MDBL). Both are smaller for COS than for CLC (the MMU of COS is 1 ha, whereas that of CLC is 25 ha, and the MDBL of COS is 20 m, whereas that of CLC is 100 m). These differences cause the surface area of the same land cover class (e.g., artificial surfaces) to be larger when calculated using COS than when calculated using CLC.

Since the Land Cover Flows of interest for our assessment were defined based on the third level of the CLC nomenclature, the two national maps (COS2007 and COS2018) were initially reclassified at the third level of the CLC nomenclature. A change matrix was then created for the period 2007–2018 by overlaying the two reclassified maps with the NUTS III boundaries of 2018. Each change between CLC classes (level 3) provided by this matrix was then associated with a Land Cover Flow (LCF level 2) and classified as to its contribution to the phenomena of interest (land take, re-naturalization, or recycling of developed land). Thus, each phenomenon totals the area of a set of changes between land cover/land use types (CLC classes), which were driven by one or several processes (LCFs).

Although the above tasks were carried out based on CLC-level 3 classes, to simplify the description of the source and destination classes to which each flow refers, we adopted the eight land cover types of the LEAC nomenclature (Table A1 in Appendix A).

3. Results

The results presented below begin by quantifying the land take, re-naturalization, and net land take in mainland Portugal and its NUTS III regions, during the period 2007–2018. Then, we describe the drivers of land take and re-naturalization and the land cover types they have transformed. Subsequently, developed land recycling and the weights of its components (densification, grey recycling, and green recycling) are estimated. Finally, urban land use efficiency is assessed by comparing the expansion of artificial areas and non-artificial land transformed by new development (i.e., land take) with population growth.

In the period under review, several spatial planning instruments were already in force in mainland Portugal to regulate changes in land use. However, it was only in 2014 that the Law for Public Policy on Soil, Land-use Planning, and Urban Planning reinforced the stance, already expressed in the previous Law of 1999, of considering land (i.e., the soil) as a scarce resource.

Although the national objectives for spatial planning in 2007 already made allowances for the profitability of existing infrastructures, to avoid unnecessary expansion of infrastructure networks and urban perimeters, the use of empty urban spaces, and the rehabilitation of historical centers, among other things, these objectives were reiterated in 2014, thereby reinforcing the need to contain urban sprawl and dispersed building, and the need to favor urban redevelopment and regeneration over new construction.

3.1. Land Take, Re-Naturalization, and Net Land Take

Land take in mainland Portugal amounted to 26.4 km^2/year between 2007 and 2018, representing a change from non-artificial to artificial areas of 7.2 ha/day. Figure 2 maps the annual land take rate by NUTS III in this period. In addition to this rate, Table 3 presents (in the second column) the annual average of these land transformations on the mainland, by NUTS III, in the same period. The following facts stand out after analysis of both:

- The land take rate shows a similar spatial distribution to that of the resident population, decreasing from the coastal regions towards the interior of the mainland. The highest rates occurred in the two metropolitan areas (M.A. LISBOA and M.A. PORTO) and their neighboring regions.
- The land take rate was highest in the Porto metropolitan area (1153 m^2/year.km^2), slightly lower in the Lisbon metropolitan area (1021 m^2/year.km^2), and reached figures of 600 m^2/year.km^2 in the Aveiro, Oeste, and Ave regions.
- The lowest land take rates refer to the interior regions of Alto Alentejo and Beira Baixa (83.8 and 107.8 m^2/year.km^2, respectively).

Table 3. Land take, land re-naturalization, and net land take indicators by NUTS III region and for mainland Portugal in 2007–2018.

NUTS III Region	Land Take		Land Re-Naturalization		Net Land Take	
	Annual Average (km^2/year)	Annual Rate (m^2/year.km^2)	Annual Average (km^2/year)	Annual Rate (m^2/year.km^2)	Annual Average (km^2/year)	Annual Rate (m^2/year.km^2)
ALTO MINHO	0.5	245.8	0.1	47.5	0.4	198.3
CÁVADO	0.7	538.6	0.1	90.4	0.6	448.2
AVE	0.9	602.0	0.1	68.1	0.8	533.9
M. A. PORTO	2.4	1153.0	0.2	93.1	2.2	1059.9
ALTO TÂMEGA	0.4	152.8	0.1	31.1	0.4	121.7
TÂMEGA E SOUSA	0.9	515.2	0.1	37.9	0.9	477.2
DOURO	1.1	281.2	0.2	45.9	0.9	235.4
TERRAS DE TRÁS-OS-MONTES	1.2	210.9	0.1	12.6	1.1	198.3
ALGARVE	1.8	353.9	0.1	28.9	1.6	325.1
OESTE	1.4	638.4	0.3	148.1	1.1	490.4
REGIÃO DE AVEIRO	1.1	663.1	0.2	89.1	1.0	574.0
REGIÃO DE COIMBRA	1.8	423.3	0.2	52.4	1.6	370.9
REGIÃO DELEIRIA	1.2	486.2	0.2	69.3	1.0	416.9
VISEU DÃO LAFÕES	0.9	281.6	0.1	39.0	0.8	242.6
BEIRA BAIXA	0.5	107.8	0.04	9.0	0.5	98.8
MÉDIO TEJO	0.8	238.3	0.04	12.2	0.8	226.1
BEIRAS E SERRA DA ESTRELA	1.0	152.4	0.2	38.4	0.7	114.0
M. A. LISBOA	3.1	1021.0	0.5	157.9	2.6	863.2
ALENTEJOLITORAL	1.2	226.6	0.1	22.8	1.1	203.8
BAIXO ALENTEJO	1.2	144.4	0.3	40.4	0.9	104.0
LEZÍRIA DO TEJO	0.8	184.3	0.2	56.5	0.5	127.8
ALTO ALENTEJO	0.5	83.8	0.1	10.9	0.4	72.9
ALENTEJOCENTRAL	0.9	126.8	0.1	9.1	0.9	117.7
MAINLAND PORTUGAL	**26.4**	**296.2**	**3.7**	**41.6**	**22.7**	**254.5**

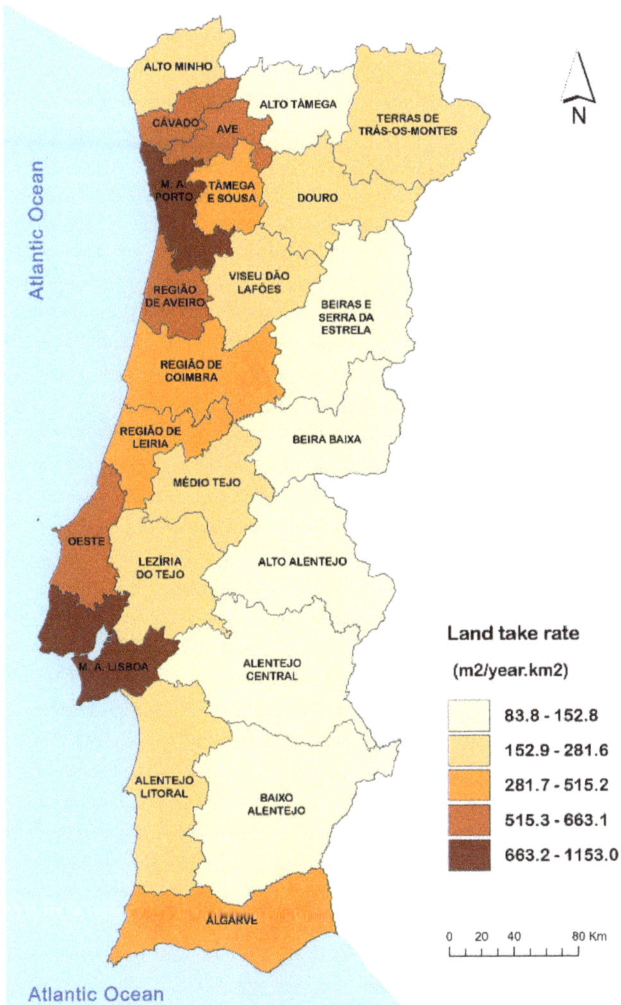

Figure 2. Land take rate by NUTS III region for mainland Portugal, 2007–2018 (m^2/year.km^2).

Land re-naturalization is, in comparison to the reverse phenomenon (land take), a rare process that covers small surface areas on the mainland. The land re-naturalized between 2007 and 2018 (40.8 km^2) represents only 0.05% of mainland Portugal and corresponds to an annual average re-naturalization of 3.7 km^2/year. Table 3 presents land re-naturalization indicators (in the fourth and fifth columns) for mainland Portugal, by NUTS III, in the above period. Figure 3 maps the land re-naturalization rate by NUTS III for the same period. The following facts stand out from the analysis of both:

- The rate of re-naturalization of developed land in mainland Portugal was 41.6 m^2/year.km^2.
- At the NUTS III level, the highest re-naturalization rates refer to the Lisbon metropolitan area (157.9 m^2/year.km^2), the Oeste region (148.1 m^2/year.km^2), the Porto metropolitan area (93.1 m^2/year.km^2), and the Cávado region (90.4 m^2/year.km^2).
- The lowest re-naturalization rates per NUTS III (<13 m^2/year.km^2) were in the inland regions of the mainland (Beira Baixa, Alto Alentejo, Alentejo Central, and Terras de Trás-os-Montes).

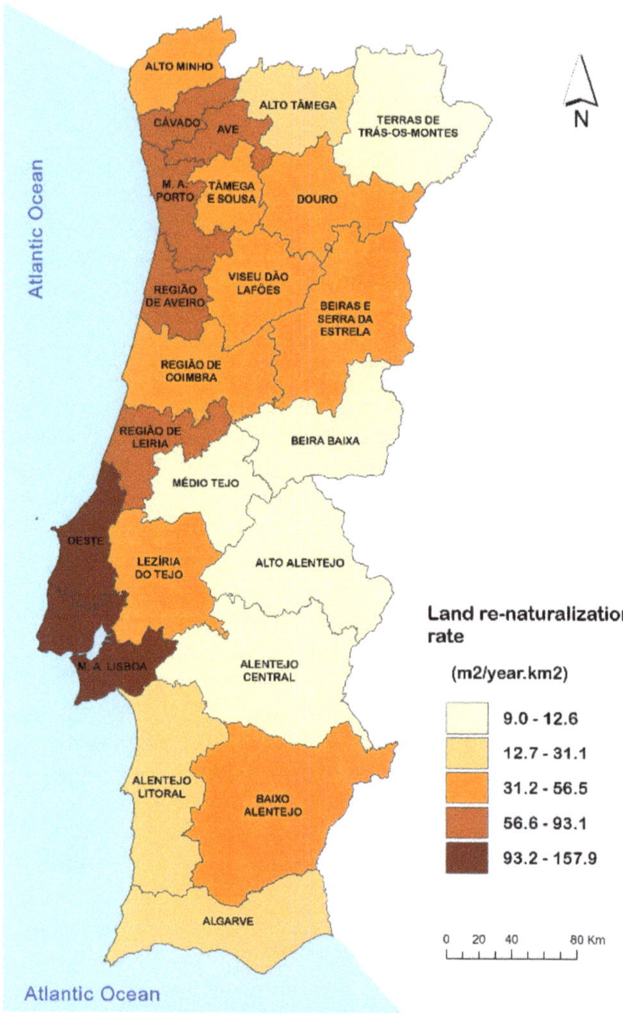

Figure 3. Land re-naturalization rate by NUTS III region for mainland Portugal, 2007–2018 (m^2/year.km^2).

A more detailed geographical analysis also revealed that 14% of the municipalities in mainland Portugal did not re-naturalize developed land in the period under review.

The balance between land take and re-naturalization shows an annual average of net land take in mainland Portugal of 22.7 km^2/year between 2007 and 2018. Figure 4 maps the annual net land take rate by NUTS III in this period, and Table 3 presents (in the 6th and 7th columns) additional net land take indicators for both NUTS III and mainland Portugal.

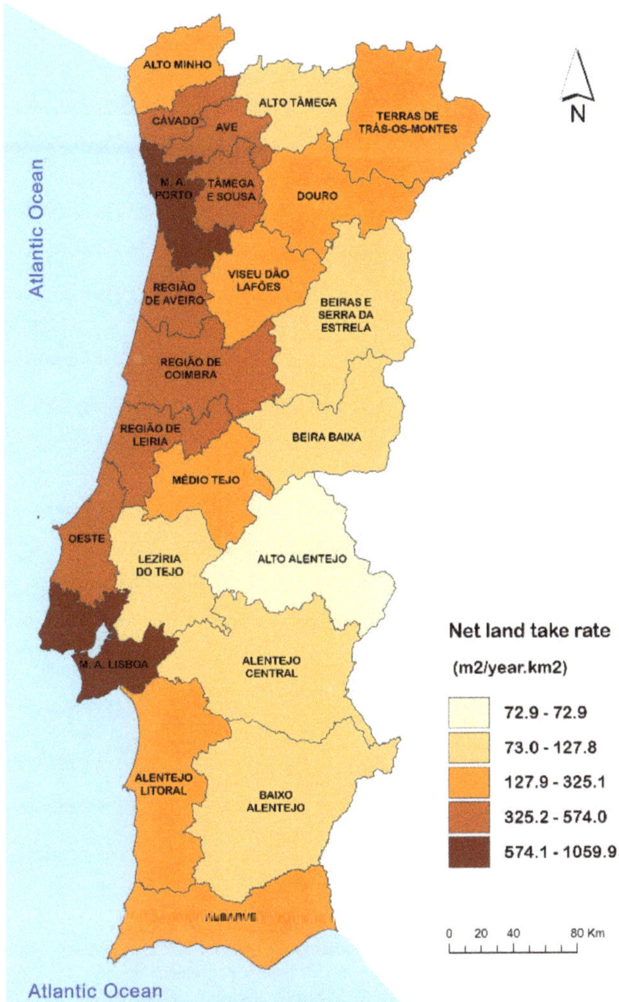

Figure 4. Net land take rate by NUTS III region for mainland Portugal, 2007–2018 (m^2/year.km^2).

As noted for the land take rates, the highest rates of net land take occurred in the two metropolitan areas and their immediate surroundings. These rates were higher in the Porto metropolitan area (1059.9 m^2/year.km^2), slightly lower in the Lisbon metropolitan area (863.2 m^2/year.km^2), and close to 500 m^2/year.km^2 in the Aveiro and Ave regions. The lowest net land take rates were in the Alto Alentejo and Beira Baixa regions (72.9 and 98.8 m^2/year.km^2, respectively).

Figure 5 identifies the processes (LCFs) that drove land take and land re-naturalization in mainland Portugal and the land cover types transformed by each LCF (in km^2) between 2007 and 2018. In this figure, the gray bars (negative values) represent the re-naturalization processes, and the bars in other colors (positive values) represent the land take processes. Table 4 identifies the surface area and land cover types transformed by the land take and land re-naturalization processes. Land cover types are shown as percentages of the surface area transformed by the process. The meanings of the processes in Table 4 are provided in Figure 5.

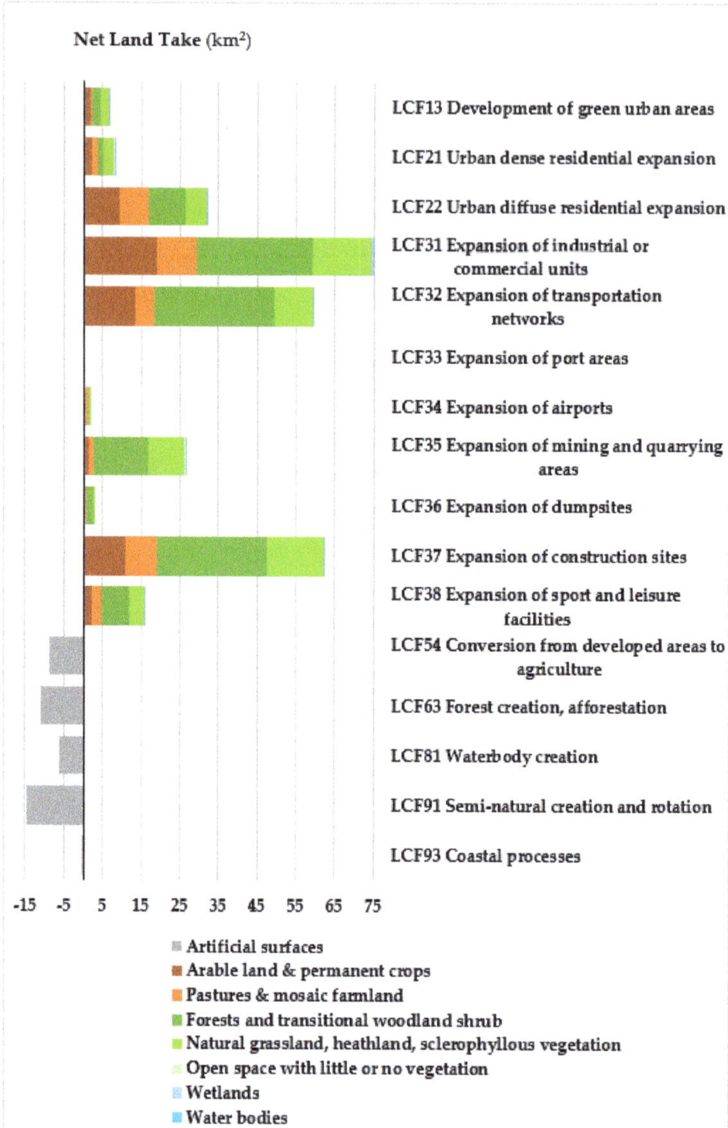

Figure 5. Land surface transformed by land take and land re-naturalization processes (LCFs) and land cover types consumed by each process for mainland Portugal, 2007–2018 (km^2).

Table 4. Land surface transformed by land take and land re-naturalization processes (LCFs) and land cover types consumed per process (% of the surface area transformed by LCF) for mainland Portugal, 2007–2018.

Land Take and Re-naturalization Processes (LCFs)	Surface Area (km^2)	Artificial Surfaces (%)	Arable Land & Permanent Crops (%)	Pastures & Mosaic Farmland (%)	Forests and Transitional Woodland Shrub (%)	Natural Grassland, Heathland, Sclerophyllous Vegetation (%)	Open Space with Little or No Vegetation (%)	Wetlands (%)	Water Bodies (%)
LCF13	6.7		24.9	10.5	29.7	34.4	0.2		0.4
LCF21	7.9		29.7	17.4	19.9	32.1	0.8	0.03	0.05
LCF22	31.8		30.2	23.2	29.4	17.1	0.1		0.03
LCF31	74.5		25.4	13.8	40.5	20.1	0.1	0.2	0.06
LCF32	59.4		23.1	7.7	52.3	16.6	0.1	0.03	0.1
LCF33	0.3				20.4	24.8	4.3		50.4
LCF34	1.9		16.2	38.5	23.7	21.5			
LCF35	26.6		6.1	3.9	53.9	33.6	2.2		0.3
LCF36	2.9		9.5	12.2	64.6	13.8			
LCF37	62.5		17.7	12.6	45.7	23.4	0.2	0.03	0.3
LCF38	15.8		13.1	17.3	45.7	23.1			0.7
Total Land Take	**290.3**	**0.0**	**21.2**	**12.8**	**43.6**	**21.8**	**0.3**	**0.1**	**0.2**
LCF54	8.8	100.0							
LCF63	11.1	100.0							
LCF81	6.3	100.0							
LCF91	14.6	100.0							
LCF93	0.004	100.0							
Total Re-naturalization	**40.8**	**100.0**							

In the period 2007–2018, five land take processes stood out for their relative importance in terms of the area of non-artificial surfaces transformed:

- Construction of new industrial and commercial sites (LCF31) accounted for 26% of the non-artificial land taken. The land cover types most used in this conversion were forests and transitional woodland shrub (41%); arable land and permanent crops (25%); natural grassland, heathland, and sclerophyllous vegetation (20%); and pastures and mosaic farmland (14%).
- Conversion from non-urban land to construction sites (LCF37) was the process with the second highest relative importance (22%) in land take. The land cover types most used by this conversion were forests and transitional woodland shrub (46%); natural grassland, heathland, and sclerophyllous vegetation (23%); arable land and permanent crops (18%), and pastures and mosaic farmland (13%)
- The expansion of transportation networks (LCF32) accounted for 20% of the non-artificial land taken. The land cover types that contributed most to the development of transportation networks were forests and transitional woodland shrub (52%); arable land and permanent crops (23%); and natural grassland, heathland, and sclerophyllous vegetation (17%).
- Diffuse residential expansion (LCF22) was responsible for 11% of the non-artificial land taken. The land cover types that contributed most to the creation of discontinuous urban fabric were arable land and permanent crops (30%), forests and transitional woodland shrub (29%), pastures and mosaic farmland (23%), and natural grassland, heathland, and sclerophyllous vegetation (17%).
- The expansion of mining and quarrying areas (LCF35) accounted for 9% of the non-artificial land taken. The land cover types that contributed most to the development of those areas were forests and transitional woodland shrub (54%) and natural grassland, heathland, and sclerophyllous vegetation (34%).

In summary, in the period under review, the consumption of non-artificial surfaces in mainland Portugal was mostly driven by the expansion of economic sites and infrastructures (LCF3). The land cover types most expended by those changes were forests and transitional woodland shrub (43.6%), natural grassland, heathland, and sclerophyllous vegetation (21.8%), and arable land and permanent crops (21.2%).

The processes with the highest significance in terms of land re-naturalization in the period 2007–2018 were the conversion of artificial areas into semi-natural areas (partial LCF91), forest and woodlands (partial LCF63), and agriculture (LCF54). These processes contributed respectively, 36%, 27%, and 22% to the creation of re-naturalized land.

An analysis based on CLC classes (level 3) showed that conversions from mineral extraction sites and construction sites (131 + 133) to non-artificial land accounted for 91% of re-naturalization. Among these transformations, the most common were the conversions from the two classes above to bushes (322) and to broad-leaved and coniferous forests (311 + 312), which had shares of 32% and 26%, respectively, in the re-naturalization of developed land. As far as the creation of agricultural areas from mineral extraction and construction sites is concerned, the development of non-irrigated arable land (211) and permanent pastures (231) stands out, which together accounted for 8% of land re-naturalization. The conversion of construction sites to inland waters (511 + 512) accounted for 14% of land re-naturalization.

3.2. *Recycling of Developed Land*

In this section, we analyze transformations between artificial surface classes, namely, urban development (LCF11) or densification, gray recycling (LCF12), and green recycling (partial LCF13). As mentioned above, these three processes are referred to (in a broad sense) as developed land recycling. Figures A1–A3 in Appendix B illustrate the conversions between CLC classes (level 3) covered by these three processes. In line with the explanation provided in Section 2, our assessment of developed land recycling (in the broadest sense) excludes conversions of sites under construction for any class (illustrated in light blue in the figures in Appendix B).

Table 5 provides indicators of recycling (in the broadest sense) of developed land for mainland Portugal, by NUTS III regions, in the period under assessment. Figure 6 maps the rates shown in the third column of this table. The following conclusions stand out from the analysis of both:

- Recycling of developed land is still rare in mainland Portugal. Between 2007 and 2018, the area recycled (6.8 km^2) was well lower than that re-cultivated (40.8 km^2). The low recycling rates are attributable to the fact that almost half of the mainland municipalities (49%) have not recycled developed land in this period.
- The rate of developed land recycling had a similar spatial distribution to that of land take, decreasing from the coast to the mainland's interior. This rate was 7 m^2/year.km^2 for mainland Portugal in the period under review.
- The highest recycling rates occurred in the Aveiro region (50.6 m^2/year.km^2); the metropolitan areas (38.4 m^2/year.km^2 in Lisbon and 27.1 m^2/year.km^2 in Porto); and the Lezíria do Tejo, Ave, Cávado, and Oeste regions, which had recycling rates of around 10 m^2/year.km^2.
- The lowest recycling rates ($\leq$0.7 m^2/year.km^2) were observed in interior Alentejo regions (Alto Alentejo, Alentejo Central, and Baixo Alentejo).
- The recycling of developed land accounted for 9.1% of the mainland's urban land management (LCF1) between 2007 and 2018. This share was much higher ($\geq$25%) in regions adjacent to the metropolitan areas, namely, Aveiro and Lezíria do Tejo.
- Developed land recycling also accounted for 1.9% of the formation of artificial surface areas on the mainland. The regions with the highest shares were also Lezíria do Tejo (7.1%) and Aveiro (6%).

Figure 7 shows the relative significance of developed land recycling processes in the period under analysis. It allows concluding that gray recycling prevailed over urban development (or densification) in mainland Portugal and most of its regions. The majority of NUTS III regions did not carry out green recycling in the 2007–2018 period. This type of land recycling only had some relative importance (less than 20%) in the following spatial units: mainland Portugal, the two metropolitan areas, and the regions of Alto Minho, Aveiro, Beira Baixa, and Coimbra.

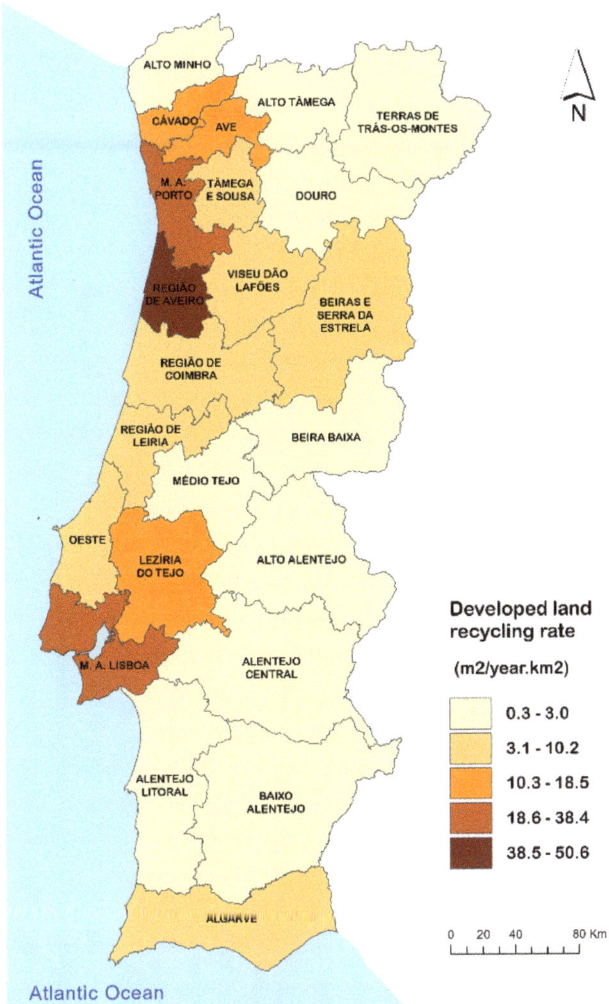

Figure 6. Developed land recycling rate by NUTS III region for mainland Portugal, 2007–2018 (m^2/year.km^2).

Table 5. Developed land recycling indicators by NUTS III region for mainland Portugal, 2007–2018.

NUTS III Region	Yearly Recycling * (km^2/year)	Land Recycling Rate * (m^2/year.km^2)	Weight of Land Recycling * in Urban Land Management -LCF1 (%)	Weight of Land Recycling * in the Formation of Artificial Areas (%)
ALTO MINHO	0.01	3.0	8.9	1.1
CÁVADO	0.02	13.6	11.3	2.1
AVE	0.02	15.7	9.7	2.2
M. A. PORTO	0.06	27.1	9.6	1.9
ALTO TÂMEGA	0.003	1.1	4.5	0.7
TÂMEGA E SOUSA	0.01	5.2	6.3	0.9
DOURO	0.003	0.8	3.1	0.3

Table 5. *Cont.*

NUTS III Region	Yearly Recycling * (km^2/year)	Land Recycling Rate * (m^2/year.km^2)	Weight of Land Recycling * in Urban Land Management -LCF1 (%)	Weight of Land Recycling * in the Formation of Artificial Areas (%)
TERRAS DE TRÁS-OS-MONTES	0.01	2.1	14.6	0.9
ALGARVE	0.04	8.9	7.4	1.6
OESTE	0.02	10.2	5.5	1.2
REGIÃO DE AVEIRO	0.09	50.6	26.9	6.0
REGIÃO DE COIMBRA	0.03	6.2	3.3	1.0
REGIÃO DE LEIRIA	0.02	8.2	9.0	1.4
VISEU DÃO LAFÕES	0.03	7.9	12.0	2.3
BEIRA BAIXA	0.01	1.5	7.5	1.2
MÉDIO TEJO	0.01	2.6	5.0	0.9
BEIRAS E SERRA DA ESTRELA	0.03	4.5	19.8	2.5
M. A. LISBOA	0.12	38.4	8.7	2.7
ALENTEJO LITORAL	0.01	2.6	6.4	1.0
BAIXO ALENTEJO	0.01	0.7	3.4	0.4
LEZÍRIA DO TEJO	0.08	18.5	24.5	7.1
ALTO ALENTEJO	0.004	0.6	1.7	0.5
ALENTEJO CENTRAL	0.002	0.3	2.1	0.2
MAINLAND PORTUGAL	**0.6**	**7.0**	**9.1**	**1.9**

* Sum of the surface areas transformed by densification (LCF11), gray recycling (LCF12), and green recycling (partial LCF13), excluding conversions from construction sites to any class.

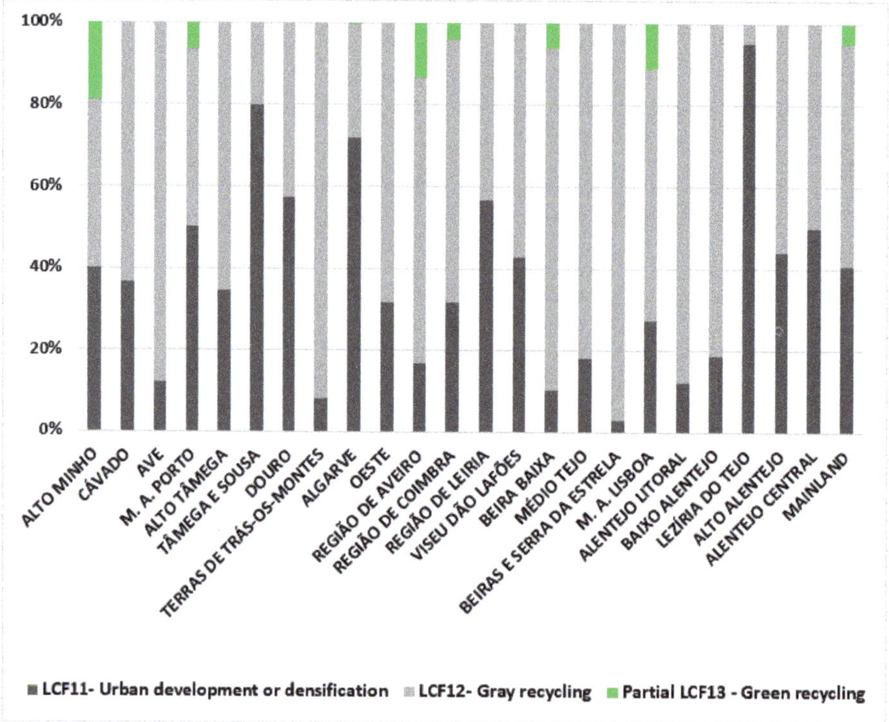

Figure 7. Developed land recycling processes per NUTS III region and for mainland Portugal (% of surface area recycled excluding conversions from construction sites to any class), 2007–2018.

3.3. Land Take and Population Growth

This section compares the population growth with the expansion of artificial areas between 2007 and 2018 in mainland Portugal and its NUTS III regions. Table 6 presents the following additional indicators to assess urban land use efficiency: land take per new inhabitant, artificial area per capita, and share of artificial land by region.

Table 6. Indicators of the growth of resident population and artificial surface areas, land take per new inhabitant, artificial surface area per capita, and share of artificial land, by NUTS III region for mainland Portugal, 2007–2018. Sources: inter-census population estimates produced by Statistics Portugal; land cover and land use classes extracted from COS2007 and CO2018 maps produced by Directorate-General for Territory.

NUTS III Region	Population Growth 2007–2018 (%)	Artificial Surface Area Growth 2007–2018 (%)	Land Take per New Inhabitant * 2007–2018 (m^2/inhab)	Artificial Surface Area per Capita 2007 (m^2/inhab)	Artificial Surface Area per Capita 2018 (m^2/inhab)	Artificial Land 2018 (%)
ALTO MINHO	−7.1	2.7	−341.6	717.9	793.5	8.3
CÁVADO	−1.1	3.2	−1631.5	471.4	491.9	15.9
AVE	−4.0	4.4	−562.1	451.0	490.4	13.9
M. A. PORTO	−2.4	4.9	−616.3	276.9	297.4	25.1
ALTO TÂMEGA	−12.1	4.6	−412.6	864.3	1028.5	3.0
TÂMEGA E SOUSA	−4.6	5.2	−515.6	425.2	468.8	10.7
DOURO	−9.7	9.8	−608.4	501.3	609.7	2.9
TERRAS DE TRÁS-OS-MONTES	−11.1	16.3	−955.0	609.6	797.8	1.6
ALGARVE	1.0	7.0	4515.4	586.6	621.6	5.5
OESTE	−0.1	5.4	−39373.2	622.1	656.3	10.5
REGIÃO DE AVEIRO	−2.2	5.2	−1484.7	559.0	601.3	12.9
REGIÃO DE COIMBRA	−6.9	6.4	−632.3	594.6	679.1	6.8
REGIÃO DE LEIRIA	−3.4	5.0	−1304.7	761.5	827.9	9.6
VISEU DÃO LAFÕES	−7.1	5.1	−520.4	626.9	709.1	5.5
BEIRA BAIXA	−11.3	8.9	−529.2	615.3	756.1	1.3
MÉDIO TEJO	−7.2	4.4	−488.6	753.7	847.5	5.9
BEIRAS E SERRA DA ESTRELA	−12.8	5.8	−339.0	553.7	671.7	2.3
M. A. LISBOA	2.5	4.6	479.3	225.2	229.7	21.7
ALENTEJO LITORAL	−5.8	15.0	−2313.7	803.5	980.4	1.7
BAIXO ALENTEJO	−10.0	9.5	−1014.5	789.8	964.2	1.3
LEZÍRIA DO TEJO	−3.8	3.5	−912.4	700.2	753.5	4.2
ALTO ALENTEJO	−13.4	7.1	−344.0	564.5	698.0	1.2
ALENTEJO CENTRAL	−10.1	8.6	−600.8	656.6	792.9	1.6
MAINLAND PORTUGAL	**−2.6**	**5.7**	**−1100.8**	**438.2**	**475.5**	**5.2**

* Land take $_{2007–2018}$ (in m^2)/(inhabitants $_{2018}$ − inhabitants $_{2007}$).

A comparison of the first two indicators in Table 6 shows that, between 2007 and 2018, in most regions and in mainland Portugal, there was an expansion of artificial land despite a population decrease. As a result, the artificial area per capita increased in all regions and the mainland. The Lisbon metropolitan area and the Algarve are the only regions that had population increases, but they were, nevertheless, lower than those for their artificial surface areas.

On the other hand, the inner regions of the mainland (Alto Alentejo, Beiras e Serra da Estrela, and Alto Tâmega) experienced the highest population reductions.

The land take per new inhabitant indicator shows the uniqueness of the Lisbon metropolitan area and the Algarve compared to the other regions of mainland Portugal, which presented negative figures for this indicator, because they had population declines between 2007 and 2018. The lowest land take per new inhabitant happened in the Oeste region (−39373.2 m^2/inhabitant). Although in this region, the area of non-artificial land transformed by urban development (15.6 km^2) is not among the highest, the decrease in its population (−396 inhabitants) was the smallest of those observed in regions that lost population.

The fact that in 2018 almost half (47%) of the resident population in mainland Portugal was concentrated in the two metropolitan areas justifies that these two regions had the highest shares of artificial land (25.1% for Porto and 21.7% for Lisbon).

Although in 2018, the Lisbon metropolitan area had about one million more inhabitants than the Porto metropolitan area, the artificial area per capita in Porto (297 m^2/inhabitant) was higher than in Lisbon (230 m^2/inhabitant), which shows that Lisbon is denser than Porto. Despite the higher shares of artificial land observed in the two metropolitan areas (see Table 6), the regions of Cávado, Ave, Aveiro, Tâmega e Sousa, and Oeste also stand out for having, in 2018, more than 10% of their surface areas occupied by artificial land. In the above regions, the lowest artificial surface area per capita value was registered in Tâmega e Sousa (469 m^2/inhabitant), and the highest was in Oeste (656 m^2/inhabitant).

4. Discussion

We would first like to point out that, like other authors, we consider that the term 'land take' is controversial, as the land is not taken but transformed by urban development [5,6,25,26]. Its use in this paper is only justified by the recognition the term has gained with the spread of the goal of "no net land take". The designation 'land consumption' is also in itself questionable, as it reflects a change in land cover/land use, and only changes of non-artificial to artificial land are accounted for by land take.

This study assessed the land take, re-naturalization, and recycling of developed land in mainland Portugal, broken down by NUTS III region, through an analysis of land cover/land use changes between 2007 and 2018.

The 2007 and 2018 national land cover/land use maps (COS) served as baseline data. Since national cadastral data is either not available or up to date for the whole of the mainland, COS maps were considered the most complete and accurate data source for quantifying and tracking the phenomena of interest due to their regular updating (every three years) and their spatial coverage (mainland Portugal).

The Land and Ecosystem Accounting system, developed by the EEA [50–52], was adopted to assess the land cover/use changes contributing to the phenomena addressed and their driving forces. In this assessment, a Minimum Mapping Unit was not defined for change identification, which may have contributed to some of the changes considered, especially those related to small areas, no longer reflecting their original land cover/land use classification. This approach was, however, justified by both the small number and small size of some of the changes of interest for this study.

The main innovative contribution of this study was the assessment of land take, re-naturalization, and recycling of developed land using national data sources and the Land and Ecosystem Accounting system. Monitoring the European target in any country or region requires a quantitative assessment of these phenomena, and their extent in mainland Portugal was previously unknown.

Since our findings depend on the data sources and assessment approach used, their comparability with results published by the EEA [7,11,17,53] for mainland Portugal, which considers different assessment periods and data sources, is limited. Regarding this issue, Decoville and Schneider [5] point out that the land take estimates provided by distinct data sources present significant variability, which questions the measurability of the 2050 target and its comparison across countries. Furthermore, land take is described by most countries using an average annual or daily surface area, which is not comparable across countries due to their uneven spatial extents, so this metric only evaluates the performance of the same spatial unit over time.

The following is a summary of the main results obtained in this study, which are compared, whenever possible, with the European trends and discussed in light of the policies in force in mainland Portugal.

The non-artificial land consumed by urbanization during the period under consideration (7.2 ha/day) represents 0.3% of mainland Portugal's surface area, which is more than double that observed between 2006 and 2018 in 39 European countries (0.14%) (percentage based on the content available at [11]).

The main drivers of land take in the mainland in the 2007–2018 period were the development of new industrial and commercial sites and the creation of construction sites, transportation networks, and residential areas. The first three accounted for 68% of the land taken, and residential extension accounted for 13.7%. Although the period available for comparison at the European level (2006–2018) was slightly longer than that taken into consideration herein (2007–2018), it was found that the main drivers of land taken in Europe (39 countries) are similar to those observed for mainland Portugal. Nevertheless, the expansion of mining and quarrying areas in the rest of Europe was more significant than the expansion of transportation networks.

The land cover types most consumed by the conversion of non-artificial areas into artificial areas in mainland Portugal were forests and transitional woodland shrub (43.6%), and natural grassland, heathland, sclerophyllous vegetation (21.8%). The consumption of arable land and permanent crops by the land take was 21.2%. In contrast to observations for mainland Portugal, arable land and permanent crops (44.6%) and pastures and mosaic farmland (26.9%) were the land cover classes most used by land take in Europe (39 countries) between 2006 and 2018 (percentages based on the content available at [11]).

Daily re-naturalization in mainland Portugal between 2007 and 2018 (1.0 ha/day) was about seven times less than the land uptake by new urbanization. The highest rates of re-naturalization were in the coastal regions, mainly those interconnecting the two metropolitan areas. Re-naturalization was even null in several municipalities in mainland Portugal (14%). Since re-naturalization makes it possible to compensate for some of the non-artificial land used by urbanization, its practice should be promoted on the mainland. Although the land lost for human development is commonly much higher than that re-naturalized, the proportion of land re-naturalized in mainland Portugal between 2007 and 2018 (0.5%) was higher than that re-naturalized in Europe (39 countries) between 2006 and 2018 (0.013%) (percentage based on the content available at [11]).

In the period under analysis, the changing of artificial areas into semi-natural areas accounted for 36% of re-naturalization in mainland Portugal; the conversion of mineral extraction sites and construction sites to bushland was the most common (32%). Given that scrub development results from land abandonment and increases the risk of forest fires, it should be noted that this is the least desirable type of re-naturalization for our territory.

The conversion of developed land into agricultural areas accounted for 22% of the land re-naturalization in mainland Portugal. This conversion was, however, the most noteworthy (at 71.5%) in 39 European countries between 2006 and 2018 (percentage based on the content available in [11]).

Regarding the reuse of developed land, the surface area recycled between 2007 and 2018 in mainland Portugal (6.8 km^2) was six times less than the surface area re-naturalized (40.8 km^2), so the annual average of land recycled (0.6 km^2/year) was very low. This low average is attributable to the fact that almost half of the mainland municipalities (49%) have not recycled developed land in this period. Of the regions that did register recycling, Aveiro stands out with a maximum annual recycling rate of 50.6 m^2/year.km^2, along with the two metropolitan areas and some regions adjacent to these.

A European study on this topic [17,53] reported that, between 2006 and 2012, the contribution of developed land recycling to the creation of artificial surface areas in 318 functional urban areas of medium to large European cities was still relatively low (13.5%). This value is, however, much higher than that observed in the mainland (1.9%) and in the Lezíria do Tejo region (7.1%), which recorded the highest figure.

As for land recycling processes, gray recycling was more relevant than urban development (or densification) in most regions, and in mainland Portugal, where it accounted for 55% of land recycled between 2007 and 2018. In contrast to mainland Portugal, densification was the most frequent recycling process in the above-mentioned European cities. This disparity between results is likely due to the use of different base data. As observed in the aforementioned European cities, green recycling was the least common recycling process in mainland Portugal (4.9%), though most regions have not implemented it.

It should be highlighted that our green recycling estimate is, by definition, lower than that of the EEA [17,53] because it only includes the development of green urban areas from artificial areas. The EEA estimate additionally covers the development of sports and leisure facilities from artificial land, based on the assumption that this creates additional green cover, and therefore increases the area of unsealed land. We have not included this option because our base maps show that more than half of the surfaces occupied by sports and leisure are is sealed.

Since the reuse of developed land prevents land take and the contribution thereof to the creation of artificial surface areas in mainland Portugal was extremely low, we conclude that practice thereof should be encouraged in Portugal.

Our results show that the daily net land take in mainland Portugal (6.2 ha/day) between 2007 and 2018 is still far from the "zero net land take" target, mainly due to the rarity or even lack of either re-naturalization (1 ha/day) or developed land recycling processes (0.2 ha/day).

Between 2007 and 2018, mainland Portugal and most of its regions experienced a decrease in inhabitants and an expansion of artificial surface areas. The Lisbon metropolitan area and Algarve were the only regions that recorded an increase in population, but these increases were less than the growth in their artificial surface areas. In the said period, the rate of land take in mainland Portugal (7.2 ha/day) decreased substantially in comparison to the 1990–2007 period (34 ha/day) [54]. However, in the period 1990–2007, the population increased in the mainland and most of its regions, whereas from 2007 to 2018, there were population declines in most of the territory. Despite the positive reduction in the land take rate between the two periods, the rate observed in the latter period did not follow population dynamics. As most of the non-artificial land taken between 2007 and 2018 served for the development of economic sites and infrastructures, and while some regions still need to invest in this to halt population loss, it is not sustainable for artificial areas to continue to grow over the next decade at the same rate.

The above results show that the current legislation in Portugal to control land use, namely, the instruments to protect agricultural and natural land (often seen by local governments as obstacles to the development of their territories), is not effective enough to promote the re-use of developed land, which is vital to reduce land take. On the other hand, the recent reinforcement in the national legislation of concerns containing urban sprawl and encouraging urban densification and regeneration may have contributed to the fact that their impact on reversing the trend of land take in the mainland is not yet noticeable.

The introduction of the European target in countries such as Portugal, which have multi-level planning systems and where land take presents significant spatial variability, makes it advisable to define differentiated guidelines or targets for the regional and local levels. Since in our country, it is up to the local authorities to authorize and issue building permits under the municipal land use plans, it is paramount that the regional guidelines allow for balance and articulation among the municipalities. Furthermore, it is necessary to make decision-makers and stakeholders aware of the existence of the European goal, inform them about the type of interventions that can help achieve it, and promote stakeholder involvement in the development and implementation of land use plans [13].

As highlighted by Build Europe [15], the target of "no net land take" ignores the different social and economic conditions in European countries and regions. Accordingly, measures aimed at achieving the target should be adapted to the specific context of each country or region. In line with this view, Decoville [33] argues that land take should not be assessed based on a single indicator, as evaluation thereof requires the identification of the main drivers and knowledge of the demographic and economic developments and policy measures in place in the same period.

The definition of interventions to reach the European goal requires a comprehensive evaluation of the phenomena addressed, such as the one provided by this study for mainland Portugal. It is recognized, however, that the study would be improved by regional and municipal assessments similar to those presented for mainland Portugal, which would allow differentiating the driving forces of net land take at these levels and support the definition of more targeted interventions. Quantifying permanent land take is another aspect that could help improve the study. Knowledge of the surface area of sealed land would reveal the most intense form of land take, which is usually irreversible [16].

Shortage of time did not allow the evaluation of the aforementioned aspects. Thus, the research we intend to undertake in the short term will include the regional analysis of both sealed land and land take drivers and the identification of inland urban areas with reuse potential.

5. Conclusions

This study discloses that the term "land take" has different significance among Member States, which prevents comparisons of its estimates across countries. For this reason, clarification of the land use changes that the term covers are needed. Despite the ambiguity of land take's definition, there is evidence that several countries are acting on the message contained in the goal, which is the protection of soils, especially those that are most fertile and productive. To this end, some countries, such as Italy, already assign economic value to soil based on the functions and ecosystems it provides, to classify land and estimate losses and gains due to its transformations [47]. Moreover, the economic valuation of soil also supports the establishment of fees for land take or rating systems for defining building rights [13].

A reduction in net land take is only feasible through an increase in the re-naturalization of developed land, a process that is undertaken in specific and rare circumstances. Accordingly, developed land recycling (or inner urban development) is the only intervention that can seriously mitigate land take. As implementation thereof may be unfeasible in already highly densified urban areas where the population continues to grow, it must be admitted that the "no net land take" target may be unattainable for some countries and regions in Europe by the year 2050. Such countries should pursue more sustainable urban development, which preserves land as a natural resource and maximizes the functions and services it provides.

The descriptions in the first section of some of the initiatives and instruments used to prevent land take also showed that the actions aimed at the 2050 target differ across countries and that the indicators monitored are equally diverse, favoring an adjustment in the target to match the land use planning objectives of the countries or regions. In line with the vision of the ESPON SUPER project, we believe that the ultimate aim of the "no net land take" target should be "to harness the potential of each territory to contribute to European sustainability" [26] (p. 97). Therefore, replication of good practices from other countries or regions (extensively described in [26]), with appropriate vertical and horizontal coordination and adapted to the political and socio-economic contexts, may be more effective in achieving this ultimate aim than policy and regulatory harmonization.

Thus, the recommendations we make regarding the implementation of the European goal are based on the RECARE project [13] and aim at the allocation of dedicated EU funding for more sustainable land use, namely, the increase in green and blue infrastructure, the reuse of brownfield sites, and the creation of binding policy instruments to reduce soil sealing. We also recommend that member states establish laws and regulations to combat land take and financially support local governments in their implementations. Along with the definition of a legal framework concerning land take and soil sealing, they must promote dialogue about the European goal, adopting communication strategies adjusted to the different stakeholders and levels of decision-making.

Regarding mainland Portugal, our findings show low efficiency in urban land use in the mainland and most of its regions (while the mainland's artificial land increased by 5.7%, its population decreased by 2.6%). The main drivers of land take were the development of new industrial and commercial sites and the expansion of construction sites, transportation networks, and residential areas. Our results also show that the recycling of land developed on the mainland is still very low because almost half of the municipalities do not undertake it. These municipalities are some of inland regions that had the most significant population reductions between 2007 and 2018. While some of these regions still need to invest in new jobs and infrastructures to halt population loss, these should be targeted to sites within the urban perimeters with the potential to be reused.

The European guidelines on land take do not seem sufficient for its mitigation in Portugal, so we consider that setting quantitative targets for different levels (national, regional, and municipal) would be more effective for reducing land take.

This study is of interest to spatial planning actors unaware of the European goal and the interventions European nations can undertake to meet it by 2050. The contents presented can serve as an example for countries that have not yet started to monitor net land take and related processes. Moreover, the study is also of interest to those concerned with sustainable urban development.

Author Contributions: Conceptualization, R.N. and B.C.; methodology, R.N.; software, R.N.; validation, R.N. and B.C.; formal analysis, R.N.; investigation, R.N. and B.C.; resources, R.N. and B.C.; data curation, R.N.; writing—original draft preparation, R.N. and B.C.; writing—review and editing, R.N. and B.C.; visualization, R.N. and B.C.; supervision, R.N. and B.C.; project administration, R.N. All authors have read and agreed to the published version of the manuscript.

Funding: This research received no external funding.

Institutional Review Board Statement: Not applicable.

Informed Consent Statement: Not applicable.

Data Availability Statement: The data supporting the presented results are Portuguese Land Cover/Use maps for 2007 and 2018 (known as "Carta de Uso e Ocupação do Solo" or COS), which are freely available on SNIG, the Portuguese National Spatial Data Infrastructure (https://snig.dgterritorio.gov.pt) (accessed on 30 June 2022).

Acknowledgments: The authors are grateful to Filipe Marcelino, from DGT, for the technical support on land cover/land use issues.

Conflicts of Interest: The authors declare no conflict of interest.

Appendix A

Table A1. -CLC land cover classes and LEAC land cover types. Source: adapted from [50,52].

CLC-Level 1	CLC-Level 2	CLC-Level 3	LEAC
1 Artificial surfaces	11 Urban fabric	111 Continuous urban fabric 112 Discontinuous urban fabric	1 Artificial surfaces
	12 Industrial, commercial and transport units	121 Industrial or commercial units 122 Road and rail networks and associated land 123 Port areas 124 Airports	
	13 Mine, dump and construction sites	131 Mineral extraction sites 132 Dump sites 133 Construction sites	
	14 Artificial, non-agricultural vegetated areas	141 Green urban areas 142 Sport and leisure facilities	

Table A1. *Cont.*

CLC-Level 1	CLC-Level 2	CLC-Level 3	LEAC
2 Agricultural areas	21 Arable land	211 Non-irrigated arable land 212 Permanently irrigated land 213 Rice fields	2A Arable land & permanent crops
	22 Permanent crops	221 Vineyards 222 Fruit trees and berry plantations 223 Olive groves	
	23 Pastures	231 Pastures	2B Pastures & mosaic farmland
	24 Heterogeneous agricultural areas	241 Annual crops associated with permanent crops 242 Complex cultivation patterns 243 Land principally occupied by agriculture, with significant areas of natural vegetation 244 Agro-forestry areas	2A Arable land & permanent crops 2B Pastures & mosaic farmland
3 Forest and semi natural areas	31 Forest	311 Broad-leaved forest 312 Coniferous forest 313 Mixed forest	3A Forests and transitional woodland shrub
	32 Shrub and/or herbaceous vegetation associations	321 Natural grassland 322 Moors and heathland 323 Sclerophyllous vegetation	3B Natural grassland, heathland, sclerophyllous vegetation
		324 Transitional woodland/shrub	3A Forests and transitional woodland shrub
	33 Open spaces with little or no vegetation	331 Beaches, dunes, sands 332 Bare rock 333 Sparsely vegetated areas 334 Burnt areas 335 Glaciers and perpetual snow	3C Open space with little or no vegetation
4 Wetlands	41 Inland wetlands	411 Inland marshes 412 Peatbogs	4 Wetlands
	42 Coastal wetlands	421 Salt marshes 422 Salines 423 Intertidal flats	
5 Water bodies	51 Inland waters	511 Water courses 512 Water bodies	5 Water bodies
	52 Marine waters	521 Coastal lagoons 522 Estuaries 523 Sea and ocean	

Appendix B

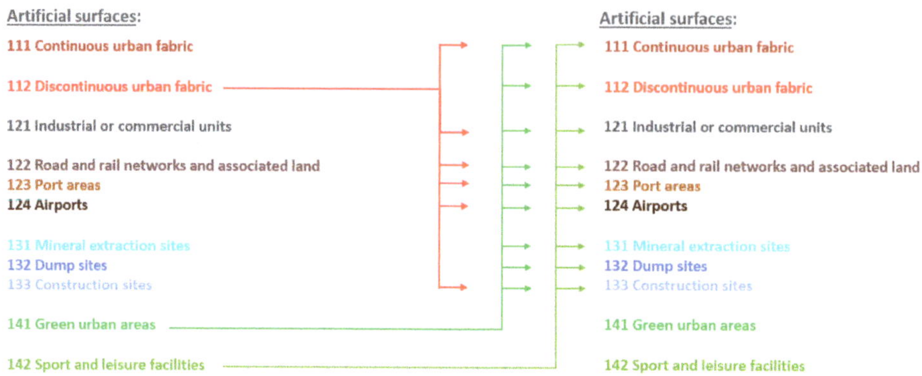

Urban development or densification (LCF11) = Conversion from discontinuous urban fabric (112) to continuous urban fabric (111), industrial, commercial and transport units (12…) and construction sites (133) + Conversion from green urban areas and sport and leisure facilities (14…) to urban fabric (11…), industrial, commercial and transport units (12…) and mineral extraction, dump and construction sites (13…).

Figure A1. Conversions between classes of artificial surfaces (CLC-level 3) covered by the urban development or densification (LCF11).

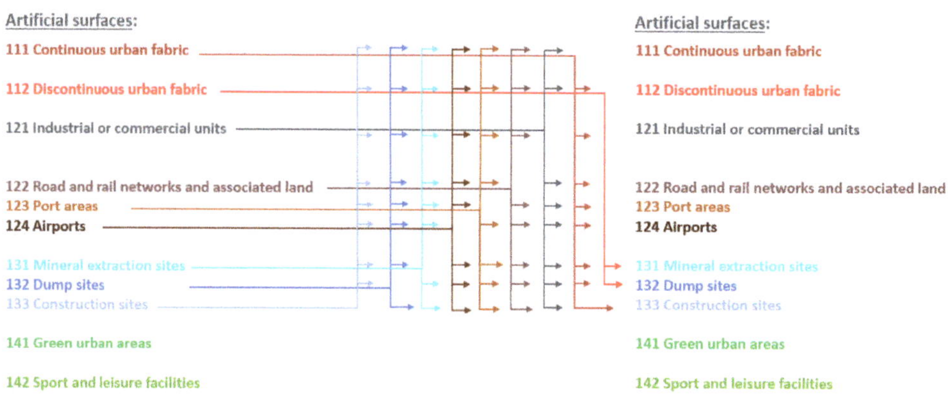

Gray recycling (LCF12) = Conversion from continuous urban fabric (111), industrial, commercial and transport units (12…) and mineral extraction, dump and construction sites (13…) to other artificial areas, except for green urban areas and sport and leisure facilities (14…) + Conversion from discontinuous urban fabric (112) to mineral extraction sites (131) and dump sites (132).

Figure A2. Conversions between classes of artificial surfaces (CLC-level 3) covered by gray recycling (LCF12).

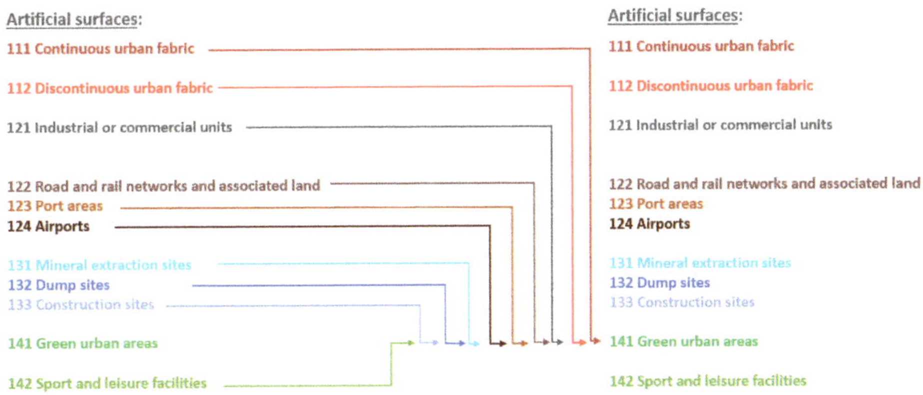

Green recycling (partial LCF13) = Creation of green urban areas (141) from developed land.

Figure A3. Conversions between classes of artificial surfaces (CLC-level 3) covered by green recycling (partial LCF13).

References

1. European Commission. Roadmap to a Resource Efficient Europe. COM/2011/571 Final; Brussels. 20 September 2011. Available online: https://eur-lex.europa.eu/legal-content/EN/TXT/PDF/?uri=CELEX:52011DC0571&from=EN (accessed on 24 May 2022).
2. European Union. 7th Environmental Action Programme of the European Union; Official Journal of the European Union, Strasbourg. 28 December 2013. Available online: https://ec.europa.eu/environment/action-programme/index.htm (accessed on 24 May 2022).
3. European Commission. EU Soil Strategy for 2030. Reaping the Benefits of Healthy Soils for People, Food, Nature and Climate. COM/2021699 Final; DG Environment, Brussels. 17 November 2021. Available online: https://ec.europa.eu/environment/document/download/ae853f10-c9a2-4665-a9f2-c29d11c49374_en (accessed on 24 May 2022).
4. European Commission. DG Environment. Science for Environment Policy. No Net Land Take by 2050? Future Brief 14; Science Communication Unit, UWE, Bristol. April 2016. Available online: http://ec.europa.eu/science-environment-policy (accessed on 24 May 2022).
5. Decoville, A.; Schneider, M. Can the 2050 zero land take objective of the EU be reliably monitored? A comparative study. *J. Land Use Sci.* **2015**, *11*, 331–349. [CrossRef]
6. Marquard, E.; Bartke, S.; Gifreu i Font, J.; Humer, A.; Jonkman, A.; Jürgenson, E.; Marot, N.; Poelmans, L.; Repe, B.; Rybski, R.; et al. Land consumption and land take: Enhancing conceptual clarity for evaluating spatial governance in the EU context. *Sustainability* **2020**, *12*, 8269. [CrossRef]
7. European Environment Agency. Land Take in Europe—Indicator Assessment (CSI 014/LSI 001). 13 December 2019. Available online: https://www.eea.europa.eu/data-and-maps/indicators/land-take-3/assessment (accessed on 24 May 2022).
8. Zambrano, L.; Aronson, M.F.J.; Fernandez, T. The Consequences of Landscape Fragmentation on Socio-Ecological Patterns in a Rapidly Developing Urban Area: A Case Study of the National Autonomous University of Mexico. *Front. Environ. Sci.* **2019**, *7*, 7–152. [CrossRef]
9. European Commission, DG Environment. *Soil Sealing. Science for Environment Policy-In-Depth Reports*; Science Communication Unit, University of the West of England (UWE): Bristol, UK, March 2012; p. 41. Available online: https://ec.europa.eu/environment/archives/soil/pdf/sealing/Soil%20Sealing%20In-depth%20Report%20March%20version_final.pdf (accessed on 18 June 2022).
10. Tobias, S.; Conen, F.; Duss, A.; Wenzel, L.M.; Buser, C.; Alewell, C. Soil sealing and unsealing: State of the art and examples. *Land Degrad. Dev.* **2018**, *29*, 2015–2026. [CrossRef]
11. European Environment Agency. Land Take and Net Land Take. 10 September 2019. Available online: https://www.eea.europa.eu/data-and-maps/dashboards/land-take-statistics (accessed on 24 May 2022).
12. Colsaet, A.; Laurans, Y.; Levrel, H. What drives land take and urban land expansion? A systematic review. *Land Use Policy* **2018**, *79*, 339–349. [CrossRef]
13. Naumann, S.; Frelih-Larsen, A.; Prokopp, G. RECARE-Preventing and Remediating Degradation of Soils in Europe through Land Care. Policy Brief on Soil Sealing and Land Take. September 2018. Available online: https://www.ecologic.eu/sites/default/files/publication/2018/2730_recare_soil-sealing_web.pdf (accessed on 24 May 2022).
14. Intergovernmental Panel on Climate Changes. IPCC Special Report on Climate Change, Desertification, Land Degradation, Sustainable Land Management, Food Security, and Greenhouse Gas Fluxes in Terrestrial Ecosystems. Summary for Policymakers.

Approved Draft. 7 August 2019. Available online: https://www.ipcc.ch/site/assets/uploads/2019/08/4.-SPM_Approved_Microsite_FINAL.pdf (accessed on 24 May 2022).
15. Build Europe. No Net Land Take by 2050 Solving the Unsolvable–10 Solutions for an Improved Management of European Land. 13 January 2022, p. 32. Available online: https://buildeurope.net/wp-content/uploads/2022/01/No-net-land-take-by-2050-Solving-the-unsolvable.pdf (accessed on 24 May 2022).
16. European Commission, Directorate-General for Environment; Prokop, G.; Jobstmann, H.; Schönbauer, A. *Overview of Best Practices for Limiting Soil Sealing or Mitigating Its Effects in EU-27: Final Report*; Publications Office: Brussels, Belgium, 2011; p. 227. Available online: https://data.europa.eu/doi/10.2779/15146 (accessed on 24 May 2022).
17. European Environment Agency. *Land Recycling in Europe-Approaches to Measuring Extent and Impacts*; EEA Report No. 31/2016; Publications Office of the European Union: Luxembourg, 2016; p. 56.
18. European Union. *European Spatial Development Perspective-Towards Balanced and Sustainable Development of the Territory of the European Union*; Official Publications of the European Communities: Luxembourg, 1999.
19. European Union. Territorial Agenda of the European Union. Towards a More Competitive and Sustainable Europe of Diverse Regions. 2007. Available online: https://ec.europa.eu/regional_policy/en/information/publications/communications/2007/territorial-agenda-of-the-european-union-towards-a-more-competitive-and-sustainable-europe-of-diverse-regions (accessed on 24 May 2022).
20. European Union. Territorial Agenda of the European Union 2020. Towards an Inclusive, Smart and Sustainable Europe of Diverse Regions. 2011. Available online: https://ec.europa.eu/regional_policy/en/information/publications/communications/2011/territorial-agenda-of-the-european-union-2020 (accessed on 24 May 2022).
21. European Union. Territorial Agenda–A Future for All Places. 2020. Available online: https://www.territorialagenda.eu/home.html (accessed on 24 May 2022).
22. United Nations. New Urban Agenda–Habitat III. 2017. Available online: https://habitat3.org/wp-content/uploads/NUA-English.pdf (accessed on 24 May 2022).
23. United Nations. Transforming Our World: The 2030 Agenda for Sustainable Development. 2015. Available online: https://sdgs.un.org/2030agenda (accessed on 24 May 2022).
24. Cowie, A.; Orr, B.; Sanchez, V.; Chasek, P.; Crossman, N.; Erlewein, A.; Louwagie, G.; Maron, M.; Metternicht, G.; Minelli, S.; et al. Land in balance: The scientific conceptual framework for Land Degradation Neutrality. *Environ. Sci. Policy* **2018**, *79*, 25–35. [CrossRef]
25. ESPON. *SUPER Project-Sustainable Urbanisation and Land-Use Practices in European Regions. Main Report*; ESPON EGTC: Luxembourg, November 2020; p. 86. Available online: https://www.espon.eu/sites/default/files/attachments/ESPON%20SUPER%20Final%20Report%20-%20Main%20report_newtemplate.pdf (accessed on 30 June 2022).
26. ESPON. SUPER Project-Sustainable Urbanisation and Land-Use Practices in European Regions. In *A Guide to Sustainable Urbanisation and Land-Use*; Cotella, G., Evers, D., Rivolin, U.J., Solly, A., Berisha, E., Eds.; ESPON EGTC: Luxembourg, November 2020; p. 108. Available online: https://www.espon.eu/sites/default/files/attachments/2020_ESPON_SUPER_Guide_final_A4_screenview.pdf (accessed on 30 June 2022).
27. Schatz, E.; Bovet, J.; Lieder, S.; Schroeter-Schlaack, C.; Strunz, S.; Marquard, E. Land take in environmental assessments: Recent advances and persisting challenges in selected EU countries. *Land Use Policy* **2021**, *111*, 105730. [CrossRef]
28. Copernicus. CORINE Land Cover. 2022. Available online: https://land.copernicus.eu/pan-european/corine-land-cover (accessed on 24 May 2022).
29. Copernicus. Urban Atlas. 2022. Available online: https://land.copernicus.eu/local/urban-atlas (accessed on 24 May 2022).
30. Eurostat Statistics Explained. LUCAS–Land Use and Land Cover Survey. 2021. Available online: https://ec.europa.eu/eurostat/statistics-explained/index.php?title=LUCAS_-_Land_use_and_land_cover_survey (accessed on 24 May 2022).
31. Copernicus. Imperviousness. 2022. Available online: https://land.copernicus.eu/pan-european/high-resolution-layers/imperviousness (accessed on 24 May 2022).
32. Wealer, R. Comment Integrér le Principe de "Zero Artificialisation Nette" dans le PDAT en Cours de Refonte? [How to Integrate the Principle of "Net Zero Artificialization" in the PDAT revision?]. ESPON Peer Learning Workshop: How to Reach no Net Land Take by 2050? Belgium. 9 February 2022. Available online: https://www.espon.eu/no-net-land-take-workshop (accessed on 24 May 2022).
33. Decoville, A. Use and misuse of indicators in spatial planning: The example of land take. *Plan. Pract. Res.* **2018**, *33*, 70–85. [CrossRef]
34. Federal Ministry Republic of Austria Climate Action, Environment, Energy, Mobility, Innovation and Technology. Flächenrecycling und Brachflächen-Dialog [Land Recycling and Brownfields Dialogue]. 2022. Available online: https://www.bmk.gv.at/themen/klima_umwelt/abfall/flr-bfd.html (accessed on 30 June 2022).
35. Umweltbundesamt. Flächeninanspruchnahme [Land Take]. 2022. Available online: https://www.umweltbundesamt.at/umweltthemen/boden/flaecheninanspruchnahme (accessed on 24 May 2022).
36. Geodynamix. Land Use Database. 2018. Available online: https://geodynamix.eu/land-use-database (accessed on 24 May 2022).
37. Bovet, J.; Marquard, E.; Schröter-Schlaack, C. Workshop Report. In Proceedings of the International Expert Workshop on Land Take, Berlin, Germany, 4–5 April 2019; Available online: https://www.ufz.de/export/data/464/239439_235934_SURFACE%20Workshop%20report_final_2019-11-07.pdf (accessed on 24 May 2022).

38. Vanongeval, S.; Vanacker, S.; Bieseman, H.; Pisman, A.; Wille, E.; Isenborghs, C. Land Take Objectives and Strategies in Flanders (Belgium). In Proceedings of the SURFACE Virtual Symposium, Virtual, 4 February 2022; Available online: https://www.ufz.de/export/data/464/262038_Land%20take%20objectives%20and%20strategies%20in%20Flanders.pdf (accessed on 24 May 2022).
39. Ruelle, C. No Net Land Take by 2050 in Wallonia? In Proceedings of the International Conference on Soils, Sediments and Water-Intersoil 2020, Brussels, Belgium, 2–3 March 2020; Available online: https://orbi.uliege.be/bitstream/2268/246065/1/Ruelle-Intersoil-3Mars2020.pdf (accessed on 24 May 2022).
40. Schorcht, M.; Krüger, T.; Meinel, G. Measuring Land Take: Usability of National Topographic Databases as Input for Land Use Change Analysis: A Case Study from Germany. *ISPRS Int. J. Geo-Inf.* **2016**, *5*, 134. [CrossRef]
41. Umwelt Bundesamt. Brownfield Redevelopment and Inner Urban Development. 2014. Available online: https://www.umweltbundesamt.de/en/topics/soil-agriculture/land-use-reduction/brownfield-redevelopment-inner-urban-development#inner-urban-development-and-brownfield-redevelopment (accessed on 24 May 2022).
42. Fina, S. Land Take Data in a European Context–Comparing Apples and Oranges. In Proceedings of the SURFACE Virtual Symposium, Virtual, 4 February 2022. Available online: https://www.ufz.de/export/data/464/262036_Land%20take%20data%20in%20a%20European%20context%20%E2%80%93%20Comparing%20apples%20and%20oranges.pdf (accessed on 24 May 2022).
43. Umwelt Bundesamt. Land Take: Make Goals Binding, UBA Recommends a Quota System and Land Certificate Trading Scheme. 2017. Available online: https://www.umweltbundesamt.de/en/press/pressinformation/land-take-make-goals-binding (accessed on 24 May 2022).
44. Leibniz Institute of Ecological Urban and Regional Development. Monitor of Settlement and Open Space Development. 2022. Available online: https://www.ioer-monitor.de/en/results/analysis-results/land-take/ (accessed on 24 May 2022).
45. République Française, France Stratégie. Objectif «zéro artificialisation nette»: Quels Leviers Pour Protéger les Sols ? [Objective "Zero Net Artificialization": Which Levers Should Be Used to Protect Soils?]. 2019. Available online: https://www.strategie.gouv.fr/publications/objectif-zero-artificialisation-nette-leviers-proteger-sols (accessed on 24 May 2022).
46. Gouvernement Francais. Portail de L'artificialisation des sols [Portal of the Artificialization of Soils]. 2021. Available online: https://artificialisation.developpement-durable.gouv.fr/suivi-consommation-espaces-naf#paragraph--2164 (accessed on 24 May 2022).
47. Sistema Nazionale per la Protezione dell'Ambiente; Munafò, M. (Eds.) Consumo di Suolo, Dinamiche Territoriali e Servizi Ecosistemici. Edizione 2021. [Land Use, Spatial Dynamics, and Ecosystem Services. 2021 Edition]. Report SNPA 22/2021; Italy. July 2021. Available online: https://www.snpambiente.it/wp-content/uploads/2021/11/Rapporto_consumo_di_suolo_2021.pdf (accessed on 24 May 2022).
48. Strollo, A.; Smiraglia, D.; Bruno, R.; Assennato, F.; Congedo, L.; Fioravante, P.; Giuliani, C.; Marinosci, I.; Riitano, N.; Munafò, M. Land consumption in Italy. *J. Maps* **2020**, *16*, 113–123. [CrossRef]
49. Wong, C.; Bäing, A.S. Brownfield Residential Redevelopment in England. What Happens to the Most Deprived Neighbourhoods? Joseph Rowntree Foundation. June 2010, p. 28. Available online: https://www.jrf.org.uk/sites/default/files/jrf/migrated/files/brownfield-residential-redevelopment-full.pdf (accessed on 24 May 2022).
50. Haines-Young, R.; Weber, J.-L. European Environment Agency. In *Land Accounts for Europe 1990–2000-Towards Integrated Land and Ecosystem Accounting*; EEA Report No. 11/2006; Office for Official Publications of the European Communities: Luxembourg, 2006; p. 107.
51. European Environment Agency. Land Cover Accounts—An Approach to Geospatial Environmental Accounting. 2020. Available online: https://www.eea.europa.eu/themes/landuse/land-accounting (accessed on 24 May 2022).
52. Ivits, E.; Milego, R.; Mancosu, E.; Gregor, M.; Petersen, J.; Büttner, G.; Löhnertz, M.; Maucha, G.; Petrik, O.; Bastrup-Birk, A.; et al. European Environment Agency & European Topic Centre on Urban, Land and Soil Systems (ETC-ULS). In *Land and Ecosystem Accounts for Europe towards Geospatial Environmental Accounting*; ETC/ULS Report 02/2020; ETC-ULS: Vienna, Austria; p. 78.
53. European Environment Agency. Land Recycling and Densification. 2018. Available online: https://www.eea.europa.eu/data-and-maps/indicators/land-recycling-and-densification/assessment-1 (accessed on 24 May 2022).
54. Nicolau, R.; Condessa, B. *Dinâmicas de Artificialização do Solo em Portugal Continental nos períodos 1990–2007 e 2007–2018 [Land Artificialization Dynamics in Mainland Portugal in the Periods 1990–2007 and 2007–2018]*; Direção Geral do Território: Lisboa, Portugal, January 2022; p. 52. Available online: https://www.dgterritorio.gov.pt/sites/default/files/projetos/Nicolau_Condessa_31janeiro2022.pdf (accessed on 24 May 2022).

land

Article

Dynamic Modeling of Land Use and Coverage Changes in the Dryland Pernambuco, Brazil

Cinthia Pereira de Oliveira [1], Robson Borges de Lima [1,*], Francisco Tarcísio Alves Junior [1], Mayara Maria de Lima Pessoa [2], Anderson Francisco da Silva [2], Nattan Adler Tavares dos Santos [2], Iran Jorge Corrêa Lopes [3], Cybelle Laís Souto-Maior Sales de Melo [2], Emanuel Araújo Silva [2], José Antônio Aleixo da Silva [2] and Rinaldo Luiz Caraciolo Ferreira [2]

1 Laboratório de Manejo Florestal, Universidade do Estado do Amapá, Macapá CEP 68901–262, Brazil; cinthia.oliveira@ueap.edu.br (C.P.d.O.); francisco.junior@ueap.edu.br (F.T.A.J.)
2 Laboratório de Manejo de Florestas Naturais "José Serafim Feitosa Ferraz", Departamento de Ciência Florestal, Universidade Federal Rural de Pernambuco, Recife CEP 52171-900, Brazil; mayara.pessoa@ifmg.edu.br (M.M.d.L.P.); engf.anderson@gmail.com (A.F.d.S.); nattantavares@gmail.com (N.A.T.d.S.); engf.cybelle@gmail.com (C.L.S.-M.S.d.M.); emanuel.araujo@ufrpe.br (E.A.S.); jaaleixo@uol.com.br (J.A.A.d.S.); rinaldo.ferreira@ufrpe.br (R.L.C.F.)
3 Departamento de Ciência Florestal, Universidade Federal do Paraná, Curitiba CEP 80210-170, Brazil; iranjorge._@hotmail.com
* Correspondence: robson.lima@ueap.edu.br

Abstract: The objective of this work was to carry out a multitemporal analysis of changes in land use and land cover in the municipality of Floresta, Pernambuco State in Brazil. Landsat images were used in the years 1985, 1989, 1993, 1997, 2001, 2005, 2009, 2014, and 2019, and the classes were broken down into areas: water, exposed soil, agriculture, and forestry, and using the Bhattacharya classifier, the classification was carried out for generating land use maps. The data was validated by the Kappa index and points collected in the field, and the projection of the dynamics of use for 2024 was constructed. The thematic maps of land use and coverage from 1985 to 2019 show more significant changes in the forest and exposed soil classes. The increase in the forest class and the consequent reduction in exposed soil are consequences of the interaction between climate and human activities and the quality of the spatial resolution of the satellite images used between the years analyzed.

Keywords: caatinga domain; digital classification; remote sensing

Citation: de Oliveira, C.P.; de Lima, R.B.; Alves Junior, F.T.; de Lima Pessoa, M.M.; da Silva, A.F.; dos Santos, N.A.T.; Lopes, I.J.C.; de Melo, C.L.S.-M.S.; Silva, E.A.; da Silva, J.A.A.; et al. Dynamic Modeling of Land Use and Coverage Changes in the Dryland Pernambuco, Brazil. *Land* **2022**, *11*, 998. https://doi.org/10.3390/land11070998

Academic Editor: Giusppe Pulighe

Received: 26 May 2022
Accepted: 23 June 2022
Published: 30 June 2022

1. Introduction

The dry forests of the Brazilian semiarid, known as Caatinga, have been going through a continuous and lengthy reduction in their coverage [1–3]. In short, changes in land use and land cover are prominent, caused mainly by the advance in agriculture and livestock farming, raising goats and cattle, etc. [4–6], exploitation of wood and non-wood products (firewood, charcoal, fodder, etc.), in addition to urban expansion [7], as well as expansion of infrastructure and changes in the land structure. Despite being responsible for meeting the demand for forest resources in the Northeast region of Brazil, multitemporal studies on changes in their use and coverage are still incipient.

Knowledge of the human and biophysical dimensions of changes in tropical dry forests and their effects is highlighted as a priority for research [8,9]. Through remote sensing tools, there is the possibility of answers that contribute to identifying problems inherent to unrestricted land use in drylands [10] and, consequently, related to the reduction of forest cover in these areas [11].

Our current understanding of the importance of this ecosystem has been generated using remote-sensing approaches that provide spatially-explicit values relating to forest area, land cover, topography, soil, and climate variables. This information is widely used in many dynamic models for generating predictive maps of land cover and land-use changes [4,6]. Although these maps have improved our understanding of the morphoclimatic characteristics of the caatinga, they currently do not address land-cover predictions, which are essential for environmental management.

Therefore, a better understanding of the spatial and temporal dynamics of land use forms and their potential drivers in recent years is needed to be projected into future scenarios as an effective way to inform environmental policy and decision-making. Importantly, spatially explicit scenarios can anticipate the magnitude and distribution of land-cover loss, thus providing valuable information to develop corresponding measures to manage, for example, deforestation and desertification and mitigate their impacts. Simulated scenarios can also be used to evaluate development policies, which involve proposals to build infrastructure in strategic natural systems, the establishment of land protection schemes [12], or the assessment of the combined effects of climate change (e.g., [13]).

The use of remote sensing, primarily orbital, as an aid to planning activities related to natural resources and the environment has facilitated, over the years, studies in different ecosystems [5,14–16] and allied to these techniques, spatial simulation models have been receiving greater attention from researchers, becoming a promising field of research [17,18].

Spatial or landscape simulation models simulate changes in environmental attributes across the geographic territory [19,20] and seek to help understand the causal mechanisms and development processes of environmental systems, and thereby determine how they evolve under a set of circumstances over time [21]. Therefore, data from remote sensing of the landscape and modeling together with field surveys become potentially relevant for disseminating sustainable forest management, especially in Pernambuco, as well as being essential tools for the formulation of public policies and environmental in the future region [11].

In order to provide information that better supports planning and land use in the medium term, the objective of this work was to carry out a multitemporal analysis of changes in land use and land cover in the municipality of Floresta in Pernambuco State in Brazil. As secondary objectives, we sought to (1) Understand the changes in land use and land cover from 1989 to 2019, based on the production of maps; (2) Analyze land use and land cover change (LULCC) conversions by investigating impacts resulting from 10 years (from 2014 to 2024) of changes (in a dry forest area from remote sensing tools.

2. Materials and Methods

2.1. Study Area

The study was conducted in the municipality of Floresta (Figure 1), located 433 km west of the city of Recife, in the São Francisco mesoregion and Itaparica microregion, Pernambuco, Brazil. The municipality covers an area of 3674.9 km^2, with an average altitude of 316 m, and is located at geographic coordinates 8°36′02″ S latitude and 38°34′05″ W longitude.

According to Köppen's climate classification, the region's climate is of the BS'h type, which reports a warm semi-arid climate. The average total annual precipitation is between 200 and 800 mm, with a concentrated rain period from January to May, with the wettest months being March and April [22]. The average annual air temperature is more significant than 26 °C. The soil in the region is classified as Chronic Luvissol, characterized as shallow and usually presenting an abrupt change in its texture [23].

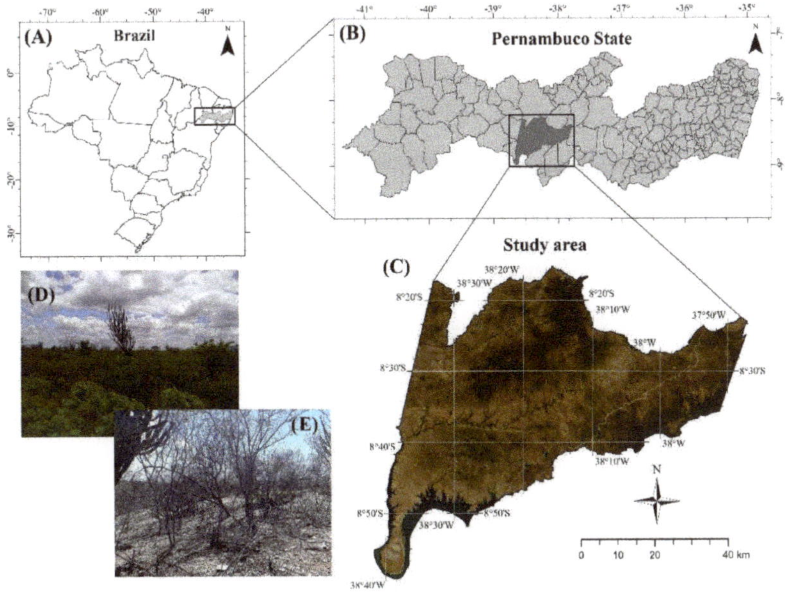

Figure 1. Coverage of the study area: (**A–C**), and photos of vegetation profiles in rainy (**D**) and dry season (**E**), in the sertão of Pernambuco, Brazil.

2.2. Classification of Land Use and Land Cover

Landsat-5 sensor TM (TematicMapper) satellite images from the years 1985, 1989, 1993, 1997, 2001, 2005, 2009, and Landsat-8 were used, with the sensor OLI (Operational Land Imager) from the years 2014 and 2019, acquired free of charge from the image catalog of the National Institute for Space Research (INPE), with cloud-cover rates of less than 30% and 30 m spatial resolution, for orbit/point 216/66, comprising a scene for each date evaluated; the images obtained from the TM sensor needed to be registered spatially. Image acquisition dates are shown in Table 1.

Table 1. Date of acquisition of the evaluated images.

Satellite	Acquisition Date	Orbit/Point	Spatial Resolution	Spectral Bands Used	Spectral Range (μm)
Landsat 5	1 October 1985 28 October 1989 7 October 1993 2 October 1997 27 September 2001 24 October 2005 20 November 2009	216/66	30	1 2 3 4 5 7	(0.45–0.52) (0.52–0.60) (0.63–0.69) (0.76–0.90) (1.55–1.75) (2.08–2.35)
Landsat 8	23 March 2014 29 October 2019	216/66	30	2 3 4 5 6 7	(0.45–0.51) (0.53–0.59) (0.64–0.67) (0.85–0.88) (1.57–1.65) (2.11–2.29)

For the digital classification, it was necessary to perform the image segmentation. For the Landsat-5 and Landsat-8 satellite images, the values of spectral similarity of 12 and 10 and the area sizes of 15 and 100 pixels were adopted, respectively. From the

Bhattacharya classifier implemented in the Spring software, the following thematic classes were identified:

- vegetation (areas covered with natural forest)
- farming (areas intended for agriculture and livestock)
- water (all watercourses present in the area of study)
- exposed soil (uncovered areas of vegetation and in the soil preparation phase and agricultural fallow)

The images generated from the classifications were quantified areas (hectares), according to thematic classes and generated maps of land use and land cover for all mapped dates.

To verify the reliability of the digital classification of land use and land cover in the municipality of Floresta, the Kappa index [24] was used, calculated from the confusion matrix, obtained during the training sample collection phase in each of the classified images. The acceptance intervals of the Kappa index (K) results followed the classification suggested by [24], in which it is categorized as "poor" when K is less than 0.4, "reasonable" with a K of 0,4 to 0.8, and "excellent" with K greater than 0.8.

The validation was carried out from georeferenced points in loco with a GPS device Garmin® GPSMAP 62sc (Chicado, IL, USA). A photographic record was carried out to compare the data from the digital classification of the year 2014.

2.3. *Dynamic Spatial Modeling*

For the input data of the model in the dynamic variables, only the thematic maps of land use and land cover for the years 2009 and 2014 were used, and the static variables were represented by the maps of hydrography, urban areas, road network, slope, altimetry, soils and geology of the study area. The urban area was vectored based on Landsat 5 and 8 images. The maps of the road network, water network, soils, and geology were generated from shapefiles of the State of Pernambuco from the Mineral Resources Research Company (CPRM) database. Altimetry and slope maps were generated using Spring software (version 5.2.6) based on NASA's Shuttle Radar Topography Mission (SRTM). The vector maps were converted to matrix format and standardized in the exact spatial resolution, the number of rows and columns, and the same coordinate system with Universal Transverse Mercator (UTM) and Datum WGS-84 projection.

Dynamic spatial modeling was performed in Dinamica EGO software, version 2.4.1. Moreover, it was divided into three stages: (1) construction and calibration of the model, (2) simulation, and (3) validation. The construction and calibration of the model were performed from the calculation of historical transition matrices, indicating the variation of land use and land-cover classes at two different time points, obtaining the transitions that occurred annually (multiple-step matrix) and the changes that happened in the total study interval (single-step matrix), in this case, five years, corresponding to the years between 2009 and 2014.

Once the transition rates were obtained, it was possible to perform, based on Bayes' conditional probability theorem, the method of weights of evidence adopted by Dinamica EGO for the definition of transition probabilities, which visualized the areas that are more favorable for possible changes. The procedure to follow was the calculation of the coefficients, using as input data the result of the weights of evidence method, initial and final land-use map, and static variables. The Weights of Evidence method assumes that the input maps must be spatially independent. The Cramer indices and the Join Information Uncertainty were used to assess this correlation between variables; with the selection requirement for the variables to remain in the model, a correlation threshold of 0.5 was adopted, and variables that presented a correlation above 0.5 were discarded.

The algorithms incorporated in Dinamica EGO (patcher and expander) and the isometry index and the variance of the changing area calculated in the change map were considered for the simulation model of the transitions of the spatial patterns of the use classes. In order to obtain the most suitable model, tests were carried out varying the input parameters. Model validation was performed by the fuzzy similarity comparison test between the 2014 simulated map and the reference map for the same date; the closer to 1, the greater the similarity between the maps; thus, the distinctions being identified between the maps of actual end and initial use, and simulated ending and natural starting.

With the validation of the model, it was possible to simulate the scenarios for 2024 with the help of the SPRING software (version 5.2.6), quantifying land use and coverage and also observing the trends in class changes (Agriculture, Exposed Soil, Water, and Vegetation) on the map of initial use (2014) and of the simulated use map (2024).

3. Results and Discussion

3.1. Land Use and Coverage

The Kappa index values obtained for the municipality of Floresta using the error matrix of classified images of the years under study showed excellent acceptance levels for the most part, except for the year 2001, which was categorized as reasonable. The thematic maps obtained by the supervised digital classification process—the Bhattacharya algorithm, in the municipality of Floresta, allowed the visualization of the spatial distribution of the thematic classes (Figure 2) and their dynamic and quantification (Figure 3).

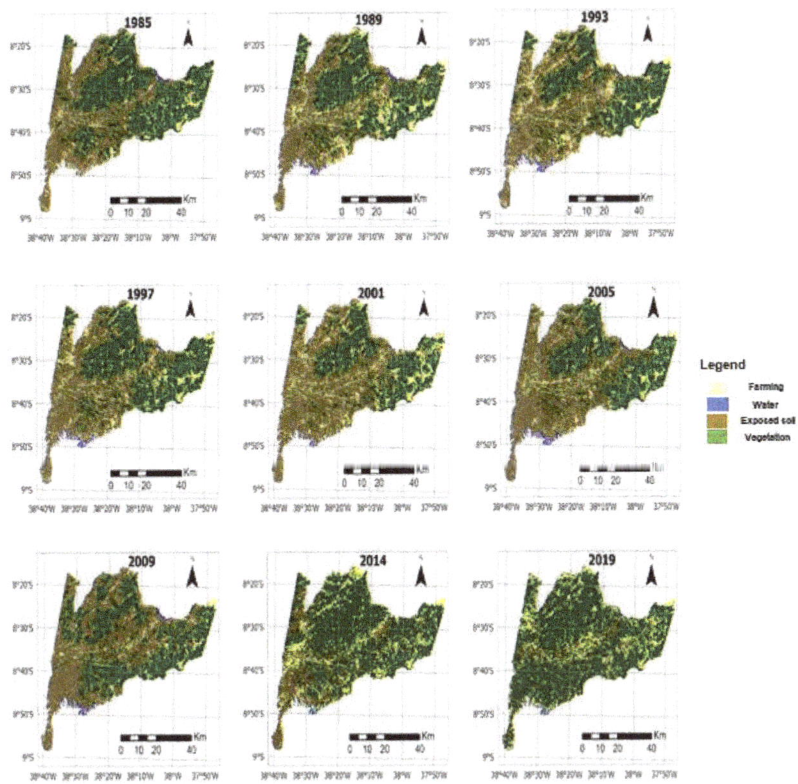

Figure 2. Thematic maps of land use in the municipality of Floresta-PE from 1985 to 2019.

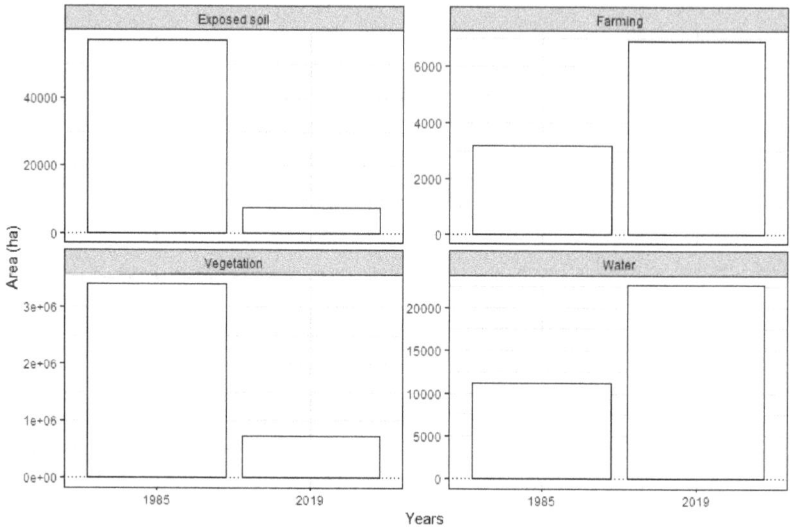

Figure 3. Dynamics and quantification of thematic classes through digital classification in the municipality of Floresta-PE.

Among the various areas where the 25 points marked in loco in the municipality of Floresta were collected, only seven did not correspond to the classification of the images, corresponding to 72% of correct answers. However, for the forest class, the correct answer was only 50%, which is associated with strong seasonality and high heterogeneity in terms of phytophysiognomy of the Caatinga, making the digital-image classification process difficult. Classifications in which the Kappa index indicates excellence can be found when working with a reduced number of classes [25].

The results corroborate [26] that for classifications involving four to seven classes, the use of the confusion matrix presents more minor variations. However, it is worth emphasizing the issues [27] raised regarding the basic assumptions underlying the accuracy assessment, such as generalization of the number of classes, mixed-pixel problems, incorrect category registrations, and sampling plan. It is also noteworthy that a problem associated with multitemporal remote-sensing data for detecting changes is that they do not have the same date (day/month), which varies between solar incidence angles, atmospheric conditions, and soil moisture [28].

The forest class presented an area in 2001 of 119,962.44 hectares, representing a smaller area compared to 2014 (218,602.62 ha.), equivalent to 61.70% of the area this year (Figure 3). However, for 2014 the classification was influenced by rainfall, as it was lower and unevenly distributed when observed in other periods (Figure 4). Thus, it is recommended to compare maps from 1985 to 2009, since rainfall is no more significant influence. Therefore, it can be observed that between 1985 and 2009, there was a reduction in the forest and agricultural classes, from 48.86 to 41.69% and from 10.0% to 7.58%, respectively. In comparison, the exposed-soil class increased from 40.58 to 49.64% and water from 0.58 to 1.09%.

According to Silva et al. [11], in a study in the municipality of Floresta, the removal of vegetation was necessary due to the works to transpose the São Francisco river from the east axis (started in 2007), which runs from the municipality of Floresta (PE) to Monteiro (PB), in a 220 km route in which 430 ha were deforested to make way for canals, reservoirs, construction sites, service roads and places for earth and stone extraction. Compared with the years between 1985 and 2009, the results found in this study reveal that it may be a reflection of these works, which are still in the execution phase.

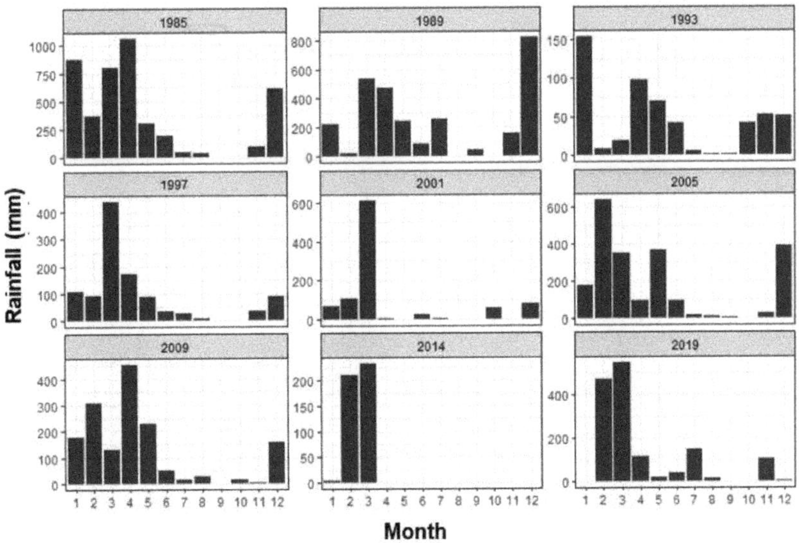

Figure 4. Monthly rainfall in the study area according to image acquisition dates, Floresta-PE.

The increase in exposed soil during the study period requires attention since the municipality of Floresta is inserted in the Cabrobó desertification nucleus [29,30], and according to [31], this class is a characteristic fundamental of this phenomenon in the semi-arid region of the Northeast, whose problem can worsen as a result of the successive droughts that devastate the Northeast.

There is more incredible difficulty in the digital classification of images in these areas, as the vegetation is of reduced size and greater spacing between woody individuals than in the other vegetation physiognomies of the study area, generally coinciding with the presence of the steppe savanna wooded and open. Still, the water class had a lower percentage share (0.44) in the study area, probably due to the long period of drought that has passed through this region since 2009, which did not allow the restoration of the most significant area observed in 2005 (4707.90 ha).

Except 2001 and 2014, the other years showed an increasing area of the water class, which can be explained by [2,11], due to the creation in 1988 of the Luiz Gonzaga Hydroelectric Power Plant (Itaparica) in Petrolândia-PE, which produced a greater flow of water for the Municipality of Floresta with the widening of the São Francisco River and also because this period had a more significant presence of public policies to alleviate the drought in construction of wells and weirs.

The reduction in agriculture observed between 1985 and 2009 can be explained by the fact that this class and exposed soils are closely linked, as they are part of agricultural areas [4,6,15,32]. In addition, exposed soils are generally fallow or under crop preparation [2]. Still, it may also reflect the prolonged period of drought that the region has been experiencing since 2009, corroborating the assertions of Soares [27] and Mariano et al. [31] that in times of drought, agriculture is seriously compromised. In addition, and among the main income-generating activities, the removal of firewood stands out, which, together with agriculture, promotes substantial changes in the caatinga vegetation and soils.

3.2. Dynamic Modeling of Land Use and Land Cover

The weight of evidence allowed us to infer what contribution a class occurred in each transition. The positive weights of evidence favor the transition's occurrence (Table 2). The highest positive values achieved in each class were considered.

Table 2. Continuous and static variables that most influenced land use and land cover transitions in the municipality of Floresta-PE.

Transition	Local Variables (0 to 500 m)	Weight of Evidence (W+)
Farming → Water	Vegetation	0.9936
Farming → exposed soil	Water	0.759
Farming → Vegetation	Water	0.8178
Water → Farming	exposed soil	0.9434
Water → exposed soil	Farming	0.9946
Water → Vegetation	Vegetation	0.891
exposed soil → Farming	Slop	0.9445
exposed soil → Water	Water	0.9163
exposed soil → Vegetation	Vegetation	0.7112
Vegetation → Farming	Farming	0.9496
Vegetation → Water	Hypsometric	0.9581
Vegetation → exposed soil	Farming	0.9551

It was observed that the transition from vegetation to agriculture and exposed soil was influenced by agriculture itself, which can be explained by the high demand for this activity in the municipality. The expansion of vegetation with agriculture and exposed soil undergoing the transition to vegetation was explained by the variable water and vegetation having to be taken into account for this result when the image was obtained. The existence of dependence in the maps tested was observed only for the variable "exposed soil," in which it presented a Cramer Index (V) greater than 0.5, as for the Joint Uncertainty Index (U), this variable presented values less than 0. 5 (Table 3). As it is an essential variable for the model, it was not excluded from it.

Table 3. Higher Cramer Index and Joint Information Uncertainty values in the model variables.

Variable	Cramer (V)	Uncertainty of Information Joint (U)
Exposed soil	0.54844644	0.285369436
Exposed soil	0.548146466	0.332466886
Exposed soil	0.547584512	0.367988266
Exposed soil	0.530289258	0.199917044
Exposed soil	0.530248771	0.26247358
Exposed soil	0.530062551	0.324499163

From the simulation performed in the Dinamica EGO software, the simulated map of the year 2014 was obtained, compared with the classified map of the same year to observe the quality of the model (Figure 5). The fuzzy similarity index values (Table 4), obtained from the constant and exponential decay functions for different sizes of windows with gradual clustering of pixels, presented good values in the literature.

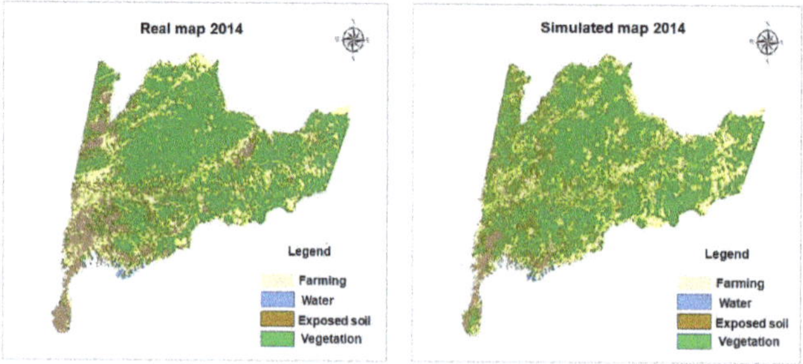

Figure 5. Simulated land use map for the municipality of Floresta, Pernambuco.

Table 4. Fuzzy similarity indices obtained from the constant and exponential decay functions for different window sizes in the periods between 2009 and 2014.

Similarity Index *Fuzzy*		
Windows (Pixels)	Decay Function	
	Constant	Exponential
1 × 1	0.4713	0.4713
3 × 3	0.5944	0.5373
5 × 5	0.6448	0.5605
7 × 7	0.6820	0.5720
9 × 9	0.7132	0.5785
11 × 11	0.7399	0.5822

Ferrari [33], for example, in an Atlantic Forest ecosystem, obtained a fuzzy similarity value for 11 × 11 windows and a constant decay function of 0.84. Macedo [34], in the border region between Cerrado and Atlantic Forest, obtained a fuzzy similarity value of 0.52 as a function of constant decay for the same windows. The generation of future scenarios, or the simulation of maps a posteriori, is illustrated in Figure 6 over ten years. The first map is presented corresponding to the 2014 map used as a reference for the comparison.

Obtaining simulated maps for ten years allowed the quantification of the conversion rates of classes between the years 2014 and 2024. Table 5 shows the modeling results for the municipality of Floresta-PE.

Table 5. Quantification of the future scenario of the municipality of Floresta-PE and comparison with 2014.

Class	Área 2014 (ha)	Área 2024 (ha)	2014–2024 (ha)	2014–2024 (%)
Vegetation	218,602.62	229,940.64	11,338.02	5.19
Farming	55,365.75	61,320.78	5955.03	10.76
Exposed soil	78,790.59	62,569.71	−16,220.88	−20.59
Water	1555.47	801.72	−753.75	−48.46
Total	354,314.43	354,632.85	-	-

It was found that there is an increase linked to the areas of vegetation and agriculture in the municipality of Floresta. The areas of exposed soil had a considerable drop. According to Benedetti [25], the trend is that if the area is maintained, the same conditions as extensive activities such as agriculture will be reduced over time. From the evolutionary analysis of land use and land cover, as well as the spatial, historical survey of the occupation of the area and its prediction of how its uses will tend to behave in the future, it is possible to understand the location of the areas of these uses and the changes to which these areas are likely to be subject [19].

However, it is essential to note that our analyzes do not specifically consider direct anthropogenic factors in the modeling. Furthermore, if the effects of indirect factors, such as feedback from the surface atmosphere, were also considered, the resulting simulated vegetation area could represent significant decreases motivated by the increased severity of droughts and fires [35,36]. The risk of triggering these processes of amplification of forest loss and, therefore, reduction of vegetation cover is possibly more significant in the scenario of currently imminent climate change [37,38]. Under current deforestation trends, not only does forest loss increase, but the remaining forest areas become more fragmented, impacting their ecological functions and the future stability of the ecosystem.

Finally, while the approaches presented here help to draw relevant links between cause and effect of changing spatial points and ecological processes in tropical, dry forest landscapes, inferring complex and dynamic land-use processes is still tricky [39] because multiple processes may account for the same pattern and may change substantially because they are geographically structured [40]. To better understand the processes that drive the observed land-cover dynamics and use [36] recommended applying dynamic models based on site-specific factors. By assessing the relative influence of different biotic and abiotic processes over longer time horizons, these models can further inform decisions about which restoration interventions will lead to spatial patterns of land use similar to those observed in reference areas. All these effects ultimately affect the ability of ecosystems to provide services to society, potentially amplifying socioeconomic inequality, which is highly documented in South America [41].

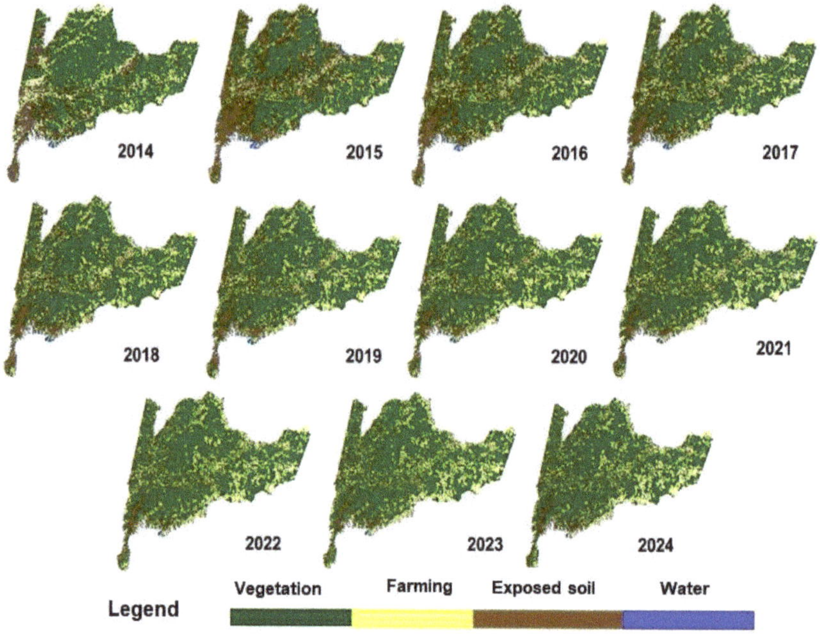

Figure 6. Result of the annual simulation between 2014 and 2024 for the municipality of Floresta—PE.

4. Conclusions

This study complements the knowledge about the direct and indirect causes of land use and land cover in tropical dry forests in Brazil. Our results indicate that from 1985 to 2014, more significant changes were observed in the forest and exposed-soil classes. The increase in forest class and the consequent reduction in exposed soil are consequences of the interaction between climate and human activities, as well as the quality of the spatial resolution of the satellite images used between the years analyzed. The low rainfall climatic conditions in the analyzed periods are primarily associated with the exposed soil throughout the municipality, as indicated by our spatially-explicit scenarios. However, their particular influences are variable in space and time and act in a complex way in combination with the other environmental drivers to produce specific trends in the transformation of the dry-forest ecosystem. These results suggest the need to complement the variables modeled in this study under the direct influence of other environmental factors inherent to the place. More specifically, our results may suggest potential future trajectories of land-cover changes, such as possible loss of vegetation area. This information is valuable for developing public policies and management strategies to combat the effects of environmental degradation and the loss of natural areas on a larger scale.

Author Contributions: C.P.d.O. and R.L.C.F. planned the study and wrote the manuscript, C.P.d.O., R.B.d.L., J.A.A.d.S. and M.M.d.L.P., F.T.A.J. participated in the processing data, calculation and modeling of the data and wrote the manuscript. E.A.S., A.F.d.S., N.A.T.d.S., C.L.S.-M.S.d.M., M.M.d.L.P.

and I.J.C.L. collected and processed the data images. They also commented on the manuscript. All authors have read and agreed to the published version of the manuscript.

Funding: This research was funded by the Coordination for the Improvement of Higher Education Personnel—Brazil (CAPES)—Financial Code 001, National Council for Scientific and Technological Development (CNPq—303991/2016-0; 400540/2016-9), Foundation for the Support of Science and Technology of the State of Pernambuco and Federal Rural University of Pernambuco (FACEPE and UFRPE—IBPG-0827-5.02/14).

Acknowledgments: The authors are grateful for the financial support granted by the National Council for Scientific and Technological Development (CNPq) for the productivity in research fellows of the tenth and eleventh co-authors, project leaders. The authors also thank the Universidade do Estado do Amapá for all the support in the publication of this manuscript.

Conflicts of Interest: The authors declare no conflict of interest.

References

1. De Almeida, A.Q.; de Mello, A.A.; Neto, A.L.D.; Ferraz, R.C. Relações empíricas entre características dendrométricas da Caatinga brasileira e dados TM Landsat 5. *Pesq. Agropec. Bras.* **2014**, *49*, 306–315. [CrossRef]
2. Monteiro Junior, J.J.; Silva, E.A.; De Amorim Reis, A.L.; Mesquita Souza Santos, J.P. Dynamical Spatial Modeling to Simulate the Forest Scenario in Brazilian Dry Forest Landscapes. *Geol. Ecol. Landsc.* **2019**, *3*, 46–52. [CrossRef]
3. De Lima, T.L.; da Silva, J.A.A.; Longhi, R.V.; Gallo, R.; Hakamada, R.E.; dos Santos, M.V.F.; Martínez Molina, J.R.; Herrera Machuca, M.Á.; Muir, J.P.; Ferreira, R.L.C. Structure, Survival, and Species Diversity in a Tropical Dry Forest Submitted to Coppicing. *For. Ecol. Manag.* **2021**, *501*, 119700. [CrossRef]
4. Foley, J.; Defries, R.; Asner, G.; Barford, C.; Bonan, G.; Carpenter, S.; Chapin, F.S., III; Coe, M.; Daily, G.; Gibbs, H.; et al. Global Consequences of Land Use. *Science* **2005**, *309*, 570–574. [CrossRef]
5. Sano, E.E.; Rosa, R.; Brito, J.L.S.; Ferreira, L.G. Land Cover Mapping of the Tropical Savanna Region in Brazil. *Environ. Monit. Assess.* **2010**, *166*, 113–124. [CrossRef]
6. Wubie, M.A.; Assen, M.; Nicolau, M.D. Patterns, Causes and Consequences of Land Use/Cover Dynamics in the Gumara Watershed of Lake Tana Basin, Northwestern Ethiopia. *Environ. Syst. Res.* **2016**, *5*, 8. [CrossRef]
7. Xavier, A.P.C. Modelagem Temporal da Dinâmica do Uso e Ocupação do Solo nas Perdas e Deposição de Sedimentos na Bacia do Rio Tapacurá (PE). Master's Thesis, Universidade Federal da Paraíba, Jardim Cidade Universitaria, PR, Brazil, 2017.
8. Blackie, R.; Baldauf, C.; Gautier, D.; Gumbo, D.; Kassa, H.; Parthasarathy, N.; Paumgarten, F.; Sola, P.; Pulla, S.; Waeber, P.; et al. Tropical Dry Forests: The State of Global Knowledge and Recommendations for Future Research. Available online: https://www.cifor.org/knowledge/publication/4408/ (accessed on 13 December 2021).
9. Sunderland, T.; Apgaua, D.; Baldauf, C.; Blackie, R.; Colfer, C.; Cunningham, A.B.; Dexter, K.; Djoudi, H.; Gautier, D.; Gumbo, D.; et al. Global Dry Forests: A Prologue. *Int. For. Rev.* **2015**, *17*, 1–9. [CrossRef]
10. Bastin, J.-F.; Berrahmouni, N.; Grainger, A.; Maniatis, D.; Mollicone, D.; Moore, R.; Patriarca, C.; Picard, N.; Sparrow, B.; Abraham, E.M.; et al. The Extent of Forest in Dryland Biomes. *Science* **2017**, *356*, 635–638. [CrossRef]
11. Silva, E.A.; Tramontina, J.; Alba, E.; de Goergen, L.C.G.; Finger, A.P.; Pereira, R.S. Forest Cover Analysis through the Weights of Evidence Method in the Campanha Ocidental Region—RS (Brazil). *Rev. Árvore* **2017**, *41*, e410105. [CrossRef]
12. Gollnow, F.; Göpel, J.; deBarros Viana Hissa, L.; Schaldach, R.; Lakes, T. Scenarios of Land-Use Change in a Deforestation Corridor in the Brazilian Amazon: Combining Two Scales of Analysis. *Reg Env. Chang.* **2018**, *18*, 143–159. [CrossRef]
13. Kim, J.; Choi, J.; Choi, C.; Park, S. Impacts of Changes in Climate and Land Use/Land Cover under IPCC RCP Scenarios on Streamflow in the Hoeya River Basin, Korea. *Sci. Total Environ.* **2013**, *452–453*, 181–195. [CrossRef] [PubMed]
14. Formigoni, M.D.H.; Xavier, A.C.; Lima, J.S.D.S. Análise temporal da vegetação na região do nordeste através de dados EVI do MODIS. *Ciênc. Florest.* **2011**, *21*, 1–8. [CrossRef]
15. Tesfaye, S.; Guyassa, E.; Joseph Raj, A.; Birhane, E.; Wondim, G.T. Land Use and Land Cover Change, and Woody Vegetation Diversity in Human Driven Landscape of Gilgel Tekeze Catchment, Northern Ethiopia. *Int. J. For. Res.* **2014**, *2014*, 614249. [CrossRef]
16. Powell, P.A.; Nanni, A.S.; Názaro, M.G.; Loto, D.; Torres, R.; Gasparri, N.I. Characterization of Forest Carbon Stocks at the Landscape Scale in the Argentine Dry Chaco. *For. Ecol. Manag.* **2018**, *424*, 21–27. [CrossRef]
17. Walsh, S.J.; Entwisle, B.; Rindfuss, R.R.; Page, P.H. Spatial Simulation Modelling of Land Use/Land Cover Change Scenarios in Northeastern Thailand: A Cellular Automata Approach. *J. Land Use Sci.* **2006**, *1*, 5–28. [CrossRef]
18. Liu, X.; Liang, X.; Li, X.; Xu, X.; Ou, J.; Chen, Y.; Li, S.; Wang, S.; Pei, F. A Future Land Use Simulation Model (FLUS) for Simulating Multiple Land Use Scenarios by Coupling Human and Natural Effects. *Landsc. Urban Plan.* **2017**, *168*, 94–116. [CrossRef]
19. Costanza, R.; Voinov, A. (Eds.) *Landscape Simulation Modeling: A Spatially Explicit, Dynamic Approach*; Modeling Dynamic Systems; Springer: New York, NY, USA, 2004; ISBN 978-0-387-00835-6.
20. Jia, Z.; Ma, B.; Zhang, J.; Zeng, W. Simulating Spatial-Temporal Changes of Land-Use Based on Ecological Redline Restrictions and Landscape Driving Factors: A Case Study in Beijing. *Sustainability* **2018**, *10*, 1299. [CrossRef]

21. Markert, K.N.; Griffin, R.E.; Limaye, A.S.; McNider, R.T. Spatial Modeling of Land Cover/Land Use Change and Its Effects on Hydrology within the Lower Mekong Basin. In *Land-Atmospheric Research Applications in South and Southeast Asia*; Vadrevu, K.P., Ohara, T., Justice, C., Eds.; Springer Remote Sensing/Photogrammetry; Springer International Publishing: Cham, Switzerland, 2018; pp. 667–698. ISBN 978-3-319-67473-5.
22. Instituto Brasileiro de Geografia e Estatística (Ed.) *Biomas e Sistema Costeiro-Marinho do Brasil: Compatível com a Escala 1:250,000*; Série Relatórios Metodológicos; IBGE: Rio de Janeiro, Brazil, 2019; ISBN 978-85-240-4510-3.
23. Dos Santos, H.G. *Sistema Brasileiro de Classificação de Solos*, 5th revised and expanded ed.; Embrapa: Brasília, DF, Brazil, 2018; ISBN 978-85-7035-800-4.
24. Congalton, R.G.; Green, K. *Assessing the Accuracy of Remotely Sensed Data: Principles and Practices*, 2nd ed.; CRC Press: Boca Raton, FL, USA, 2008; ISBN 978-0-429-14397-7.
25. Benedetti, A.C.P.; Lippert, D.B.; Pereira, R.S.; de Almeida, C.M.; Cardoso, C.D.V.; Hendges, E.R. Uso do produto MOD13Q1 do sensor Modis para análise temporal e mapeamento das florestas nas Serras do Sudeste e Campanha Meridional do Rio Grande do Sul. *Rev. Árvore* **2013**, *37*, 459–467. [CrossRef]
26. Comber, A.; Fisher, P.; Brunsdon, C.; Khmag, A. Spatial Analysis of Remote Sensing Image Classification Accuracy. *Remote Sens. Environ.* **2012**, *127*, 237–246. [CrossRef]
27. Foody, G.M. Status of Land Cover Classification Accuracy Assessment. *Remote Sens. Environ.* **2002**, *80*, 185–201. [CrossRef]
28. Eckhardt, D.W. Automated Update of an Irrigated Lands GIS Using SPOT HRV Imagery. *Photogramm. Eng.* **1990**, *8*, 1515–1522.
29. Soares, D.B. Sobre o Processo de Desertificação. *Rev. Bras. Geogr. Física* **2011**, *15*, 174–188. [CrossRef]
30. De Lopes, L.S.O.; Soares, R.C. Suscetibilidade à Desertificação das Terras Secas de Gilbués (Estado do Piauí) e Cabrobó (Estado do Pernambuco), Nordeste do Brasil. *Rev. Geogr. Am. Cent.* **2016**, *1*, 307–323.
31. Gariglio, M.A.; Brazil Serviço Florestal Brasileiro (Eds.) *Uso Sustentável e Conservação dos Recursos Florestais da Caatinga*; Serviço Florestal Brasileiro: Brasília, DF, Brazil, 2010; ISBN 978-85-63269-04-1.
32. Dupin, M.G.V.; Espírito-Santo, M.M.; Leite, M.E.; Silva, J.O.; Rocha, A.M.; Barbosa, R.S.; Anaya, F.C. Land Use Policies and Deforestation in Brazilian Tropical Dry Forests between 2000 and 2015. *Environ. Res. Lett.* **2018**, *13*, 035008. [CrossRef]
33. Ferrari, R. Modelagem Dinâmica do Uso e Cobertura da Terra da Quarta Colônia. Master's Thesis, Universidade Federal de Santa Maria, Santa Maria, RS, Brazil, 2008; 131p.
34. De Macedo, R.C.; de Almeida, C.M.; dos Santos, J.R.; Rudorff, B.F.T. Modelagem dinâmica espacial das alterações de cobertura e uso da terra relacionadas à expansão canavieira. *Bol. Ciênc. Geod.* **2013**, *19*, 313–337. [CrossRef]
35. Salazar, A.; Sanchez, A.; Villegas, J.C.; Salazar, J.F.; Ruiz Carrascal, D.; Sitch, S.; Restrepo, J.D.; Poveda, G.; Feeley, K.J.; Mercado, L.M.; et al. The Ecology of Peace: Preparing Colombia for New Political and Planetary Climates. *Front. Ecol. Environ.* **2018**, *16*, 525–531. [CrossRef]
36. González-González, A.; Villegas, J.C.; Clerici, N.; Salazar, J.F. Spatial-Temporal Dynamics of Deforestation and Its Drivers Indicate Need for Locally-Adapted Environmental Governance in Colombia. *Ecol. Indic.* **2021**, *126*, 107695. [CrossRef]
37. Mariano, D.A.; dos Santos, C.A.C.; Wardlow, B.D.; Anderson, M.C.; Schiltmeyer, A.V.; Tadesse, T.; Svoboda, M.D. Use of Remote Sensing Indicators to Assess Effects of Drought and Human-Induced Land Degradation on Ecosystem Health in Northeastern Brazil. *Remote Sens. Environ.* **2018**, *213*, 129–143. [CrossRef]
38. Siyum, Z.G. Tropical Dry Forest Dynamics in the Context of Climate Change: Syntheses of Drivers, Gaps, and Management Perspectives. *Ecol Process* **2020**, *9*, 25. [CrossRef]
39. Griscom, H.P.; Ashton, M.S. Restoration of Dry Tropical Forests in Central America: A Review of Pattern and Process. *For. Ecol. Manag.* **2011**, *261*, 1564–1579. [CrossRef]
40. Salazar, A.A.; Arellano, E.C.; Muñoz-Sáez, A.; Miranda, M.D.; Oliveira da Silva, F.; Zielonka, N.B.; Crowther, L.P.; Silva-Ferreira, V.; Oliveira-Reboucas, P.; Dicks, L.V. Restoration and Conservation of Priority Areas of Caatinga's Semi-Arid Forest Remnants Can Support Connectivity within an Agricultural Landscape. *Land* **2021**, *10*, 550. [CrossRef]
41. Laterra, P.; Nahuelhual, L.; Vallejos, M.; Berrouet, L.; Arroyo Pérez, E.; Enrico, L.; Jiménez-Sierra, C.; Mejía, K.; Meli, P.; Rincón-Ruíz, A.; et al. Linking Inequalities and Ecosystem Services in Latin America. *Ecosyst. Serv.* **2019**, *36*, 100875. [CrossRef]

land

MDPI

Article

Dynamic Landscapes in the UK Driven by Pressures from Energy Production and Forestry—Results of the CORINE Land Cover Map 2018

Beth Cole [1,*], Geoff Smith [2], Bestabé de la Barreda-Bautista [3], Alexandra Hamer [4], Matthew Payne [1], Thomas Codd [5], Sarah C. M. Johnson [1,6], Lok Yung Chan [1] and Heiko Balzter [1,7]

1 Centre for Landscape and Climate Research, School of Geography, Geology and Environment, University of Leicester, University Road, Leicester LE1 7RH, UK; mjp80@leicester.ac.uk (M.P.); sj239@leicester.ac.uk (S.C.M.J.); syrena.gym@gmail.com (L.Y.C.); hb91@leicester.ac.uk (H.B.)
2 Specto Natura, College Road, Impington, Cambridge CB24 9PL, UK; geoffsmith@specto-natura.co.uk
3 School of Geography, University of Nottingham, Sir Clive Granger Building, Nottingham NG7 2RD, UK; betsabe.delabarreda@nottingham.ac.uk
4 Department of Geography, School of Environment Education and Development, University of Manchester, Manchester M13 9PL, UK; alexandra.hamer@manchester.ac.uk
5 Division of Agricultural and Environmental Sciences, School of Biosciences, University of Nottingham, Leicestershire LE12 5RD, UK; thomas.codd@nottingham.ac.uk
6 Centre for Environmental Health and Sustainability, University of Leicester, University Road, Leicester LE1 7RH, UK
7 National Centre for Earth Observation, University of Leicester, University Road, Leicester LE1 7RH, UK
* Correspondence: bc132@leicester.ac.uk

Citation: Cole, B.; Smith, G.; de la Barreda-Bautista, B.; Hamer, A.; Payne, M.; Codd, T.; Johnson, S.C.M.; Chan, L.Y.; Balzter, H. Dynamic Landscapes in the UK Driven by Pressures from Energy Production and Forestry—Results of the CORINE Land Cover Map 2018. *Land* 2022, 11, 192. https://doi.org/10.3390/land11020192

Academic Editor: Giuseppe Pulighe

Received: 14 December 2021
Accepted: 24 January 2022
Published: 26 January 2022

Abstract: The CORINE Land Cover (CLC) map was established in 1985 and is now one of the most widely used products from the Copernicus Land Monitoring Service. As the world's longest consistent operational land cover monitoring product, CLC maps have been produced for reference years 1990, 2000, 2006, 2012 and now for 2018. This paper presents the results from the CLC2018 mapping project in the UK and analyses the results of the land cover status layer and the change layer from the period 2012–2018. It sets this change in context with the change results from the period 2006–2012 and finds that the rate of change between the subsequent CORINE land cover maps is continuing to increase. Changes mapped for the period 2012–2018 covered 76,032 ha greater than the change mapped between 2006 and 2012, an increase of 26% of mapped change. The area of changes mapped covered an area equivalent to 1.16% of the total land area of the UK. The number of different types of changes also continue to diversify; however, the dominance of rotational forestry is consistent with the previous map. The process of urban land take has been highlighted in the results between 2012 and 2018 and is a trend identified in previous iterations of the CLC inventories. The largest gain is in industrial or commercial units (an increase of 14.4%). This growth is mainly attributed to renewable energy infrastructure. As well as the descriptive analysis, the results have been analysed to identify the likely pressures being experienced on the land in the UK. Although the CLC mapping approach is consistent, there have been improvements to the input EO data used to map the changes. For 2018, the Copernicus Sentinel-2 system offered a consistent and reliable image source for the first time. This increased the spatial resolution of the source datasets to 10 m, allowing for more accurate identification of small features and those with fine spatial textures such as suburban, road networks and windfarms. We also look forward to the development of CLC+, the new generation of CORINE land mapping, and the improvements it could make.

Keywords: land cover; land use; change mapping; land use pressures; energy production; forestry

1. Introduction

Over decadal time scales, land cover and land use in the UK have undergone significant changes as a result of multiple policy drivers, economic shifts and now, increasingly, environmental impacts. The primary policy drivers of change in the coming decade are the new Agriculture Bill that replaces the Common Agricultural Policy of the European Union in the UK after Brexit, and the Defra 25-year Environment Plan [1], with its ambitious goals for environmental improvements and the directive to achieve net zero carbon emissions by 2050 to address climate change. Global policy options towards climate-friendly and sustainable lifestyle changes that are being discussed include bans on advertising high-carbon foods, reducing food waste and prioritising the distribution of food to undernourished people [2]. If higher costs were placed on emissions-intensive foods such as beef and lamb, whilst encouraging fruit and vegetable consumption as a means to mitigate climate change, substantial land use changes would likely be the consequence. The impacts of climate change such as sea-level rise and the flooding caused by the increase in the intensity of storms are likely to see more land being handed over to mitigation schemes.

A recent policy report by the British House of Commons Committee on Climate Change concluded that land use in the UK must change to meet its net-zero greenhouse gas emissions target by 2050: "Fundamental change in the use of land across the UK is needed to maintain a strong agriculture sector that also delivers climate mitigation, adaptation and wider environmental objectives" [3]. The report sets out ambitions for a high uptake of low carbon farming practices and suggests releasing 22% of land out of traditional agricultural production for long-term carbon sequestration. Other policy drivers include the drive to stimulate housing supply, with a pledge of 300,000 new homes a year in The Housing White Paper [4]. Land use change on this scale is unprecedented in the UK since the operational monitoring of land cover and land use from satellites began. Monitoring of land cover and land use (LCLU) has been one of the major uses of new satellite data products, which have continued to increase in coverage, quality and accessibility in recent decades [5]. There are many approaches to the mapping of LULC change, with different methods suited to different contexts. Land cover and land use maps have been produced at a range of scales from local to continental levels to address particular requirements, political drivers and funding regimes, and as a baseline for simulations of future LULC change under different scenarios [6].

LULC change mapping is conventionally based on satellite images from two reference periods sufficiently separated in time to allow the changes to be identified reliably, given the constraints of the input data and the mapping approach. However, satellite data cannot always be captured at regular intervals due to weather conditions or technical issues such as variation in sensor orbits [7]. This has been a challenge within the remote sensing community since its inception, and as such there are numerous methods for interpolating missing images for planned timeseries for analysis while maintaining accurate results, such as pixel-based temporal composites [8] and spatiotemporal data fusion [9]. Frequent return periods of recent satellite missions such as the European Space Agency's Sentinel satellites are valuable for addressing this challenge [10].

Automated methods for identifying and mapping LULC change are a large research focus, as the volume of geospatial data collected outstrips our capacity for manual interpretation and analysis [11], and as high-performance computing and machine learning methods become more advanced [12,13]. These approaches are highly suited to certain mapping applications, for example, where the focus is on land cover as opposed to land use, or there are a small number of highly contrasting classes in the mapping specification [9]. Complex classifications such as CORINE, which characterise land cover and land use using 44 classes in total, incorporate contextual information for interpretation of changes, for example, an awareness of the policy pressures discussed above, for which visual interpretation by skilled analysts remains unparalleled. Despite this, some elements of automation have been incorporated into the CORINE methodology [14].

The CORINE Land Cover (CLC) methodology was established in the mid-1980s and is now one of the most widely used products from the Copernicus Land Monitoring Service [15]. As the world's longest consistent operational land cover monitoring product, CLC maps have been produced for reference years 1990, 2000, 2006, 2012 and now for 2018.

This paper presents results from the CLC2018 mapping exercise in the UK and analyses the results of the land cover status layer with a 25 ha minimum mapping unit (MMU) and the change layers from 2006 to 2012 and 2012 to 2018 with a 5 ha MMU [16].

Over its long history, more than 870 publications have made use of the CLC data [17], and a large body of work has been summarised in [18]. CLC data have been used for research in geography, remote sensing, ecology, forestry, agriculture, engineering, optics and computer science [17], as well as for many operational applications in businesses and policy contexts. The CLC data addresses many common challenges identified in remote sensing of land cover, because it is free to access and comprises a long-standing data timeseries, covering a large, multiregional scope, with consistent methodology, data processing, validation and verification, the details for which are publicly available [19].

The CLC2018 is therefore a further step in delivering a powerful and rich time-series of landscape dynamics in Europe. The UK component thus provides an important stocktake of the primary land cover and land use types before all of the new drivers described above manifest themselves as landscape changes. There are some limitations in the CLC methodology, namely the spatial resolution [20] and some nomenclature definitions, for detailed assessment of the implementation of environmental objectives. However, there is huge value in the large-scale consistent mapping approach. Recent improvements in the use of higher resolution input data from Sentinel and future developments of CLC+ are addressing some of these limitations. This paper aims to present an assessment of the large-scale land cover and land use types in the UK in 2018, examine the primary changes since 2006, and their likely drivers, and analyse change patterns and land cover transitions quantitatively.

2. Methods

The CLC is now part of the Copernicus Land Monitoring Service (CLMS) and integrated within a portfolio of products that provide a range of spatial and temporal detail for thematic or biophysical properties for either selected hotspot locations or wall-to-wall coverage at a pan-European level. The CLC could be described as a classical parcel-based land cover map that covers the EEA-38 countries plus the UK. Selected products from the CLMS can be combined to produce further monitoring and assessment information that is beyond the capabilities of a single dataset. The CLMS is now producing the second generation of CLC, or CLC+, giving improved spatial detail and an advanced thematic data model, but it will still be able to generate the traditional CLC for long term monitoring purposes.

The CLC classes are defined in a 3-level hierarchical structure grouped into artificial surfaces, agricultural areas, forest and semi-natural areas, wetlands and water bodies at Level 1. The detailed class descriptions as set out in the nomenclature [21] were followed during the interpretation processes of this work. The status and changes in CLC are mapped based on the 44 individual classes at Level 3, but during reporting and analysis these can be aggregated up to 15 classes at Level 2, and just the 5 broad classes at Level 1. Table 1 shows an overview of the CLC class nomenclature. In this paper, we focus on the 44 classes at Level 3.

Table 1. CORINE class nomenclature of the 44 land cover/land use classes at Levels 1, 2 and 3.

Level 1		Level 2		Level 3	
1	Artificial Surfaces	1.1	Urban fabric	1.1.1	Continuous urban fabric
				1.1.2	Discontinuous urban fabric
		1.2	Industrial, commercial and transport units	1.2.1	Industrial or commercial units
				1.2.2	Road and rail networks and associated land
				1.2.3	Port areas
				1.2.4	Airports
		1.3	Mine, dump and construction sites	1.3.1	Mineral extraction sites
				1.3.2	Dump sites
				1.3.3	Construction sites
		1.4	Artificial, non-agricultural vegetated areas	1.4.1	Green urban areas
				1.4.2	Sport and leisure facilities
2	Agricultural areas	2.1	Arable land	2.1.1	Non-irrigated arable land
				2.1.2	Permanently irrigated land
				2.1.3	Rice fields
		2.2	Permanent crops	2.2.1	Vineyards
				2.2.2	Fruit trees and berry plantations
				2.2.3	Olive groves
		2.3	Pastures	2.3.1	Pastures
		2.4	Heterogeneous agricultural areas	2.4.1	Annual crops associated with permanent crops
				2.4.2	Complex cultivation patterns
				2.4.3	Land principally occupied by agriculture with significant areas of natural vegetation
				2.4.4	Agro-forestry areas
3	Forests and semi-natural areas	3.1	Forests	3.1.1	Broad-leaved forest
				3.1.2	Coniferous forest
				3.1.3	Mixed forest
		3.2	Shrub and/or herbaceous vegetation associations	3.2.1	Natural grassland
				3.2.2	Moors and heathland
				3.2.3	Sclerophyllous vegetation
				3.2.4	Transitional woodland scrub
		3.3	Open spaces with little or no vegetation	3.3.1	Beaches, dunes, sand plains
				3.3.2	Bare rock
				3.3.3	Sparsely vegetated areas
				3.3.4	Burnt areas
				3.3.5	Glaciers and perpetual snow
4	Wetlands	4.1	Inland wetlands	4.1.1	Inland marshes
				4.1.2	Peat bogs
		4.2	Coastal wetlands	4.2.1	Salt marshes
				4.2.2	Salines
				4.2.3	Intertidal flats
5	Water bodies	5.1	Continental waters	5.1.1	Water courses
				5.1.2	Water bodies
		5.2	Marine waters	5.2.1	Coastal lagoons
				5.2.2	Estuaries
				5.2.3	Sea and ocean

The production of CLC2018 in the UK followed the same production methodology as CLC2012, with some modifications and improvements. Following the technical guidelines set out by the European Environment Agency (EEA) [22], the method applied is the 'change mapping approach' based on the CLC2012 product and using visual interpretation of satellite imagery. The change mapping approach aims to map real land cover/land use change, representing the change process on the ground, and also technical changes (errors in CLC2012 that were missed during the 2012 update). This is achieved by interpreting change based on a comparison of multi-date satellite imagery with direct delineation of change polygons relative to the 2012 status map. This produces a CLC $\text{Change}_{2012\text{–}2018}$ layer with an MMU of 5 ha. The process also produces a revised CLC2012 dataset ($\text{CLC2012}_{\text{revised}}$) in which the technical changes are corrected. The CLC2018 status product is then produced by combining the CLC change$_{2012–2018}$ polygons and the CLC2012$_{revised}$ polygons and is represented by the equation:

$$CLC2018 = CLC2012_{revised} + CLC\text{-}Changes_{2012\text{-}2018}$$

As the CLC2012$_{revised}$ and CLC2018 status layers are mapped at a 25 ha MMU and the change layer is mapped at 5 ha MMU after intersection and unification, any small polygons are generalised with their neighbours according to a priority table [23].

As a consequence of the two different MMUs between the CLC status layers and the change layer, 'Technical change polygons' are produced as auxiliary change polygons to avoid major inaccuracies in the CLC2018 database, but do not represent a real change in land cover/land use. These technical change polygons have been removed from the 2012–2018 change database for analysis in this paper.

There are some limitations to the current CLC methodology for monitoring the implementation of environmental interventions or the impact of land cover/land use change on climate change drivers. As the MMUs of 25 ha for the status layer map and 5 ha for the change layer map are considerably larger than some other monitoring and mapping in the UK, this can lead to different figures if compared directly between data sources (e.g., the natural capital accounts [24]). The differing results are not incorrect but caused by variations in scale and nomenclature. For example, the construction of renewable energy infrastructure is often captured in the CLC MMU, but does not always fill the land parcel, and the definition of the nomenclature translates this change to industrial development. Further analysis at a finer resolution would require additional field or land inventory data at a much higher spatial resolution. The improvements made by incorporating higher resolution input data, and the development of the CLC+ and the full integration of the CLMS data product suite, go a long way to improving these limitations. The power of the CLC time series of products is its length and the consistency of implementation so that multi-decadal landscape changes and trends of change are identified easily and reliably using these data

Input Data

The available satellite data for interpretation of changes between 2012 and 2018 consisted of two slightly different datasets. IMAGE2012, the imagery from the previous production, was reused and mainly consisted of images from 2011 through 2013 (a mixture of IRS-P6 LISS III, IRS-R2 LISS IV and RapidEye systems). IMAGE2018 consisted of images from 2017 only (aiming to document the situation at the beginning of the reference period) acquired by the European Copernicus Sentinel-2 satellites, with the US Landsat 8 satellite used for gap filling. The move to Sentinel-2 data as the input for the CLC2018 represents a big change in the ability to interpret land cover types and change as the spatial resolution is improved from approximately 25 m down to 10 m. This increase in spatial resolving power of over 6 enabled clearer identification of features. Furthermore, images were available within a shorter reference period due to the more frequent revisit of Sentinel-2 compared to other similar class systems.

Each set of image data contained images from two separate acquisition windows, coverage 1 and 2, preferably for the same year and with an optimum date difference of 6 weeks to allow for relative phenological change of different land covers to aid their discrimination (Table 2). The aim was to have at least two full coverages of the UK. However, due to cloud cover, it was necessary to select some images that only conformed to the minimum time difference of 4 weeks. The image availability in IMAGE2018 was greatly improved to that in IMAGE2012, where availability was limited and had to be combined over several years to piece together the coverage required [25]. In 2017, 64.6% of the UK had between 3 and 6 images available and only 10% was interpreted by a single image.

Table 2. UK acquisition windows for IMAGE2018, for the North and South of the UK.

		Coverage 1		
UK_N	01/05/17	15/06/17	01/09/17	30/09/17
UK_S	15/04/17	15/06/17	01/09/17	31/10/17
	Coverage 2			Coverage 2

In-situ and online data was also used in the CLC2018 production in areas where satellite imagery was not sufficient for the interpretation of particular classes and/or changes. Examples include ordnance survey open data, the national forestry inventory and very-high spatial resolution satellite imagery. All data was open access and sourced online from commercial or government department websites. The satellite imagery can be viewed freely in Google Earth and greatly aided visual interpretation of IMAGE2012 and IMAGE2018.

The results presented in this paper are from a map representing the whole of the United Kingdom. The Channel Islands and Northern Ireland were reprojected from their native coordinate systems and stitched together with the UK data. The map has a 25 km buffer around the coastline to ensure all islands, estuaries, tidal flats and ports and harbours are included. The buffer is clipped to the border between Northern Ireland and the Republic of Ireland. The status layer statistics used in each set of results are the up-to-date revisions created during each production, i.e., CLC2018 and $CLC2012_{revised}$ for 2018 and CLC2012 and $CLC2006_{revised}$ for 2012. Class 523 (sea and ocean) has been removed from the statistics for the status layers to avoid skewing the results.

3. Analysis of the CORINE Land Cover Map

3.1. Status Layer

The UK CLC2018 status map is shown in Figure 1; it provides a comparable representation of the UK in the context of the EEA-38 countries, our immediate geographical neighbours. The CLC2018 map includes 36 of the 44 Level 3 classes in the CORINE nomenclature (Table 1); 8 classes were not present in the UK, such as olive groves and rice fields. The area covered by each class is shown in Figure 2; this has been aggregated up to Level 1 nomenclature to see the overall proportions of land cover types in Figure 3. Agriculture remains the most dominant land cover, accounting for 55% coverage of the UK. There is an approximately even split between 6,660,235 ha of non-irrigated arable land (211) and 6,926,447 ha of pasture (231), covering 26.7% and 27.7% of land, respectively. There is also an east-west split, with arable land predominantly occupying the east of the country and pastures in the west, due to environmental conditions. However, at a local scale, there is often a more complex mosaic between these classes due to topography, soils and farm management practices.

The second most dominant Level 1 class is forest and semi-natural areas, representing 24.2% of the territory. The natural vegetation classes 321 (natural grassland) and 322 (moors and heaths) account for the largest part of this, with 13.3% of the coverage. When the peatland class, 412, is added, which covers another 9.2%, the natural open moorland landscapes occupy nearly a quarter of the country, concentrated in the upland areas of Scotland, Wales and the Pennines. The forestry classes cover a combined area of 2,437,987 ha, or 9.8% of the country if you include class 324, transitional woodland scrub. This class can represent woodland degradation, forest regeneration or natural succession. It also includes clear cut areas in forests, regeneration areas in the transitional stage or regrowth lasting 5–8 years or until the trees reach 5 m in height [21]. In the UK, the majority of this class is attributed to the clear cut or regrowth of harvested coniferous woodland. There are

556,176 ha of broad-leaved forest, 1,198,586 ha of coniferous forest and 300,353 ha of mixed forest.

Artificial surfaces occupy 8.6% of the country and are widely distributed, with a greater density in the south. The majority of this class is made up of urban settlements, containing class 112 discontinuous urban fabric (5.4%) and 142 sport and leisure facilities (1.2%), which can be large areas of open land including, for example, golf courses. To put this into context, 83.4% of the UK population lived in urban areas in 2018 [26].

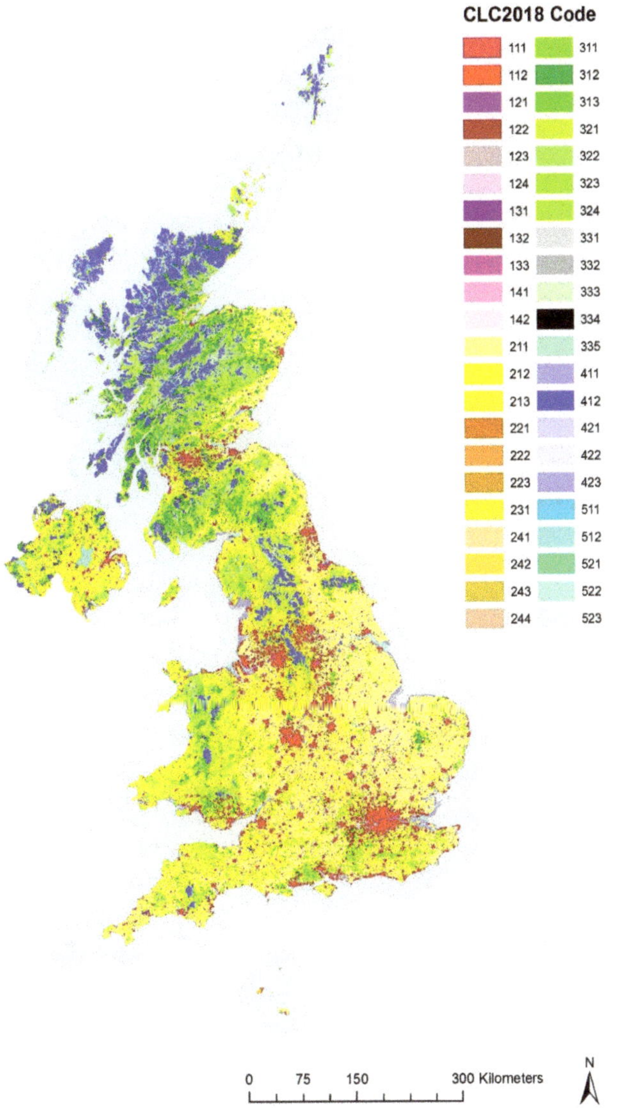

Figure 1. CORINE Land Cover (CLC) map for the UK 2018. See Table 1 for class descriptions.

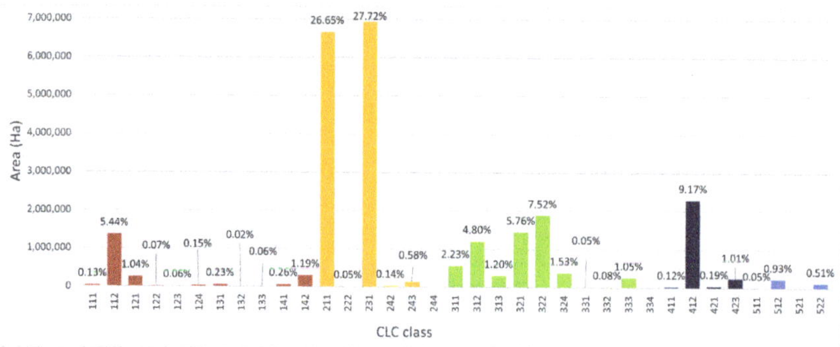

Figure 2. Area and proportion of UK land in each class for 2018.

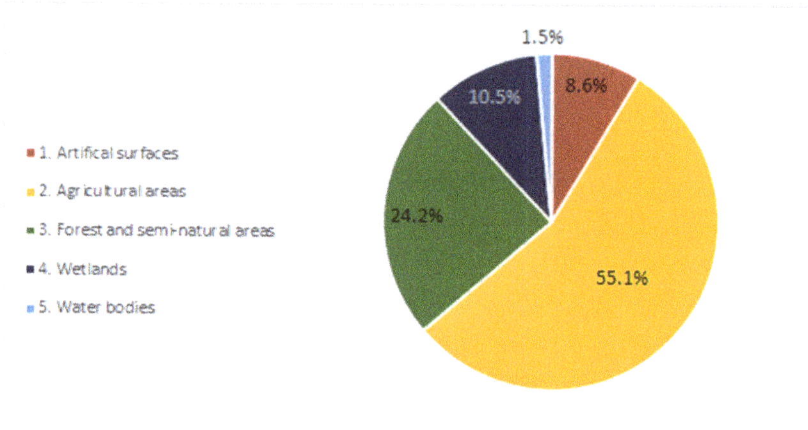

Figure 3. Proportion of UK land area aggregated to Level 1 class for 2018.

3.2. *Main Land Cover/Land Use Changes between 2012 and 2018*

The total area of land cover/land use changes between 2012 and 2018 amount to 290,368 ha, which corresponds to 1.16% of the total land area of the UK. Most of the changes occurred around the English–Scottish border, the southwest of Scotland and in Wales, and are predominantly related to forest management, i.e., clear-cutting and replanting of coniferous woodland. The spatial distribution of changes between 2012 and 2018 is similar to those detected from 2006 to 2012. The amount and the spatial distribution of changes between CLC classes are shown in Figure 4.

There were 205 different types of land cover/land use change in the CLC change layer during the period 2012–2018. The complete dataset of all changes at the national scale is summarised in the change matrix shown in Table S1 in the Supplementary Material. Table 3 shows a slimmed-down version, showing classes with over 1000 ha of change. The class transitions with the largest area of change between 2012 and 2018, which represents almost 90% of the area, are shown in Table 4.

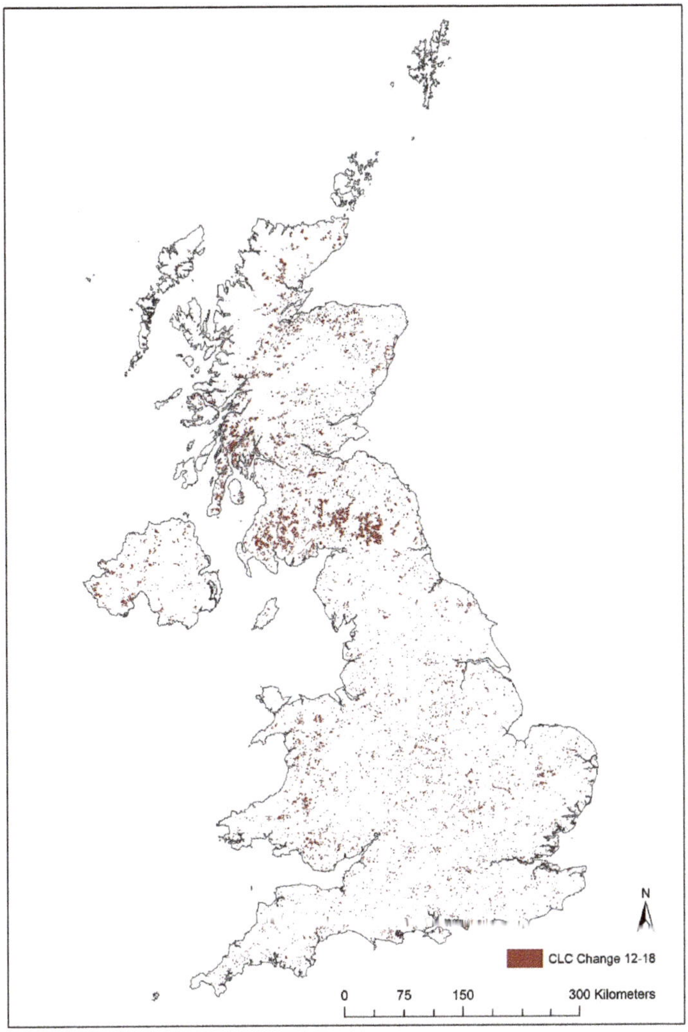

Figure 4. Overview map of all observed CORINE land cover/land use changes between 2012–2018 for the UK.

Table 3. Change matrix, 2012 to 2018, showing the total area (ha) of each change. Only classes with more than 1000 ha are shown in this matrix for size purposes. Totals are for all classes. The full change matrix can be found in Supplementary Material Table S1.

		2018															
	CLC Code	**112**	**121**	**122**	**131**	**133**	**142**	**211**	**231**	**311**	**312**	**313**	**321**	**322**	**324**	**512**	**Total**
	121	190				1289			152				10				1640
	131	134	186			143	66	682	3186				1383	573	208	778	7819
	132		26			81	56		910				436	116	85		1742
	133	5867	3314	215			473	130	575				137			125	11,473
	141	362	179	40	41	525	11										1165
	142	134	75			576		90	73	17			17		25		1017
	211	2805	12,827	486	3094	10,874	260		1118	21	23				518	209	33,498
2012	**231**	2495	5655	878	1345	7073	107	2989		16	9		25	125	299	39	21,524
	311	59	25	23	48	80			27						781		1044
	312	14	3528	18	143	250			10			400			126,935		131,317
	313	27	8	12	63	40									2482		2633
	321	11	1762	38	248	168	14	65	437		54				111		3000
	322	9	2496	107	231	125	14	35	308		64		52		346		3979
	324		1019		76	124		8		1418	55,133	2342	220	345			60,716
	334												1018	322			1340
	412		3448	22	96	149	11		41		16		7			113	3915
	Total	**12,199**	**35,203**	**1840**	**5401**	**22,206**	**1033**	**4094**	**7161**	**1521**	**55,299**	**2743**	**3466**	**1498**	**131,843**	**1263**	

Table 4. The 20 most common CLC land cover/land use changes by total area between 2012 and 2018 detected in the change layer with a minimum mapping unit of 5 ha. The change code shows the transition from one Level 3 class to another. The impacts of these land cover/land use changes in relation to climate change are noted as positive or negative.

Change Code	Change Description	Impact for Climate Change	Area (ha)	% Changed Area in UK
312-324	Clear-cutting of coniferous forest	-ve	126,935	43.7
324-312	Regrowth of coniferous forest	+ve	55,133	19.0
211-121	Arable land to industrial and commercial development	-ve	12,827	4.4
211-133	Arable land converted to construction sites	-ve	10,874	3.7
231-133	Pastureland converted to construction sites	-ve	7073	2.4
133-112	Completion of construction sites to urban areas	-ve	5867	2.0
231-121	Pastureland to industrial and commercial development	-ve	5655	1.9
312-121	Coniferous forest to industrial and commercial development	-ve	3528	1.2
412-121	Peatland to industrial and commercial development	-ve	3448	1.2
133-121	Completion of construction sites to industrial and commercial developments	-ve	3314	1.1
131-231	Mineral extraction sites converted to pastureland	+ve	3186	1.1
211-131	Arable land to mineral extraction sites	-ve	3094	1.1
231-211	Pastureland converted to arable land (intensification of agriculture)	-ve	2989	1.0
211-112	Arable land to urban areas	-ve	2805	1.0
322-121	Moors and heath to industrial and commercial development	-ve	2496	0.9
231-112	Pastureland to urban areas	-ve	2495	0.9
313-324	Clearing of mixed forest	-ve	2482	0.9
324-313	Growth/replanting of mixed forest	+ve	2342	0.8
321-121	Natural grassland to industrial and commercial development	-ve	1762	0.6
324-311	Regrowth of broad-leaved forest	+ve	1418	0.5
	Total		259,724	89

The felling and planting of coniferous woodland (changes 312-324 and 324-312) accounted for the greatest amount of change; 62.7% to 64.4% of all change in the UK can be attributed to this driver of change (the uncertainty is introduced when considering the mixed wood class, 313). This forestry change was most common in Scotland, particularly focused around both sides of the Scottish Border with some occurring in Wales and other parts of England.

Seven of the top twenty changes are from a vegetated CLC class to industrial and commercial development (121). This type of change may initially seem unusual, but this change can, in some cases, be a predominantly land use rather than a land cover change. The majority of these areas were not completely denuded of vegetation but converted to a land use primarily aimed at renewable energy generation. New windfarm developments in vegetated areas necessitate access road building and small areas without vegetation in the immediate vicinity of the windfarm, but they generally leave the majority of the CLC polygon unchanged in terms of its land cover. However, because the primary land use changes from forestry or agriculture to energy generation, under the CLC technical guidelines, such alterations are mapped as a class change. The agricultural classes that changed from 2xx to industrial and commercial development (121) were often converted to solar energy farms (72%), while the forest and semi-natural classes in the uplands that changed to 121 were mainly converted to windfarm developments (92%). The agricultural classes changing to the industrial or commercial class covered 9241 ha, accounting for a combined 6.4% of all changes. This is an increase from previous CLC inventories [25] and can be mainly attributed to the increase in the number and scale of solar energy farm developments. The overall increase in the 121 industrial class is 33,562 ha, a greater increase than that of the urban fabric, class 112, with 11,896 ha. The net decrease of both the arable class by 29,404 ha and the pasture class by 14,363 ha represents a combined total of 15.1% of the total UK change.

Figure 5a shows the ten CLC classes with the largest amount of land use/land cover change between 2012 and 2018 in terms of change area in hectares, and Figure 5b shows the net change (gains and losses) for those classes. These ten classes showed over 2000 ha of changed area per class.

The artificial surfaces classes 111 (continuous urban fabric), 121 (industrial or commercial units) and 133 (construction sites) all gained more area than they lost, which is indicative of continued urban land take and in particular large housing developments and construction sites over 5 ha in area. Only class 131 (mineral extraction sites) showed a net decrease. More new construction site areas (133) were started than completed from 2012 to 2018.

The classes 211 (non-irrigated arable land) and 231 (pastures) both lost more in area than they gained. Urban expansion often occurs on former agricultural land around cities and towns in the UK, and also farmland has been given over to renewable energy supply. The advent of government solar incentives for farmland since 2010 [27] is likely to have contributed to this apparent industrial expansion.

More than twice as much land of class 312 (coniferous forest) in 2012 was lost than gained. Mirroring that change, class 324 (transitional woodland/shrub) gained approximately twice as much land as it lost over the 6 years. Transitions between these two classes are representative of typical rotation forestry practices in the UK, where mature forest stands are clear-cut and then replanted. Until the replanted forest stand is sufficiently established, the land is classed as transitional woodland/shrub in the CORINE nomenclature. What is significant about the change statistics between these two classes is that coniferous forest was harvested much faster than it was re-establishing on harvested land. Even though this does not change the primary land use of coniferous forest land, it does change the total coniferous forest cover and the carbon stock stored in UK forests. Wood production has increased across the UK since 1975 [28], although since 2014 it has reached a plateau (Figure 6a). The increase in wood production is driven mainly by softwood production. Since the Kyoto Protocol baseline year of 1990, new tree planting in the UK has declined, mainly because of a drop in conifer planting, due to the ending of the tax breaks that fuelled conifer planting in the 1970s and 1980s [29,30], which has not been compensated for by the increase in planting of deciduous trees (Figure 6b). We can see this in the CLC change data, with class 312 showing an overall loss of 76,018 ha (Figure 5).

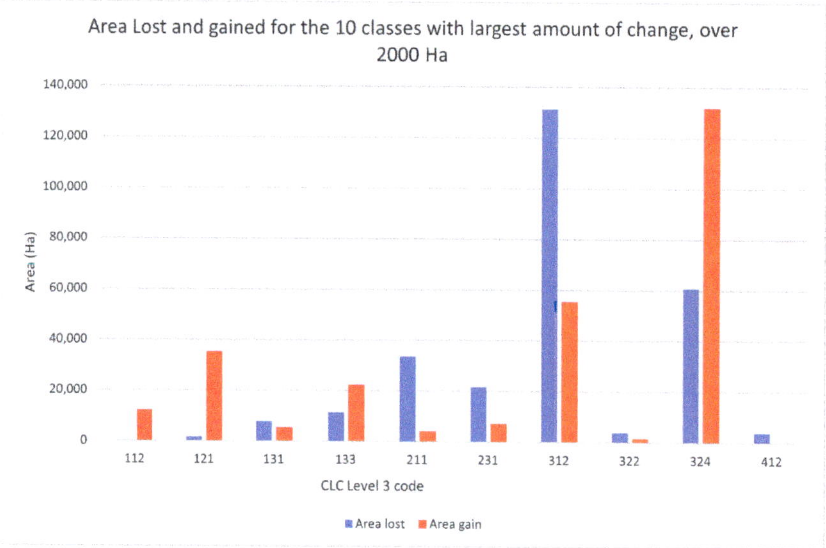

(**a**)

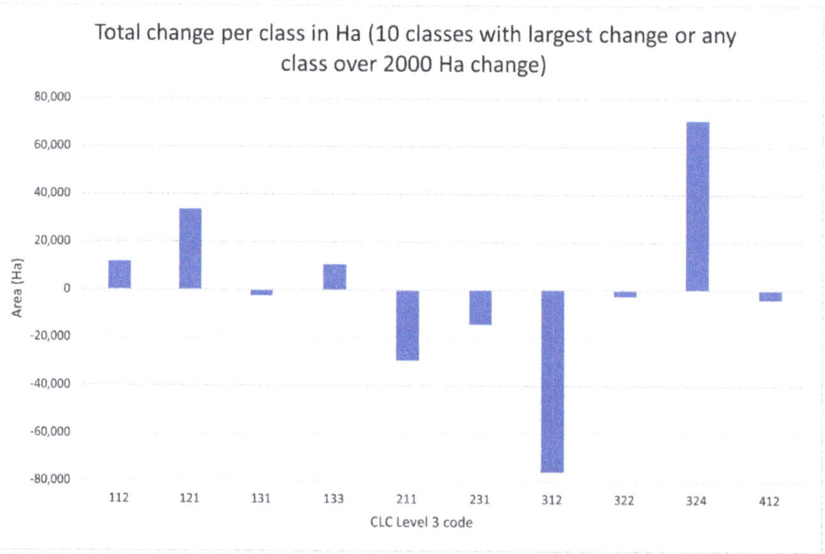

(**b**)

Figure 5. CLC land cover/land use area gained and lost (ha) between 2012 and 2018 (**a**) and total area of overall change (**b**) per class for the 10 largest changes.

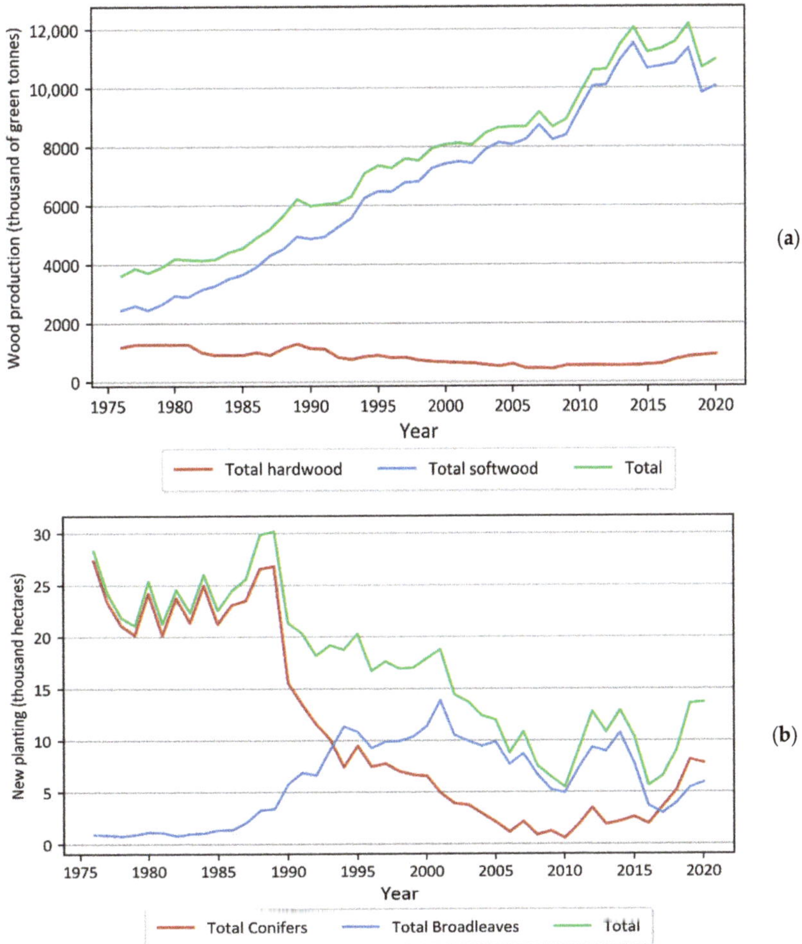

Figure 6. UK Forestry statistics. (**a**) Wood production statistics in thousands of green tonnes. Data for 2020 are provisional. (**b**) New tree planting in thousands of hectares. Data for 2020 are revised. Source: [28].

Classes 322 (moors and heathland) and 412 (peat bogs) also lost more than they gained, which is concerning given that these land cover types are generally thought of as deserving and receiving, in many cases, protected status. The change that is occurring in these classes is mainly to 121 (industrial), which can be attributed to renewable energy and the building of windfarms. As discussed above, these occupy a smaller area than is often classified on the map because of the CLC minimum mapping width of 100 m, and that the surface provisions in windfarms still leave vegetated areas.

3.3. Land Cover/Land Use Classes That Have Experienced the Most Gains or Losses

To put the amount of land cover/land use changes observed in the CLC change layer 2012 to 2018 into the context of the nature of UK landscapes, Figure 7 shows the gains and losses of CLC classes as a percentage of their area extent in the UK in 2012.

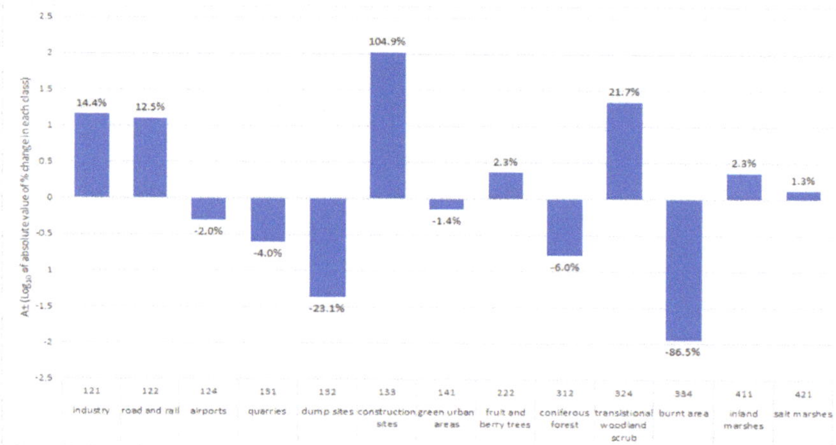

Figure 7. Percentage of the area gained and lost between 2012 and 2018 per CLC class as a proportion of the area covered by that class in 2012. Graph shows A ± x the log of the absolute value of % change. Only the classes with more than 1% of the change in their area are shown.

In contrast to Figure 5 above, Figure 7 emphasises the relative change that affected the CLC classes over the six years. The artificial surface classes 121 (industrial or commercial units) and 122 (road and rail networks and associated land) expanded their cover substantially compared to their area extent in 2012. Industrial or commercial units increases in cover by 14.4% and road and rail networks is only slightly less at 12.5%.

Classes 124 (airports), 131 (quarries) and 132 (dump sites) decreased in area. There is a significant redevelopment of retired military airbases into residential areas, with dozens of projects currently underway showing changes to other urban classes, and the ceasing of operations of some smaller airports. Similarly, there is a regeneration of quarry and landfill sites as brownfield areas for residential and industrial expansion. Class 133 (construction sites) showed the largest relative increase in area, which is indicative of the large areas that were made available for housing development, road building and industrial development. Class 141 (green urban areas) decreased in coverage, although urban green spaces are often protected against development. It is possible, however, that developments on brownfield sites and losses of greenbelt land in cities occurred over the reporting period. The total loss was only 929 ha; this is distributed fairly evenly across the country, with one or two polygons of change to either construction sites or discontinuous urban fabric in each of the larger cities.

Out of the agricultural areas, Class 222 (fruit trees and berry plantations) has increased significantly relative to the small absolute area that they covered in 2012. The forest and semi-natural area class 312 (coniferous forest) has decreased significantly since 2012, while class 324 (transitional woodland-shrub) increased even more relative to its cover in 2012. These two large relative changes show an intensification of timber harvesting in the UK. Class 334 (burnt areas) decreased considerably in comparison to its original extent in 2012, which was driven by large-scale wildfires between 2006 and 2012 that were not matched in area by new fires from 2012 to 2018. The extent of burnt areas was very low in the UK, so a small number of changes can make a big difference to the relative size of the class.

The wetland classes 411 (inland marshes) and 421 (salt marshes) saw an increase in relative extent compared to the baseline of 2012, representing the development of new sites for nature conservation and natural flood management, typically by conservation charities in partnership with the UK Environment Agency, e.g., Steart Coastal Management Project [31].

3.4. Changing Trends—2006–2012 Change Layer Compared to 2012–2018 Change Layer

Figure 8 shows a visualisation of the main land cover/land use transitions for both time periods as connected graphs. The two graphs show at a glance that, from 2012 to 2018, there were a substantially greater number of transitions in land cover/land use than from 2006 to 2012 (thicker arrows of darker colour and red hue). In both time periods, the largest transitions in terms of change area were between 312 (coniferous forest) and 324 (transitional woodland-shrub) in both directions, which indicate rotation forestry practices for softwood production and tree planting after harvest. As already mentioned, the transition from 312 (coniferous forest) to 121 (Industrial or commercial units) are largely indicative of new wind turbines in forested landscapes for the primary purpose of renewable energy generation. This type of change is classed as 121 under the technical methodology of CORINE because the primary purpose of the land use changes from forestry to energy production, even though in the vast majority of cases the wind turbines only require small clearings in the forest and most trees remain intact (Figure 9).

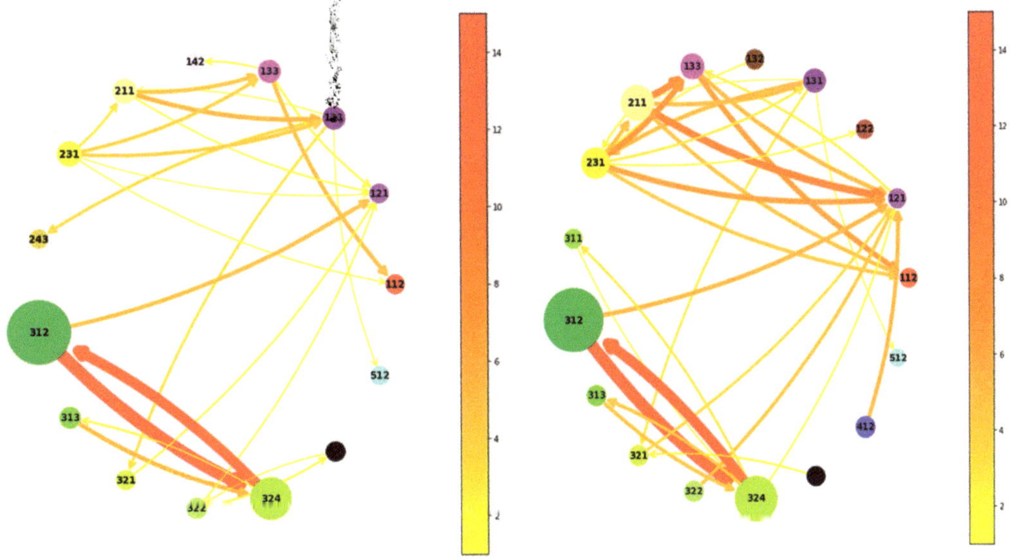

(**a**) (**b**)

Figure 8. Transitions over 600 ha in magnitude between CORINE land cover classes at Level 3 for 2006–2012 (**a**) and 2012–2018 (**b**). Arrow widths and colours are proportional to the logarithm of the area of change (ha) from the CORINE change layers. Only changes over 500 ha are shown.

Between 2010 and 2019, wind power capacity in the UK more than quadrupled to 24 GW, with greater production capacity on land (up from 4.1 GW in 2010 to 14.2 GW in 2019) and offshore (up from 1.3 GW to 9.8 GW) (BEIS, 2020). This trend is set to continue. Figure 8 shows an increasing transition from 211 (non-irrigated arable land) to 131 (mineral extraction sites) and 133 (construction sites), with some changes from 231 (pastures) to 131 in 2006–2012 and 133 in 2012–2018. The transitions from 211 (non-irrigated arable land) and 231 (pastures) to 112 (discontinuous urban fabric) also increased in magnitude. Class 112 describes types of housing development with associated gardens and green spaces that are typical for urban expansion around the fringes of towns and cities. Urban expansion is often taking place on agricultural land previously used for crop production or as pastures, and the CORINE statistics suggest a much greater degree of urban expansion and new settlements over 5 ha in area since 2012 compared to 2006–2012. Additionally, the transitions from agricultural land (211) and pastures (231) to industrial or commercial land

(121) increased substantially from 2012; it increased from a change of 1477 ha in 2006–2012 to 18,462 ha in 2012–2018. From 2012–2018 there is also a new transition from 412 (peat bogs) to 121 (industrial or commercial units), which was caused by the creation of wind turbines in the uplands where the main peatland cover still remains but the primary land use is now for energy production.

Figure 9. Land use change from coniferous forest to renewable energy (industrial or commercial units), showing the exaggeration in area caused by the mapping guidelines.

3.5. Analysis of Pressures on Land Cover/Land Use Change in the UK

Beyond the descriptive analysis of large-scale land cover/land use change, the question arises of which pressures have acted on land cover/land use in the UK that led to the observed transitions between CORINE classes over these two 6-year periods. Most land use and land cover changes are caused by a combination of social, economic and natural processes which operate at all scales, from the local to the global level. For example, agricultural policies combined with varying local employment opportunities lead to intensified use of land in some areas and abandonment in other areas. These changes can therefore affect the environment and its condition and biodiversity in either a positive or a negative manner, depending on context.

In the early 2000s, the BIOPRESS project [32] aimed to provide quantitative information on how changes in land cover and land use affected the environment and biodiversity in Europe in terms of pressures. The project produced consistent and coherent sets of historical (1950–1990–2000) land cover/land use change information for selected sites located from the Boreal to the Mediterranean, and from the Atlantic to the Continental regions of Europe. BIOPRESS focused on Pressures, State and Impact parts of the DPSIR framework (D = Drivers; R = Response) and the Pressure State Impact (PSI) model MIRABEL [33]. This model was originally designed as a tool for predicting pressures, impacts and scenarios of change, so the MIRABEL approach was not only being used to identify (and quantify) pressures based on observed land cover changes over the preceding 50 years, but also aimed to look at predicting future impacts on biodiversity. The land cover change statistics were converted into quantitative measures of pressures on biodiversity through the integration of socio-economic indicators.

In the BIOPRESS project, pressures have been defined as the processes that can be determined by the spatial patterns of land cover changes that are related to habitat fragmentation at the local scale. In other words, the processes determine how land cover changes may affect local environmental conditions. It is well known that land cover change is not a unidirectional process (e.g., forests being converted to agriculture). In the land cover change–pressure matrix, each land cover conversion was associated with a unique and particular process, which was meant to represent a specific anthropogenic pressure on biodiversity. This assumption was far from being satisfactory. Firstly, the same land cover change might be associated with two different processes at the same time, depending on where the land cover change took place in the first period. Secondly, it is not advisable to assume that any type of land cover change is always a pressure on a habitat. BIOPRESS dealt with these issues by applying expert knowledge to carry out the final mapping of land cover conversions.

The change statistics in BIOPRESS were produced by the backdating of CORINE land cover for Level 3 of the nomenclature at the selected sites. A land cover change–pressure association matrix was developed which can be considered as a compact format for representing the pressures that have resulted in different transitions between all possible land cover categories. Six main pressures were selected in BIOPRESS for the statistical analysis of land cover change patterns in combination with economic development, technology, and other social factors:

- Agricultural Intensification (I): agricultural conversions as well as transformations to more intensive practices.
- Land Abandonment (Ab): cropping cessation and conversion into early successional, herbaceous habitats. The transition to woody, later-successional habitats was considered as a Mediterranean extension of Afforestation.
- Afforestation (A): conversion of open (more or less natural) habitats into forests or macchias.
- Deforestation (D): conversion of forest to non-forest classes.
- Drainage (Dr): All changes affecting aquatic habitats that are transformed into more terrestrial ones, including land gain from intertidal and sea areas and the loss of peatlands drained due to agricultural practices or forests.
- Urbanisation (U): transformation to urban covers but also to related covers (road system, leisure areas, construction sites, etc.).

The work here considered three further pressures/processes because it was beyond the scope of this paper to analyse local context, specific interpretations and fragmentation which had been undertaken in the BIOPRESS project. The additional pressures were therefore:

- Urban greening (Ug): conversion of urban classes to more vegetated classes.
- Extensification (Ex): conversion of intensive agricultural classes to more extensive management.
- Re-wetting (Rw): conversion of 'dry' classes to wetlands and intertidal cover types.
- The resulting adapted land cover change–pressure matrix is given in Figure 10, where the colour of the land cover change combination gives the type of pressure.

In this paper, we have therefore compared the UK CLC changes from 2006 to 2012 and from 2012 to 2018 to the adapted land cover change–pressure matrix to summarise the likely pressures being experienced in the UK.

As expected from the Level 3 change results reported above, the dominant pressures are related to afforestation (A) and deforestation (D), Figure 11, but this is only the manifestation of rotational planting in commercial forestry. The BIOPRESS analysis shows that deforestation had a higher percentage of the total changes than afforestation over both periods, but deforestation declined more between the two periods than afforestation (Figure 11). These changes are also reflected by the Forest Research wood production data (Figure 6). According to these official statistics, the average annual rate of increase in wood production (slope of the trend line) reduced between 2012 and 2018 to 64,000 green tonnes per year from 397,000 tonnes per year between 2006 and 2012 (Figure 6). At the same time, the average annual rate of tree planting across the UK slowed down substantially from +380 ha per year between 2006 and 2012 to a declining trend of −930 ha per year from 2012–2018 (Figure 6). These trends are problematic in the context of the UK's legally binding commitment to achieve a net zero greenhouse gas budget by 2050. In an assessment of pathways to achieve this commitment, the Sixth Carbon Budget for the UK by the Climate Change Committee made four key recommendations in 2020, one of which states that 460,000 ha of new mixed woodland need to be planted by 2035, increasing woodland cover from 13% of UK land in 2020 to 15% by 2035, and 18% by 2050 to remove CO_2 and deliver wider environmental benefits. Additionally, 260,000 ha of farmland would need to switch to producing energy crops [3].

Figure 10. Adapted land cover change–pressure matrix. The colours identify the pressures that the land cover change represents: Brown—Agricultural Intensification, Light green—Land Abandonment, Dark green—Afforestation, Yellow—Deforestation, Purple—Drainage, Grey—Urbanisation, Mid green—Urban greening, Orange—Extensification, Blue—Rewetting.

When the likely changes related to commercial forestry are removed, the pressures of afforestation and deforestation are equal at around 0.5% of the changes and stable between the two change periods. This would suggest that the impacts of any government tree planting schemes are not being detected; however, it is reported that the government are falling short of their targets. The CLMS High spatial Resolution Layer (HRL) for forests records changes in forest area in its dominant leaf type change sublayer. It records forest changes on a 20 m grid (0.04 ha) rather than the 25 ha MMU of CLC, but it gives a similar order of magnitude for the changes and a similar split between afforestation and deforestation.

The next largest pressure is related to urbanisation or urban land take through a range of developments, from housing and industrial to infrastructure and construction. Given the MMU of CLC-Changes, only the large urbanisation projects will be detected and the considerable number of small brownfield infilling developments will not be recorded. Of particular interest, as they were identified by the interpreters during the production of both CLC2012 and CLC2018, were the creation of renewable energy sites related to onshore wind and solar. In CLC terms, these sites would be represented by conversion to industrial and commercial (121). Between the 2006–2012 and 2012–2018 periods, urbanisation doubled, but this conversion to industrial and commercial actually trebled. As onshore wind dominates in the uplands and solar tends to be in the lowlands, a rough breakdown of the split between onshore wind and solar can be made by considering the source class for the change. From a sample of the 500 largest change objects going to 121, it was found that 92% of the changes from forest, semi-natural and bog classes were related to onshore wind, and 72% of the changes from arable and pasture were to solar, so the assumption holds. In fact, of the changes in the 2012–2018 period to 121, 79% were related to renewable energy. In the 2006–2012 period, the split in the percentages of change was 2.60 and 0.69 for onshore wind and solar, respectively. By the 2012–2018 period, the changes had both increased and the split altered to 3.88 and 6.37, respectively, giving increases by a factor of 1.5 and 9.2. These increases and the change in the split between onshore wind and solar are comparable with the BEIS electrical generating capacity figures for renewable sources, which increased factors between the two periods of 1.7 and 6.5, respectively. The CLC-Change results are very realistic, even considering there will be some commission of changes that are not related to renewable energy.

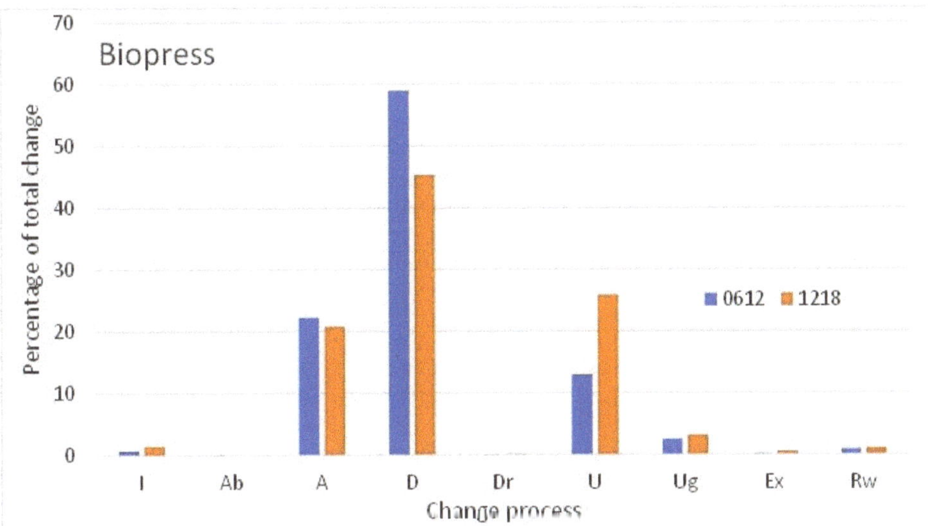

Figure 11. Pressures being experienced in the UK based on the BIOPRESS approach. Key: Agricultural Intensification (I), Land Abandonment (Ab), Afforestation (A), Deforestation (D), Drainage (Dr), Urbanisation (U), Urban greening (Ug), Extensification (Ex) and Re-wetting (Rw).

The remaining pressures are relatively minor compared to forestry and urban development. However, there appear to be increases in agricultural intensity which goes against the need for increased biodiversity in the agricultural environment and the promotion of regenerative farming in recent years. It will be interesting to see if this has changed by the next CLC update. More promising is the increase in urban greening, a trend which must continue in order to keep established urban areas habitable given the likely impacts of climate change. This result may appear to contradict those in Figure 7, but in that case only changes to the CLC class 141 (green urban areas) were considered. In this pressure analysis, all changes from an urban to a vegetated class, e.g., 131 (mineral extraction sites) to 231 (pastures), associated with the restoration of sites were included. It is hoped that this pressure will increase in future and be more strongly detectable in CLC updates.

4. Discussion

The change results described above raise a number of issues and important conclusions for land cover/land use mapping and changes to the UK landscape. Firstly, the rate of change between the subsequent CORINE land cover maps is continuing to increase. As reported in [25], between the 2000–2006 map and the 2006–2012 map, there was an increase in the amount of change mapped by 21,854 ha, or 11%, and the variety of types of change also increased. This increase is continuing with changes between 2012 and 2018, covering an area of 76,032 ha greater than 2006–2012, which represents 26% of the total change. The changes mapped in 2006–2012 covered an area equivalent to 0.86% of the total land area of the UK, and in 2012–2018 this increased to 1.16%. The number of different types of changes is also continuing to diversify, as seen in Figure 8, the diagram of directions of changes between the two time periods. The results of land cover/land use change presented in the paper are broadly indicative of the important processes happening in the UK landscape. However, developments in both the CORINE mapping process and the pressures on the UK landscape have resulted in some changes being represented accurately, or enhanced, while others are less well-captured. It is important to consider these while moving forwards into the next generation of CORINE.

Because the CORINE class descriptions and the technical manual have stayed essentially unchanged for decades, some new land use forms have to be mapped into an existing class. This is most prominent in the mapping of renewable energy infrastructure, which is classed as 'artificial surfaces' if the primary land use of an area has changed from, say, forestry or agriculture to energy production. However, the dominant land cover of that land parcel may not have changed that much. Often, wind turbines are erected that use up very small areas of the parcel and that are connected by narrow roads or bridleways. Similarly, fields which host solar farms can and do continue to be used as pasture for small livestock such as sheep and poultry in a multi-purpose land use arrangement [34]. However, under the CORINE guidelines, these parcels are still classified as an artificial surface. This represents a trade-off between continuity and contemporary relevance; on the one hand, mapping to the same specification over decadal timescales enables long-term trends to be analysed, which is valuable in landscape studies [35]. On the other hand, newly developing trends in the dynamics of the landscape may not be best captured by older specifications and, as a result, the interpretation of the results of the changes requires an in-depth understanding of the mapping guidelines [21] in order not to misinterpret the data. From the perspective of the map producer, this means that the mapping guidelines must be clearly communicated as part of the data product.

Although the CLC mapping guidelines have remained relatively stable from the first products in 1990, there has been continuous variability in the specifications of the input EO data against which the changes are mapped. Originally, the 30 m spatial resolution Landsat series of satellites was the preferred choice, but due to the loss of Landsat 6, instrument problems on Landsat 7 and the shortening repeat frequency of the product from 10 to 6 years, other systems had to be used to plug the gap to give a relatively heterogeneous image data source. For 2018, the Copernicus Sentinel-2 system offered a consistent and reliable image source with similar capabilities to Landsat. However, a key change was the reduction of the spatial resolution down to 10 m, giving a resolving power six times better than Landsat (Figure 12). As can be seen, this allowed more accurate identification of small features and those with fine spatial textures such as suburban, road networks and windfarms. Although there was no change to the method, the improved 2018 data allowed the identification of technical changes (e.g., small suburban areas) that were not visible in IMAGE2012. Now that these technical changes have been made to CLC2018, future updates should be easier and more accurate, and Sentinel-2 will continue to be available over the next few decades at least.

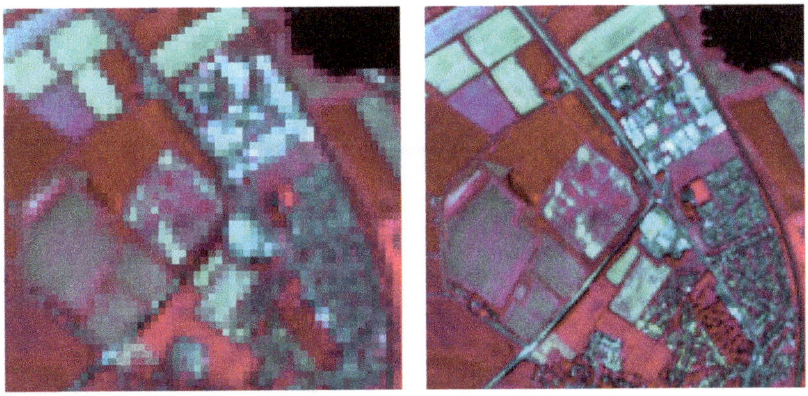

(a) (b)

Figure 12. A comparison of Landsat 8 (**a**) and Sentinel-2 (**b**) data (fcc) on the same day for an area in northeast England, showing the increased resolving power of 10 m spatial resolution data (right), particularly the traffic roundabout in the centre of the image and the suburban areas in the lower right.

The dominant landscape change in the UK is clearly the cutting of coniferous forests, and their subsequent regrowth as part of the rotation forestry for timber production. This has not changed since the last analysis between 2006 and 2012; however, what has been identified this time is the net reduction of the replanting and regrowth. A significant number of land parcels that were 'coniferous forest' in 2012 changed to 'transitional woodland/scrub' in 2018, leading to a net reduction in the coniferous forest cover over this period in the CORINE map. This is also reflected in Deforestation being one of the main pressures on the UK habitats according to the BIOPRESS methodology (Figure 11). Although over the same timeframe, the UK wood harvesting statistics are fairly stable at a very high level (Figure 6), the new tree planting statistics (Figure 6) appear to be insufficient to compensate for the high harvesting intensity. This is particularly an issue of concern in the context of the UK's pathway towards net zero carbon emissions by 2050, which was turned into law in June 2019. The UK's Sixth Carbon Budget [3] recommended that the UK's woodland cover should increase from 13% to 15% by 2035. To achieve this goal, a further 440,000 ha of mixed woodland will have to be planted to remove CO_2. Over the same timescale, 260,000 ha of agricultural land should switch to bioenergy production, including short-rotation forestry. The CORINE class definitions in their current form would not be able to adequately quantify the changes towards these new forms of land use. Bioenergy production would be classed as agricultural cropland or, if covered by young trees, as transitional woodland/scrub. In terms of assessing the UK's habitats and carbon balance, it is important to know whether an area of transitional woodland/scrub is actively growing new trees as part of commercial forestry and bioenergy production or if it has been abandoned; there is an opportunity to incorporate this into the next generation of landscape mapping. Peatland restoration is also part of ongoing climate mitigation efforts in the UK, which has substantial peatland areas, most of which are degraded to some degree. The CORINE classes do not currently distinguish between restored and degraded peatland, although that will become a major issue for policy.

The process of urbanisation or urban land take has been highlighted in the results between 2012 and 2018 and is a trend identified in previous iterations of the CLC inventories. The major headline to note is the dramatic increase in the amount of renewable energy infrastructure in the UK over the last two decades. Due to the specifications of CLC, it may not always precisely represent the area of land that is being taken, but it is clearly picking up the trends in this sector and even the split between onshore wind and solar projects due to their context. In the case of onshore wind, the visual interpretation of satellite images and the exaggeration of this class with the mapping rules have been an advantage. Future developments of CLC, described below, will more accurately map windfarms as turbines and access roads, but the use of a characterisation rather than classification approach will allow both land cover and land use attributes to be attached to these features.

Other growths in impervious land such as the reported 12.5% rise in transport infrastructure and 105% increase in construction sites, although significant, need to be viewed with caution as the total areas these land uses cover are relatively low across the whole of the UK. Urban fabric, in the sense of the built environment, is captured in the CLC classes 111 and 112. These classes have both increased between each iteration of the CLC maps; however, the rise between 2012 and 2018 is 11,925 ha, accounting for an increase in just below 1% of their cover in the UK.

Moving forward the CLMS will continue to expand its portfolio to provide improved land cover/land-use products and provide more information on surface characteristics and dynamic behaviours. Central to the CLMS development at the pan-European scale will be the CLC+, or 2nd generation CLC, initiative, which will provide base datasets with improved spatial resolution and thematic content. The CLC+ Backbone will provide a wall-to-wall land cover map with a 0.5 ha MMU, but a relatively simple nomenclature compared to current CLC standards. The CLC+ Core will be a corresponding grid-based (1 ha) thematic information engine holding detailed land cover, land-use and additional characteristics for each cell. CLC+ Core will adopt the EAGLE data model [36], which is specifically designed to characterise rather than classify landscape features, thus allowing improved descriptions, greater flexibility and better representations in different application contexts. CLC+ Core will be drawn from a broad range of CLMS, EEA Member State and open-source datasets. By combining CLC+ Backbone and CLC+ Core via appropriate mapping specification and rule sets, an almost infinite number of derived CLC+ Instances can be produced, based on a common underlying information source. The first CLC+ Instances to be produced will be related to Land Use, Land-Use Change and Forestry (LULUCF), an inventory for emissions and removals of greenhouse gases resulting from direct human-induced land use, and CLC+ Legacy, the conventional CLC that we have today. Together, these products will form a powerful tool for tracking and understanding our environment in increasing detail while being compatible with the long time series of CLC.

5. Conclusions

There has been an increase in the rate of change in UK land cover/land use, both in area and type of change. An area covering 1.16% of the total UK has changed between 2012 and 2018. The urban fabric has increased by 1% and the industrial or commercial units by 14%, which has been driven by the increase in renewable energy infrastructure, accounting for 79% of this change. The dominant landscape change in the UK remains the clearcutting and regrowth of coniferous forests; however, there seems to be a reduction in replanting and regrowth leading to net reduction of forest cover.

The two largest changes in the UK landscape during the period 2012–2018—forestry and renewable energy—while represented clearly in the change mapping are both somewhat vulnerable to misinterpretation under the current CORINE methodology. The producers of this map emphasise the need for careful review of mapping guidelines before use of the data.

The nomenclature has been stable for decades, which enables long-term changes and trends to be identified; however, some more recent types of changes, such as the renewable energy types, are not as well captured. In its current form, CORINE is not well suited to mapping the classes required for the UK net zero assessment because the class definitions would not adequately quantify the change to the new forms of land use required to meet the UK climate change targets. However, the new methodology based on CLC+, which enables the distinction of land use and land cover, and the increase in spatial resolution will be able to more accurately capture some of the most important contemporary land changes.

Consistent long-term monitoring of land change is extremely important in the changing environment of the 21st century and should be maintained in the UK to support progress towards climate commitments, which will require a component of significant land-use change in the coming decades.

However, at the time of writing, it had not been confirmed whether the UK will continue to be part of the Copernicus programme and therefore whether the CLMS products will be updated for the UK following Brexit. As environmental time-series grow, their value increases, and at this time of climate emergency and biodiversity crisis it is important that CLC and other CLMS products continue to be regularly produced for the UK in a manner consistent with the complete European coverage.

Supplementary Materials: The following are available online at https://www.mdpi.com/article/10.3390/land11020192/s1, Table S1: full change matrix showing total area (Ha) of each change 2012–2018.

Author Contributions: B.C., writing—original draft, investigation, validation, formal analysis, supervision, project administration; G.S., writing—original draft, validation, formal analysis; B.d.l.B.-B., writing—review and editing, investigation, validation, formal analysis, supervision, project administration; S.C.M.J., writing—review and editing, supervision, project administration; A.H., writing—review and editing, investigation, validation; M.P., writing—review and editing, investigation, validation; T.C., investigation, validation; L.Y.C., investigation, validation; H.B., writing—original draft, formal analysis, supervision. All authors have read and agreed to the published version of the manuscript.

Funding: Supported by the European Environment Agency under Framework service contract No EEA/IDM/R0/16/009/United Kingdom with funding by the European Commission. IMAGE 2012 and IMAGE2018 data were provided by the Copernicus land monitoring service with financial support from the European Space Agency and the European Commission. H.B. was supported by the NERC National Centre for Earth Observation. The APC was funded by Leicester Research Archive.

Institutional Review Board Statement: Not applicable.

Data Availability Statement: The data presented in this study are openly available from the Environmental Information Data under the Open Government License and is available at: https://doi.org/10.5285/027b6432-f44d-41ab-b349-89fae673d5f5 (CLC12-18 change), https://doi.org/10.5285/084e0bc6-e67f-4dad-9de6-0c698f60e34d (CLC2018), https://doi.org/10.5285/9bb7caab-764d-407b-9a81-0d758722d900 (CLC2012 revised).

Conflicts of Interest: The authors declare no conflict of interest. The funders (The EEA) specified the method for the CLC mapping to ensure consistence across Europe. They had no role in the analyses or interpretation of data, in the writing of the manuscript, or in the decision to publish the results.

References

1. Rodgers, C. Delivering a better natural environment? The Agriculture Bill and future agri-environment policy. *Environ. Law Rev.* **2019**, *21*, 38–48. [CrossRef]
2. Capstick, S.; Wang, S.; Khosla, R. Bridging the gap–the role of equitable low-carbon lifestyles. In *UNEP Emissions Gap Report 2020*; United Nations Environment Programme: Nairobi, Kenya, 2020.
3. Committee on Climate Change. *The Sixth Carbon Budget The UKs path to Net-Zero*; Committee on Climate Change: London, UK, 2020; p. 448.
4. Department for Communities and Local Government. *Fixing Our Broken Housing Market*; Department for Communities and Local Government: London, UK, 2017; p. 106.
5. Wulder, M.A.; Coops, N.C.; Roy, D.P.; White, J.C.; Hermosilla, T. Land cover 2.0. *Int. J. Remote Sens.* **2018**, *39*, 4254–4284. [CrossRef]
6. Ren, Y.; Lü, Y.; Comber, A.; Fu, B.; Harris, P.; Wu, L. Spatially explicit simulation of land use/land cover changes: Current coverage and future prospects. *Earth-Sci. Rev.* **2019**, *190*, 398–415. [CrossRef]
7. Zhang, H.K.; Roy, D.P. Landsat 5 Thematic Mapper reflectance and NDVI 27-year time series inconsistencies due to satellite orbit change. *Remote Sens. Environ.* **2016**, *186*, 217–233. [CrossRef]
8. White, J.C.; Wulder, M.A.; Hobart, G.W.; Luther, J.E.; Hermosilla, T.; Griffiths, P.; Coops, N.C.; Hall, R.J.; Hostert, P.; Dyk, A.; et al. Pixel-Based Image Compositing for Large-Area Dense Time Series Applications and Science. *Can. J. Remote Sens.* **2014**, *40*, 192–212. [CrossRef]

9. Lu, Y.; Wu, P.; Ma, X.; Li, X. Detection and prediction of land use/land cover change using spatiotemporal data fusion and the Cellular Automata–Markov model. *Environ. Monit. Assess.* **2019**, *191*, 68. [CrossRef] [PubMed]
10. Caballero, I.; Ruiz, J.; Navarro, G. Sentinel-2 Satellites Provide Near-Real Time Evaluation of Catastrophic Floods in the West Mediterranean. *Water* **2019**, *11*, 2499. [CrossRef]
11. Hussain, M.; Chen, D.; Cheng, A.; Wei, H.; Stanley, D. Change detection from remotely sensed images: From pixel-based to object-based approaches. *ISPRS J. Photogramm. Remote Sens.* **2013**, *80*, 91–106. [CrossRef]
12. Khatami, R.; Mountrakis, G.; Stehman, S.V. A meta-analysis of remote sensing research on supervised pixel-based land-cover image classification processes: General guidelines for practitioners and future research. *Remote Sens. Environ.* **2016**, *177*, 89–100. [CrossRef]
13. Carranza-García, M.; García-Gutiérrez, J.; Riquelme, J.C. A Framework for Evaluating Land Use and Land Cover Classification Using Convolutional Neural Networks. *Remote Sens.* **2019**, *11*, 274. [CrossRef]
14. Balzter, H.; Cole, B.; Thiel, C.; Schmullius, C. Mapping CORINE Land Cover from Sentinel-1A SAR and SRTM Digital Elevation Model Data using Random Forests. *Remote Sens.* **2015**, *7*, 14876–14898. [CrossRef]
15. Büttner, G. CORINE Land Cover and Land Cover Change Products. In *Land Use and Land Cover Mapping in Europe: Practices & Trends*; Manakos, I., Braun, M., Eds.; Springer Netherlands: Dordrecht, The Netherlands, 2014; pp. 55–74, ISBN 978-94-007-7969-3.
16. Bossard, M.; Feranec, J.; Ot'ahel', J. *Corine Land Cover Technical Guide-Addendum 2000*; European Environment Agency: København, Denmark, 2000; p. 105.
17. Bielecka, E.; Jenerowicz, A. Intellectual Structure of CORINE Land Cover Research Applications in Web of Science: A Europe-Wide Review. *Remote Sens.* **2019**, *11*, 2017. [CrossRef]
18. Feranec, J.; Soukup, T.; Hazeu, G.; Jaffrain, G. (Eds.) *European Landscape Dynamics: CORINE Land Cover Data*, 1st ed.; CRC Press: Boca Raton, FL, USA, 2016; ISBN 978-1-315-37286-0.
19. Nedd, R.; Light, K.; Owens, M.; James, N.; Johnson, E.; Anandhi, A. A Synthesis of Land Use/Land Cover Studies: Definitions, Classification Systems, Meta-Studies, Challenges and Knowledge Gaps on a Global Landscape. *Land* **2021**, *10*, 994. [CrossRef]
20. Diaz-Pacheco, J.; Gutiérrez, J. Exploring the limitations of CORINE Land Cover for monitoring urban land-use dynamics in metropolitan areas. *J. Land Use Sci.* **2014**, *9*, 243–259. [CrossRef]
21. Kosztra, B.; Büttner, G.; Hazeu, G.; Arnold, S. *Updated CLC Illustrated Nomenclature Guidelines*; European Environment Agency: København, Denmark, 2017; p. 124.
22. Büttner, G.; Kosztra, B. *CLC2018 Technical Guidelines*; European Environment Agency: Wien, Austria, 2017; p. 61.
23. Pataki, R. *ArcGIS 10.5-10.3 Toolbox for Creation of CLC2018 Database. User Guide*; European Topic Centre on Urban, Land and Soil Systems: Wageningen, The Netherlands, 2018; p. 13.
24. ONS Woodland Natural Capital Accounts, UK-Office for National Statistics. Available online: https://www.ons.gov.uk/economy/environmentalaccounts/bulletins/woodlandnaturalcapitalaccountsuk/2020 (accessed on 17 June 2021).
25. Cole, B.; Smith, G.; Balzter, H. Acceleration and fragmentation of CORINE land cover changes in the United Kingdom from 2006–2012 detected by Copernicus IMAGE2012 satellite data. *Int. J. Appl. Earth Obs. Geoinf.* **2018**, *73*, 107–122. [CrossRef]
26. The World Bank World Bank Open Data. Urban Population-United Kingdom. Available online: https://data.worldbank.org/indicator/SP.URB.TOTL.IN.ZS?locations=GB (accessed on 14 July 2021).
27. DECC. *UK Solar PV Strategy Part 1: Roadmap to a Brighter Future*; Department of Energy and Climate Change: London, UK, 2013; p. 38.
28. Forest Research Forestry Statistics 2020. Available online: https://www.forestresearch.gov.uk/tools-and-resources/statistics/data-downloads/ (accessed on 1 June 2021).
29. Gabbatiss, J. *In-Depth Q&A: How Will Tree Planting Help the UK Meet Its Climate Goals?* CarbonBrief: London, UK, 2020.
30. Barkham, P. *Fir's Fair: UK must Embrace Conifers in Climate Fight, Says Forestry Chief*; The Guardian: London, UK, 2020.
31. McGrath, T. To enable communities to engage with the biggest European tidal flat and saltmarsh restoration project. In *Coastal Wetlands Restoration*; Routledge: London, UK, 2021; p. 13, ISBN 978-0-367-86309-8.
32. Gerard, F.; Petit, S.; Smith, G.; Thomson, A.; Brown, N.; Manchester, S.; Wadsworth, R.; Bugar, G.; Halada, L.; Bezák, P.; et al. Land cover change in Europe between 1950 and 2000 determined employing aerial photography. *Prog. Phys. Geogr. Earth Environ.* **2010**, *34*, 183–205. [CrossRef]
33. Petit, S.; Firbank, L.; Wyatt, B.; Howard, D. MIRABEL: Models for Integrated Review and Assessment of Biodiversity in European Landscapes. *AMBIO J. Hum. Environ.* **2001**, *30*, 81–88. [CrossRef] [PubMed]
34. BRE. *Agricultural Good Practice Guidance for Solar Farms*; Scurlock, J., Ed.; National Farmers Union: Washington, DC, USA, 2014; p. 8.
35. Crumley, C.L.; Kolen, J.C.A.; de Kleijn, M.; van Manen, N. Studying long-term changes in cultural landscapes: Outlines of a research framework and protocol. *Landsc. Res.* **2017**, *42*, 880–890. [CrossRef]
36. Arnold, S.; Smith, G.; Hazeu, G.; Kosztra, B.; Perger, C.; Banko, G.; Soukup, T.; Strand, G.-H.; Valcarcel Sanz, N.; Bock, M. The eagle concept-A paradigm shift in land monitoring. In *Land Use and Land Cover Semantics-Principles, Best Practices and Prospects*; CRC Press: Boca Raton, FL, USA, 2015; pp. 107–144, ISBN 978-1-4822-3739-9.

land

Perspective

Native Trees as a Provider of Vital Urban Ecosystem Services in Urbanizing New Zealand: Status Quo, Challenges and Prospects

Jihwi Jang [1] and Su-Young Woo [2,*]

[1] School of Biological Sciences, University of Canterbury, Christchurch 8041, New Zealand; jihwi.jang@pg.canterbury.ac.nz
[2] Urban Environment and Plant Sciences Laboratory, Department of Environmental Horticulture, University of Seoul, Seoul 02504, Korea
* Correspondence: wsy@uos.ac.kr; Tel.: +82-2-6490-2691

Abstract: In New Zealand, over 87% of the population currently resides in cities. Urban trees can face a myriad of complex challenges including loss of green space, public health issues, and harm to the existence of urban dwellers and trees, along with domestic greenhouse gas (GHG) and air pollutant emissions. Despite New Zealand being a biodiversity hotspot in terms of natural environments, there is a lack of knowledge about native tree species' regulating service (i.e., tree development and eco-physiological responses to low air quality, GHG, rising air temperatures, and drought) and how they grow in built-up environments such as cities. Therefore, we argue for the value of these native species in terms of ecosystem services and insist that they need to be viewed in relation to how they will respond to urban abiotic extremes and climate change. We propose to diversify planted forests for several reasons: (1) to improve awareness of the benefits of diverse planted urban forests; (2) to foster native tree research in urban environments, finding new keystone species; and (3) to improve the evidence of urban ecosystem resilience based on New Zealand native trees' regulating services. This article aims to re-evaluate our understanding of whether New Zealand's native trees can deal with environmental stress conditions similarly to more commonly planted alien species.

Keywords: tree diversity; ecosystem resilience; native tree; urban environment; urbanization

Citation: Jang, J.; Woo, S.-Y. Native Trees as a Provider of Vital Urban Ecosystem Services in Urbanizing New Zealand: Status Quo, Challenges and Prospects. *Land* **2022**, *11*, 92. https://doi.org/10.3390/land11010092

Academic Editor: Giuseppe Pulighe

Received: 26 November 2021
Accepted: 5 January 2022
Published: 6 January 2022

Publisher's Note: MDPI stays neutral with regard to jurisdictional claims in published maps and institutional affiliations.

1. Introduction

1.1. Effects of Urbanization on Tree Growth and Development

Urbanization is a worldwide phenomenon and a key driver of environmental degradation and climate change [1,2]. An urban environment can generally be defined as an area containing an aggregation of infrastructure, buildings, and open spaces that provide for the urban community's socio-economic functions [3]. Currently, over half of the global human population lives in urban and metropolitan areas [4], and this proportion is expected to increase to 70% by 2050 [5].

Trees in urban areas can suffer from chronic abiotic stresses, such as changes in the growing season and circadian rhythm due to urban thermal discomfort, disorders caused by air pollution, and droughts, which are typically enhanced by increasing urbanization [6]. The drastic changes in the urban landscape and environment have negatively affected urban tree and ecosystem health as many plant species have been moved from their provenance to cities (i.e., new environment) [7]. For instance, soil moisture, atmospheric temperature, relative humidity (RH), and vapor pressure deficit (VPD) are often less favorable for urban trees than for their rural environmental counterparts. This is because they result in different tree growth rates (i.e., slower or faster), lower density root systems, and higher leaf temperature, showing different relative tree growth rates until final tree development [8,9]. Another environmental feature for an urban area is a specific airborne chemical composition produced by emissions from traffic, households, and industries, which results in higher CO_2 concentration and more air pollution, raising the atmospheric temperature through GHG. Hence, urban trees growing in built-up areas are subjected to a microenvironment characterized by higher pollution and GHG emission levels due to traffic volume, additional soil drought, and contamination by the input of heavy metals or high salinity [10], as well as a restricted area for root extension which in turn decreases water availability (i.e., cover plate of a tree disc, tree pit covers, and road pavement) [11,12].

This article examined case studies currently used for carbon sequestration and air pollutant removal of urban trees native to New Zealand and compiled currently available results in New Zealand native trees in cities. However, the currently available results are related to alien tree studies and a large degree of uncertainty due to the limitation of applied studies on New Zealand native trees. For a better understanding of New Zealand native trees for urban ecosystem services, it is proposed that planted forests should be diversified for several reasons in this paper: (1) to improve awareness of the benefits of diverse planted urban forests; (2) to foster native tree research in urban environments, finding new keystone species; and (3) to improve the evidence of urban ecosystem resilience based on New Zealand native trees' regulating services. This article aims to re-evaluate our understanding of whether New Zealand's native trees can deal with environmental stress conditions similarly to the more commonly planted alien species. We compiled 146 publications that reported existing data, literature, and opinion on urban forestry and ecology. This perspective article explored and discussed whether New Zealand native trees can provide urban ecosystem services and confirmed that the existing literature can support the advantages of having native trees in cities.

1.2. The Decline of Native Forests after Human Settlement in New Zealand

As shown in Figure 1, the decline of New Zealand's native forests began with the arrival of Māori pioneers in AD 1000, who began deforestation for land-use conversion [13–16]. With the arrival and establishment of the first European settlers around 1840, more natural forests were lost as more towns were developed and agricultural activity increased. By 2000, nationally forest cover in New Zealand had been reduced to only 25% of its pre-settlement level ([15,17]; see Figure 1A). The decline of native trees has also been consistent with urban sprawl and the urbanization trend of New Zealand chronologically by the early 20th century (1920s) ([18]; see Figure 1B). Currently, many introduced species (approximately 2264 species: 30 mammals, 34 birds, and 2200 plants), including in urban areas, are reported in New Zealand [19].

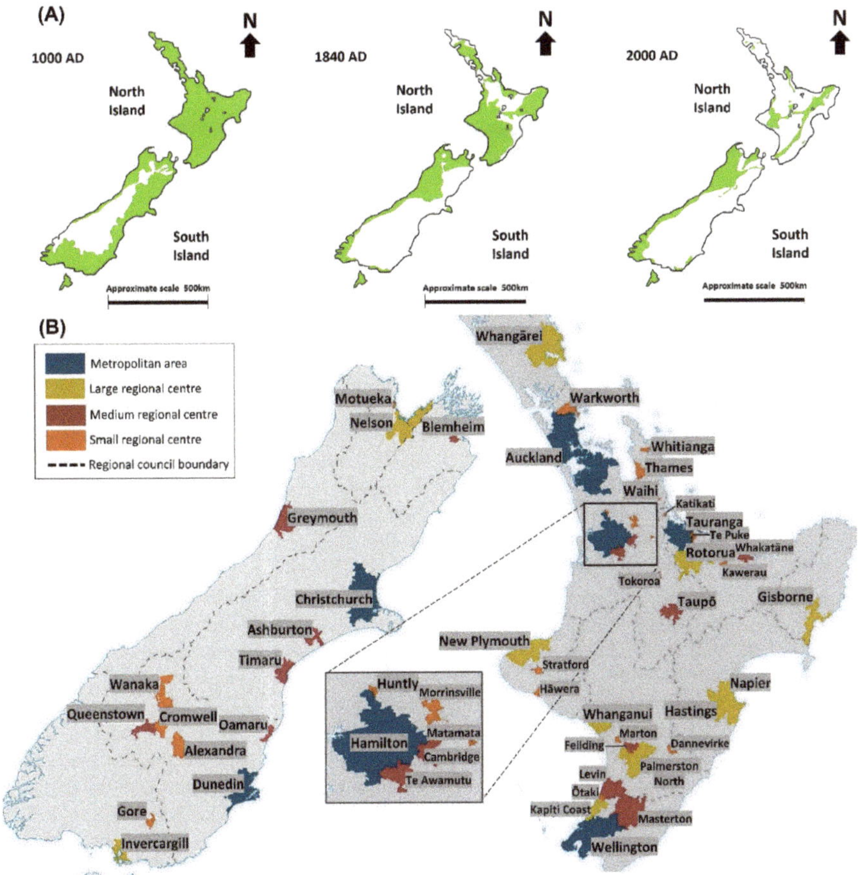

Figure 1. (**A**) Changes in native tree/forest coverage (green color) over time in New Zealand, adapted from Stevens et al. [15] and Nomura et al. [20]; (**B**) functional urban areas by type, New Zealand [21].

With agriculture, dairy farming, and township settlement, forestry activities (i.e., establishing plantation forests, logging, and timber yield) have contributed to the decline of native forests [22–24]. As a result of these activities, many native New Zealand tree species that were once common are now classified as threatened or protected in both rural and urban areas [25,26]. However, New Zealand is currently host to a wide range of alien species, defined as species non-native to New Zealand [24–30]. Since the 1990s, alien (non-native) tree species have had a significantly higher afforestation rates than native species in New Zealand [31,32]. Historically, the use of alien tree species, such as *Quercus* spp. and *Fraxinus* spp., has been preferred in urban green spaces and for garden planning [33,34]. The invasion of alien species in New Zealand cities has contributed to a severe decline in native clusters (indigenous trees clusters) over time [27–29]. In Hamilton, currently the fourth largest and second fastest-growing city in New Zealand, the distribution of native trees in the city is only 2.1%, which is the lowest among New Zealand's six main cities—namely, Auckland, Wellington, Christchurch, Hamilton, Tauranga, and Dunedin [18]. The remnants or patches of native dominated vegetation in each of the cities are very small (2.1–8.9% in the urban boundary) [18,19] and most native trees, except for nature heritage parks in cities, have been planted through urban restoration projects since the 1990s [19]. Although the resilience and flexibility of all trees to abiotic stress caused by human settlement and urbanization require further study, the physiological adaptation of urban trees that are native to New Zealand has been investigated less (especially in urban settings) than that of species indigenous to other countries, such as Central Europe [6,35,36], North America [37], East Asia [38], and Australia [39]. Since the 2000s, urban restoration, including native tree planting in cities, has continued to grow in New Zealand, but relevant research effort is required to overcome a lack of interdisciplinary breadth (i.e., environmental science, plant physiology and biochemistry, forest science, and urban ecology) [19].

2. Urbanization in New Zealand, Its Consequences, and the Role of Tree Diversity

2.1. Urbanization and CO_2 Emission Rate Increment in New Zealand

In the case of New Zealand, as much as 87% of the population currently reside in urban environments and cities [40], and urbanization is increasing especially in the Auckland region, reaching suburban areas such as Tauranga and Hamilton [41]. Urbanization is a strong influencer of population growth (including internal and international migration), building and infrastructure construction, and the spreading of residential areas and fragmentation of urban forests [41,42]. Half of New Zealand's population is expected to live in the Auckland metropolitan area by 2050, and the country's population is expected to reach 8 million by 2073 [41]. New Zealand is heading toward the upper end of urbanization, defined as rapid population growth with new infrastructure, based on Auckland [43]. In total, 76.5% of New Zealanders reside on the country's North Island, which has four major cities: Auckland, Wellington, Hamilton, and Tauranga [44].

Anthropogenic impacts are likely to accelerate abrupt changes in tree growth conditions (i.e., atmospheric temperature, humidity, CO_2 concentration, air quality, and drought extent) in cities. Even though New Zealand has among the highest air quality in the world [45], the amount of domestic anthropogenic greenhouse gas (GHG) emissions has increased over time. This increase in GHG emissions (mainly CO_2, SF_6, and HFCs) is highly related to urbanization in New Zealand [46,47]. However, since the adoption of the Climate Change Response (Zero-Carbon) Amendment Act 2019 of New Zealand, it is expected that New Zealand's government will focus on the reduction of GHG emissions [46].

The GHG inventory report of New Zealand's government largely attributed the increases in GHGs to the energy/transport sectors, determining that these sectors are responsible for 38.2% of the net increase in CO_2 emissions since 1990. In addition, land use, land-use change, and forestry activities (LULUCF) have not shown a decreased rate (i.e., carbon sequestration process in plants and soils) against the build-up of atmospheric CO_2 over time (+7203.3 Kt(CO_2) increment between 1990 and 2017) ([46]; see Appendix A). Approximately 20% of New Zealand's annual energy consumption is from road transportation in urban areas [48], and the emission rate from road transportation steadily grew during the last two decades with an increased rate of private vehicle ownership [49]. Consequently, 47% of New Zealand's total domestic CO_2 emissions come from the road transportation sector and these emissions have tripled over the past three decades [50].

2.2. High Private Vehicle Usage and Deterioration Extent in New Zealand

Over time, road transportation and the use of fossil fuel-dependent vehicles have dramatically increased. They are consistent with the 2020 population growth rate per year (2.1–2.8%). Although Auckland has 52.4 km of bicycle routes [51], private vehicle usage is still the most common form of daily transport [41,52]. In Christchurch, the second most populated city in New Zealand [53], the proportion of CO_2 emissions from vehicles has increased over the last two decades [48]. The use of private vehicles is very dominant in Christchurch, being used for the daily commute by 84% of commuters, which is similar to Auckland (85%). However, the proportion of public transportation use in many cities is still low (2–8%), except for Wellington (21%) ([52]; see Appendix B). This dominance of private vehicles is likely to affect New Zealand's urban environment and contribute to global climate change (GCC), especially as the population growth rate of Christchurch has been 13.5% for five years since 2013 [52], and therefore, the population is predicted to continue to rise.

The deterioration of private vehicles is likely to have a profound effect on New Zealand's GHG emissions. The average age of New Zealand's vehicle fleet is estimated at 14.2 years [54], which is older than that of most OECD countries; the average private vehicle ages in USA, Canada, and Australia are lower than 12 years [55]. Between 2000 and 2017, the proportion of vehicles over 15 years old in New Zealand increased from 24.5% to 42.3% [55,56], and this trend is likely to continue [55]. In addition, over the last 15 years, the proportion of 0–4-year-old vehicles remained under 20% [55,56]. Kjellström and Mercado [57] reported that the average age of vehicles is an important indicator of urban environmental health; old vehicles are likely to be less energy efficient than newer vehicles, have lower fuel efficiency, and their exhaust fumes have stronger links to GHG emissions in cities, including CO_2, CO, NO, NO_2, and particulate matter of less than 10 or 2.5 μm (PM_{10} and $PM_{2.5}$, respectively) [58]. In the case of Auckland, the concentration of multiple air pollutants ($PM_{2.5}$, black carbon (BC), and NO_2) is highly related to a high density of vehicular traffic, showing 2.5-fold ($PM_{2.5}$) and 2.9-fold (NO_2) higher concentrations in the city center (central business district) than other suburban areas in Auckland [59]. Consequently, it might affect human health and tree disservices issues to urban dwellers (87% of New Zealand population). Therefore, it is important to find proper urban tree species among various genetic diversity pools for effective GHG mitigation and air pollutant removal in the changing environment of New Zealand cities.

2.3. The Role of Tree Diversity in Ecosystem Resilience

Ecosystem services are the varied benefits to people provided by the natural environment and healthy ecosystems [60,61]. Ecosystem resilience can generally be defined as the ability to absorb disturbance and provide a stable condition for the ecosystem without loss of ecological function or ecosystem service [62]. Therefore, understanding the role of species diversity of native trees in ecosystem resilience can be vital for strategic ecosystem management tactics to combat anthropogenic disturbances, because it supports functional diversity based on species interaction under interwoven abiotic factors [63]. As a sufficient level of species diversity affects the maintenance of resilience-based management [64], native species can constitute an important proportion of resilience. Species richness is empirical evidence of plant biodiversity [65] and can contribute to effective ecosystem resilience [66]. Species diversity can improve ecosystem stability and act as an environmental buffer [66]. Moreover, increased diversity of trees in an ecosystem can mitigate the disturbance of carbon cycling through trees' species-specific eco-physiological functions and different spreading extent of root systems [67]. Thus, understanding tree diversity is important for climate change regulations [63]. One example of the vital role of native species in imparting resilience is that they attract more pollinator species than alien species [68]. The flowering and fruit production of trees are significantly increased when the monotonous alien proportion decreases [68]. In addition, native trees provide diverse faunal biodiversity habitats [69–71].

Tree species diversity in cities can provide a characterized tree population for improved species structure, function, and value [72]. However, native trees tend to be underutilized in cities [73,74]. Relying excessively on a small number of species threatens urban forest resilience and reduces ecosystem services [75]. Urban tree species are generally removed and/or replanted once they are regarded as having disservice and/or no use in urban forest management [2,76]. Complex interactions between biotic and abiotic factors can affect species imbalances and/or deletions in resilience [77]. Hence, it is important to conduct strategic management of urban ecosystems and vegetation to create a sustainable urban forest that is resilient to environmental disturbances (e.g., fragmentation and imbalances caused by invasive species) [78].

During the last five years, in the Auckland area, there was a net increase of 226 ha in tree canopy cover in built-up areas, 46 ha in urban parkland/open space, and 4 ha in transport infrastructure. However, owing to limited information on the effectiveness of native trees in urban ecosystems and environmental services in New Zealand, there remains an imbalance between alien and native species in this new tree canopy cover, possibly hindering the long-term environmental, cultural, and socioeconomic impacts on urban areas [71].

Planting alien trees in cities can be suitable for environmental regulating, different cultural or heritage purposes, and ecosystem services in some cases, especially for deciduous trees required to enlarge the canopy, or to establish community orchards [65]. Previous studies have noted that alien species on diversity can foster soil nutrients by increasing nitrogen cycling and the composition of soil microbial communities [79]. However, alien species can affect local native plant communities and diversity by minimizing species richness [80] and by affecting pollinators and soil carbon-degrading enzymes of native species [81,82]. In addition, native trees might provide better ecosystem services (with beneficial environmental regulating services) than alien trees in cities. Rahman et al. [2] reported that Central European native tree species planted in cities showed better regulating services (i.e., cooling effect) with 2.8 °C air temperature reduction (ΔAT) and higher transpiration rate than that of alien tree species in a case study of an urban area in Munich, Germany. Many urban trees (alien and native) have differing wood anatomies that highly affect trees' strategies under urban environmental stress such as drought and urban heat island (UHI). Moser-Reischl et al. [83] reported that diffuse-porous and anisohydric trees have a higher cooling effect with high canopy-scale transportation rates amid thermal discomforts in cities, whereas ring-porous and isohydric trees provide higher water potential with high survival rate (low maintenance) by affecting urban hydrology over time. Sonti et al. [37] reported that North American native trees planted in cities showed higher or equivalent stress tolerances with alien trees (i.e., increased air temperature stress, air pollution, and drought) by showing higher chlorophyll fluorescence parameters (e.g., F_v/F_m) than those of alien trees in cities (case studies of New York, NY; Philadelphia, PA; Baltimore, MD). Therefore, it is likely important to find native trees' characteristics (e.g., benefits on regulating services) [37,83], control the number of alien species, and reject uniformity [68]. According to previous studies, however, species diversity showed positive or negative impacts on ecosystem resilience to environmental stresses in many case studies. For instance, Mulder et al. [84] and Steiner et al. [85] reported that diversity enhances plant communities with species interactions by reducing drought impact. However, Wardle et al. [86], Griffiths et al. [87], and Caldeira et al. [88] reported that species diversity did not affect ecosystem resilience, resistance, or mitigating effects on drought. Hence, further studies of the impact of tree diversity on ecosystem resilience to abiotic stressors in urban areas are required.

3. Lower Proportion of Native Trees That Live in New Zealand's Cities

In this article, we define urban forest as a collection of trees that grow in a city and/or town that encompasses green space in a developed (built-up) area, yards and corridors, and park/roadside trees [89]. New Zealand's urban forests are dominated by alien tree species [90]. There are no well-documented reports on whether trees native to New Zealand have prominent ecosystem functions and increase resilience to abiotic stressors in the city. Because of their well-known benefits (environmental regulating services, e.g., carbon storage and air pollutant removal) [91], alien trees are often planted in urban forests and streets, leading to an imbalance in the ratio between alien and native tree species particularly in Auckland [92] and Christchurch [34]. Despite Christchurch being named the "garden city" of New Zealand (due to an urban botanic garden area and the number of urban parks), native species vegetation, clusters, and forests have become increasingly fragmented and insignificant in size [24,34], with the trend being toward small numbers of alien species, leading to a small genetic pool of native trees [34,90]. There are several reasons for the unequal distribution of tree cover across the region in New Zealand cities, such as land ownership (public/private greenspace), land use (urban/industrial/agricultural), geography, and natural heritage for legal protection. For instance, except for natural heritage sites in urban areas and some public/private green spaces, mostly alien species have planted and grown with higher coverages in the cities [90]. Historically, the types of tree planting and development, street trees, and urban vegetation are influenced by municipal urban planning manuals, funding resources, available space, urban dwellers' species preferences, practitioners' preferences based on alien species well known for their environmental regulating services and physiological functions for tree planting rather than genetic diversity, cultural services, and provenance [2,92–94]. Consequently, alien tree species (mainly *Betula pendula*, *Fraxinus ornus*, *Quercus palustris*, *Prunus yedoensis*, *Liquidambar styraciflua*, and *Quercus robur*) have become more dominant than native trees (mainly *Plagianthus regius*, *Sophora tetraptera*, *Cordyline australis*, and *Sophora microphylla*) in parklands and on streets in Christchurch [34,95]. Previous case studies of other countries' cities reported that increasing tree diversity and enlarging green spaces through planting native trees may increase physiological resistance to environmental stressors (regulating service), including those caused by urbanization [91,93] with the fulfillment of cultural services (i.e., cultural identity (e.g., Māori culture, local history) and aesthetic inspiration in New Zealand cities). This means that high genetic diversity with native trees might improve ecosystem resilience to miscellaneous abiotic extremes in cities. Native species can, therefore, constitute an important proportion of resilience [94].

Native trees in cities are generally planted in private greenspaces, where they have moderate to high canopy cover rates but offer a low level of protection to biotic/abiotic stressors and management [94]. Many native tree species are statistically highly distributed across housing estates with a high New Zealand Social Deprivation Index (NZDep) [92]. Huang [92] reported that alien street trees were higher in species richness (75.76% of total species) and abundance (68.51% of total individuals) than native trees in many urban forests and street trees in Auckland. A previous study in Christchurch also found that 84.1% of street trees were alien species, and found a similar array of alien street tree species in Auckland (i.e., *Acer* spp., *Betula* spp., *Quercus* spp., *Prunus* spp., *Ulmus* spp., and *Fraxinus* spp.) Recent data also show that tree cover canopy of all the land in Christchurch is 15.59%, and alien street trees are more dominant than native tree species in Christchurch ([33,34,96,97]; see Table 1).

Table 1. List of the main street trees of Christchurch and planting status in 2020 [33,34,96–98].

Species	Common Name	Provenance	Species Abundance [††]
Betula pendula	Silver birch	Europe	4642
Fraxinus ornus	Manna ash	southern Europe, southwestern Asia	4384
Quercus palustris	Swamp Spanish oak	United States	4241
Plagianthus regius [†]	Lowland ribbonwood	New Zealand	3340
Prunus yedoensis	Yoshino Cherry	Japan	2722
Liquidambar styraciflua	Sweetgum	North America, Asia	2594
Sophora tetraptera [†]	Large-leaved Kōwhai	New Zealand	2472
Cordyline australis [†]	New Zealand cabbage tree	New Zealand	2411
Sophora microphylla [†]	Kōwhai	New Zealand	2291
Quercus robur	English oak	Britain	2242
	Sum of main trees		31,339
Others (mixed with small numbers of numerous alien species)			81,547
Total (total abundance of street trees)			112,886

[†] native tree species; [††] number of trees.

4. Current Roles and Further Research Direction for Urban Ecosystem Services Provided by New Zealand Native Trees

4.1. *Definition of Ecosystem Service in New Zealand*

Urban forests have a history of providing ecosystem services (i.e., cultural, provisioning, supporting, and regulating services) and increasing resilience to abiotic stresses in cities and can help to mitigate GHG emissions and GCC caused by urbanization, road transport, and prolonged exhaust exposure [89,99,100]. The planting and management of trees in urban forests offer effective ecosystem services [101] (e.g., pleasing esthetic values [102], shade/shelter functions against thermal disservices [2], and cooling effects [6]). New Zealand focuses on six major ecosystem services (benefits) for forestry: (1) carbon storage, (2) soil erosion control, (3) biodiversity for threatened species, (4) water purification, (5) provision of agroforestry/understory crops, and (6) recreation [103]. However, there are certain ecosystem services that are more likely to affect urban dwellers in New Zealand's cities, that present many opportunities to support ecosystem services in urban areas, which is not possible in rural landscapes [60]. Meurk et al. [61] reported that ecosystem services in New Zealand's urban areas can be classified as (1) provisioning services, (2) environmental regulating services, and (3) cultural services. They noted that regulating services have more direct benefits for human health, well-being, and environmental rehabilitation for urban dwellers.

For urban ecosystem services, such as the conversion of land use and biodiversity conservation, tree species selection can substantially contribute to developing biosphere reserves. Urban forests have a wide spectrum of environmental ecosystem services, such as air, water, soil, and climate regulation, as well as ecological habitat quality through the function of various tree species and their assemblage [104]. Trees are crucial for carbon reduction and GHG elimination in cities as part of New Zealand's Zero-Carbon Act (Climate Change Response Act) in the post-Paris Agreement era [46,105]. Under Article 7 of the Rio Earth Summit ratified in 1992, New Zealand is required to submit an annual inventory of GHG emissions to the UNFCCC [106]. With these regulating services, planting native trees in cities can also contribute to diversity conservation and educating society about native species (e.g., cultural and supporting services). Clarkson [107] reported that native tree species such as *Cordyline*, *Sophora*, and *Carex* can be important for the restoration of native vegetation in New Zealand's urban areas. However, there is less understanding of native trees in New Zealand's cities than in cities of other countries [108], as there is less preference for native trees [70,109,110]. Therefore, developing an understanding of native trees for suitable species selection and utilization is likely to contribute to improved urban ecology and urban ecosystem services provision.

4.2. *Urban Trees' General Environmental Regulating Service: Carbon Sequestration*

Proper tree species selection and management contribute to carbon storage and act as urban ecosystem services. Various perspectives and approaches to species selection for urban ecosystem services have been proposed to reduce GHG emissions in many cities [99]. For instance, in an urban forestry context, "carbon-neutral carbon commonly involves measuring carbon emissions through emission reduction actions and carbon offsets" [111]. Moreover, urban forests can contribute to carbon neutrality and sequestration through urban tree management with updated tree inventories [72]. In addition, carbon management and urban ecosystem service functions are strongly influenced by the level of urbanization, knowledge of carbon sequestration management, and education levels, such as management skills and environmental awareness, and familiarity with the ecosystem services and carbon storage functions of urban trees [100]. For example, Akbari et al. [112] reported that atmospheric temperature reduction by vegetation in cities has an equivalent effect of 7 kg of total CO_2 emission reduction. Urban areas can contribute to long-term carbon storage for carbon emission mitigation through absorbing CO_2 with urban trees and forest resources through alternative methods such as chemical carbon substitution [2,43,113]. Forests are non-artificial terrestrial carbon sinks that account for approximately 45% of global land surface [114,115]. Moreover, forests account for 80% of the global above ground and 40% of the global below-ground carbon storage in terrestrial ecosystems [116]. During the last two decades, the carbon sink in temperate forests increased by more than 10%. However, the carbon sink of tropical forests decreased and that of boreal forests showed insignificant changes [114]. The decrease in tropical forests was driven by decreases in tree size, shifts in tree species distribution, and elevated tree respiration rates under high temperatures due to GCC [117,118]. Conversely, in boreal forests, the significant change was due to vulnerability to GCC and the very low nutrient-absorption ability of trees [119]. New Zealand belongs to the temperate region, except for some subtropical parts of North Island [120]. Therefore, focusing on the role of temperate forest trees and urban forests in GHG mitigation is important for New Zealand.

The importance of forest conservation in global efforts to fight climate change was recognized by Article 5 of the Paris Agreement, on Forests, which endorsed the role forests play in mitigating GHG emissions [115]. Unlike natural forest (non-urban forest), urban forests generally include green space/infrastructure and roadside trees located within or close to cities, namely population centers of building aggregation, such as commercial, residential, and industrial areas [89,121,122]. Therefore, scientists are debating how to use native trees as "green infrastructure for climate change adaptation" and for mitigation in an urban forestry context [69,107] to try to ameliorate environmental problems that threaten ecosystems and human health [107,123]. There are also discussions about the role of urban forests in ecosystem services for urban dwellers in the post-COVID-19 era [124]. Social restrictions and changes in lifestyle paradigms may fundamentally alter the relationship between urban dwellers and urban green spaces [125]. Hence, it is crucial to study and determine the roles of native trees in tackling current challenges such as climate change, water scarcity, after-effects of COVID-19, and plant biodiversity loss [109,123]. Each tree species has different climate change-adaptation strategies and responds with different mechanisms and/or resistances to these changes [126,127].

4.3. Unexploited Potential of Native Trees' Regulating Service in New Zealand's Cities

Past studies have demonstrated that trees native to New Zealand (see Appendix C for pictures of a sample of common native trees) are valuable for urban ecosystem services. By adopting selective native tree planting, afforestation in built-up environments might have similar effects as those of natural native forests on carbon storage potential and absorption rates in New Zealand [128–130].

Huang [92] reported that the mean diameter growth rate of abundant street trees managed by the city council in Auckland was 13.54 $\pm$ 1.04 mm y^{-1}. Even though the average growth rate of native trees (9.59 $\pm$ 4.76 mm y^{-1}) is slower than alien trees (13.15 $\pm$ 7.08 mm y^{-1}) in urban areas, several scientists have suggested that Auckland's urban forests/street trees composed of native trees have equivalent or better climate change mitigation potential than alien trees and can support enhanced provision of ecosystem services through eco-assessment and carbon sequestration [131–133]. By studying the carbon sequestration potential of native trees, Carswell et al. [134] found that the sequestration rate of Kānuka (*Kunzea ericoides*) was approximately 2.3 MgC ha^{-1} y^{-1} (slower sequestration rate than average of alien trees through comparison study). In addition, Schwendenmann and Mitchell [133] reported that the carbon sequestration values of native trees ranged from 69.8 to 290.9 kgC, with carbon concentration values of 44.9–49.6%. This is based on a case study of native tree species widely planted in Auckland for urban revegetation and restoration project fulfilment: Kānuka, Karaka (New Zealand laurel; *Corynocarpus laevigatus*), Lemonwood (Tarata; *Pittosporum eugenioides*), and Kōhūhū (*Pittosporum tenuifolium*). Compared with the sequestration rate of New Zealand's common alien tree *Pinus radiata* (8 MgC ha^{-1} y^{-1}) and the common trees of U.S. cities (2.8 MgC ha^{-1} y^{-1}) (Nowak et al., 2013, as cited in [133]), the average value of these four native trees in urban areas (2.1 MgC ha^{-1} y^{-1}) was significantly lower (Maclaren 2000, as cited in [133]). Nevertheless, Carswell et al. [134] stated that native trees have significant potential to mitigate

GHG emissions, providing that they have success in long-term woody succession. They reported that Kānuka and red beech (*Nothofagus fusca*) showed notable carbon storage potential after 50 years of succession with values of 148 ± 13 MgC ha^{-1} 50 $years^{-1}$ and 145 ± 19 MgC ha^{-1} 50 $years^{-1}$ with biodiversity fulfilment, respectively.

Marden et al. [130] reported that the eight most distributed native trees in New Zealand are conifers—Matai (*Prumnopitys taxifolia*), Kauri (*Agathis australis*), Miro (*Prumnopitys ferruginea*), Totara (*Podocarpus totara*), Kahikatea (*Dacrycarpus dacrydioides*), and Rimu (*Dacrydium cupressinum*)—and broadleaved species: Tītoki (*Alectryon excelsus*) and Puriri (*Vitex lucens*). Native conifers collectively contribute 90% of New Zealand's total live-plant carbon by volume, with the softwoods Rimu, Totara, Miro, and Kahikatea being the most abundant species (Peltzer and Payton, 2006, as cited in [130]). Among them, only Tītoki and Totara trees are relatively dominant in proportion to the Auckland urban area [131]. However, the potential for carbon storage and sequestration of large native trees is scarcely reported in urban areas.

Nīkau (*Rhopalostylis sapida*) and Pōhutukawa (*Metrosideros excelsa*) are the most common native tree species in New Zealand cities. In particular, Pōhutukawa is the most numerous street tree in the Wellington urban area, and it has the highest air pollutant (PM_{10} and O_3) removal efficiency (75 g (PM_{10}) $tree^{-1}$ y^{-1}, 61 g (O_3) $tree^{-1}$ y^{-1}) in the Auckland urban area [132]. Dale [131] investigated the carbon sequestration potential of seven native species (Nīkau, Pōhutukawa, Northern rata, Pōhutukawa × Northern rata hybrid, Taraire, Puriri, and Karaka) in the Wynyard Quarter area, Auckland, and estimated the total tree carbon storage potential for the sample street trees to be 1.5 MgC y^{-1}, which is equivalent to the carbon emissions from driving 30,000 km in a private vehicle (57 tree samples of 7 native species). Dale [131] also reported that Pōhutukawa trees had the highest average storage potential (0.099–0.11 MgC $tree^{-1}$ y^{-1}) due to higher wood density and tree maturity. In addition, in a case study of the Wynyard Quarter area, Findlay [132] determined Nīkau and Pōhutukawa as having the highest carbon removal efficiencies with higher canopy values and biomass. These findings can have significant implications for the debate over diversity needs and ecosystem services along with environmental acclimation through the provenance of trees in cities, but more information is still required (i.e., carbon storage, physiological responses, and long-term assessment) for various types of native tree species in urban settings.

4.4. Further Research Direction of Urban Ecosystem Services in New Zealand

Most studies on tree responses in urban areas to GCC have focused on species alien to New Zealand, and there is a lack of knowledge regarding how native urban trees will respond to the changing climate in New Zealand's cities. The annual precipitation in New Zealand is predicted to be strongly affected by changing patterns of evaporation, which are influenced by higher surface temperatures [135]. Moreover, intensification of the El Niño cycle is likely to enhance the regularity, severity, and duration of droughts in New Zealand [136,137]. Indeed, recent New Zealand climate change projections indicate that droughts are likely to increase in both intensity and duration in many cities on the North Island [138]. Currently, drought in New Zealand is not a serious issue, despite a drought occurring in Huapai, Auckland during the summer season of 2013. During this drought, the soil volumetric moisture content was recorded in the range of 29–51% at 10 cm depth, compared with 43–60% in 2012 [139]. The threat of drought leading to urban water shortages has been raised as a severe issue on the Kapiti Coast and in Wellington. This is because climate change can lower the water level and yield of the Waikanae River, leading to water shortage in surrounding urban areas that rely on the river for water [140,141]. More frequent drought events are therefore likely to lead to water shortages from the river to the built-up environment in the Kapiti Coast/Wellington and Wairarapa regions. In the Auckland and Northland regions, the frequency and intensity of El Niño events are associated with periods of drought [138]. Changes in the physiological responses and carbon and nitrogen budgets of New Zealand native trees in response to climatic conditions, such as drought, higher temperatures, and elevated CO_2, especially in urban environments, have seldom been explored in New Zealand [95]. There is also very little information available on fluxes of nutrients (e.g., carbon allocation) in New Zealand's native trees [139]. Consequently, there is a poor understanding of native tree growth and responses in New Zealand, as most research and management in New Zealand urban forestry has focused on alien tree species [95,107].

Species diversity contributes to a better provision of urban ecosystem services [142]. It affects ecosystem resilience in terms of urban forest protection from pests and plant diseases, climate change, warmer (higher) temperatures, and abiotic extremes [142,143]. Therefore, tree diversity is an important buffer against catastrophic tree loss in managed forests, including urban forests [144]. Generally, monocultures are more vulnerable to biotic and/or abiotic stressors [75,144]. Urban forests with low tree diversity and biotic homogenization may be vulnerable to ecological disturbances and are at greater risk from local/regional climate changes [145]. Therefore, it is necessary to confirm whether these findings are consistent with the large body of evidence that shows that most urban trees grow better with a diverse mixture of species rather than in a monoculture or with less diversity.

5. Conclusions

Urban trees grow under extreme/harsh/difficult and complex conditions. There has been continuous debate and controversy regarding whether native trees are resilient to urban abiotic stresses and should be planted in cities instead of alien trees [70,107,109,110]. In the case study that explored the carbon sequestration potential of native trees growing in an Auckland urban park, the potential sequestration of native trees was estimated to be in the range of 69.8–290.9 kgC, with a carbon concentration of 44.9–49.6%. Even if these carbon sequestration rates are lower than those of alien trees such as *Pinus radiata*, New Zealand native trees may have significant potential in mitigating GHGs if they are competitive in long-term woody succession.

The stress resistance of native tree species in New Zealand cities to GCC and air pollution has received less attention [47,95,108]. This is due to the relatively short history of anthropogenic environmental changes in the growth of trees in urban settings. Therefore, further investigations are needed on the growth and physiological changes in response to future GCC projections, including high temperatures, elevated O_3, $PM_{2.5}$, and CO_2 levels, and increased drought severity. Previous studies have considered the effects of individual components of GCC on tree species. However, few studies have assessed the interactive effects of stress factors, such as higher temperatures, drought stress, and elevated CO_2 [146]. Therefore, these must be assessed together more in future studies. In particular, intensive tree physiological studies during drought and the combined effects of more than two factors on species tolerance to GCC will aid in proper tree species selection and environmental policy in New Zealand's cities. Research on the adaptability to urban abiotic extreme conditions would improve the current poor understanding of native trees' responses in the urban areas of New Zealand. Therefore, it is necessary to pay attention to the role of native trees in cities to develop novel ideas that can positively affect New Zealand's climate policy in the post-Paris Agreement era.

Author Contributions: Conceptualization, J.J. and S.-Y.W.; writing—original draft preparation, J.J.; writing—review and editing, S.-Y.W.; visualization, J.J.; supervision, S.-Y.W. All authors have read and agreed to the published version of the manuscript.

Funding: This work was supported by the 2021 Research Fund of the University of Seoul.

Data Availability Statement: https://www.stats.govt.nz/information-releases/national-population-estimates-at-31-march-2020-infoshare-tables (accessed on 20 November 2021); https://environment.govt.nz/acts-and-regulations/acts/climate-change-response-amendment-act-2019/ (accessed on 20 November 2021); https://www.stats.govt.nz/information-releases/2018-census-population-and-dwelling-counts (accessed on 20 November 2021); https://www.nzpcn.org.nz/flora/species/ (accessed on 20 November 2021).

Conflicts of Interest: The authors declare no conflict of interest.

Appendix A

Table A1. New Zealand's CO_2 emissions by sector between 1990 and 2017.

Sectors	Kt (CO_2)-Equivalent		Change from 1990 (Kt (CO_2)-Equivalent)	Change from 1990 (%)
	1990	2017		
Energy and road transport	23,785.7	32,876.6	+9090.9	+38.2
Industrial processes and product use	3579.9	4968.6	+1388.7	+38.8
Agriculture	34,257.2	38,880.7	+4623.5	+13.5
Waste	4041.9	4124.7	+82.9	+2.1
Gross (excluding LULUCF †)	65,668.3	80,853.5	+15,185.2	+23.1
LULUCF	−31,161.8	−23,958.4	+7203.3	+23.1
Net (including LULUCF)	34,506.5	56,895.0	+22,388.5	+64.9

Source: MfE [46]. † LULUCF refers to land use, land-use change, and forestry sector under the United Nations Framework Convention on Climate Change (UNFCCC) and Kyoto Protocol.

Appendix B

Table A2. Comparison of the 'means of transportation for daily commute' among New Zealand's six biggest cities [43,44,52].

Private Vehicle Fleet Usage			Region	2020 Population Growth Rate (%)	
City (Population in 2020)	Percentage (%)	Rank			
Tauranga (136,700)	91	1	Bay of Plenty	2.8	
Hamilton (160,900)	87	2	Northland	2.6	
Auckland (1,571,700)	85	3	Waikato	2.3	
Christchurch (369,000)	84	4	Auckland	2.2	
Dunedin (126,300)	82	5	Canterbury †	2.2	
Wellington (202,700) ††	54	6	National wide	2.1	
Bus/Train Usage			**Walk**		
City	**Percentage (%)**	**Rank**	**City**	**Percentage (%)**	**Rank**
Wellington	21	1	Wellington	21	1
Auckland	8	2	Dunedin	12	2
Christchurch	4	3	Hamilton	7	3
Hamilton	3	4	Christchurch	5	4
Dunedin	3	5	Auckland	5	5
Tauranga	2	6	Tauranga	4	6

† Selwyn's growth rate is 5.2%, which means the largest net internal migration, followed by Tauranga city and Waikato. †† This value has excluded the population of Upper Hutt and Lower Hutt.

Appendix C

Figure A1. Sample pictures of New Zealand native trees grown and investigated environmental regulating services in domestic cities. (**A**) Kānuka (**B**) Kōhūhū (**C**) Lemonwood (Tarata) (**D**) Nīkau (**E**) Karaka (**F**) Pōhutukawa (**G**) Puriri (Retrieved from New Zealand Plant Conservation Network (NZPCN), 2021). Image credit by Mike Wilcox (Kānuka), John Barkla (Kōhūhū), Peter J. de Lange (Lemonwood), Colin C. Ogle (Nīkau), Simon Walls (Karaka), Gillian M. Crowcroft (Pōhutukawa), and John E. Braggins (Puriri), CC BY 4.0 [147].

References

1. Grimmond, S.U. Urbanization and global environmental change: Local effects of urban warming. *Geogr. J.* **2007**, *173*, 83–88 [CrossRef]
2. Rahman, M.A.; Hartmann, C.; Moser-Reischl, A.; von Strachwitz, M.F.; Paeth, H.; Pretzsch, H.; Rötzer, T. Tree cooling effects and human thermal comfort under contrasting species and sites. *Agric. For. Meteorol.* **2020**, *287*, 107947. [CrossRef]
3. Živković, J. Urban Form and Function. In *Climate Action. Encyclopedia of the UN Sustainable Development Goals*, 1st ed.; Leal Filho, W., Azul, A.M., Brandli, L., Özuyar, P.G., Wall, T., Eds.; Springer: Berlin, Germany, 2020; pp. 862–871.
4. United Nations. World Urbanization Prospects: The 2018 Revision, Custom Data Acquired via Website. 2018. Available online: https://population.un.org/wup/DataQuery/ (accessed on 24 February 2020).
5. Edmondson, J.L.; Davies, Z.G.; McHugh, N.; Gaston, K.J.; Leake, J.R. Organic carbon hidden in urban ecosystems. *Sci. Rep.* **2012**, *2*, 963. [CrossRef]
6. Rahman, M.A.; Armson, D.; Ennos, A.R. Effect of urbanization and climate change in the rooting zone on the growth and physiology of *Pyrus calleryana*. *Urban For. Urban Green.* **2014**, *13*, 325–335. [CrossRef]
7. Berry, B.J. A new urban ecology? *Urban Geogr.* **2001**, *22*, 699–701. [CrossRef]
8. Close, T.J. Dehydrins: Emergence of a biochemical role of a family of plant dehydration proteins. *Physiol. Plant.* **1996**, *97*, 795–803. [CrossRef]
9. Jeong, S.J.; Park, H.; Ho, C.H.; Kim, J. Impact of urbanization on spring and autumn phenology of deciduous trees in the Seoul Capital Area, South Korea. *Int. J. Biometeorol.* **2019**, *63*, 627–637. [CrossRef] [PubMed]
10. Ugolini, F.; Tognetti, R.; Raschi, A.; Bacci, L. Quercus ilex L. as bioaccumulator for heavy metals in urban areas: Effectiveness of leaf washing with distilled water and considerations on the trees distance from traffic. *Urban For. Urban Green.* **2013**, *12*, 576–584. [CrossRef]
11. Craul, P.J. Urban soil: Problems and promise. *Arnoldia* **1991**, *51*, 23–32.
12. Morgenroth, J.; Buchan, G.D. Soil moisture and aeration beneath pervious and impervious pavements. *Arboric. Urban For.* **2009**, *35*, 135–141.

13. McGlone, M.S. The Polynesian settlement of New Zealand in relation to environmental and biotic changes. *N. Z. J. Ecol.* **1989**, *12*, 115–129.
14. McGlone, M.S.; Wilmshurst, J.M. Dating initial Maori environmental impact in New Zealand. *Quatern. Int.* **1999**, *59*, 5–16. [CrossRef]
15. Stevens, M.I.; Winter, D.J.; Morris, R.; McCartney, J.; Greenslade, P. New Zealand's giant Collembola: New information on distribution and morphology for *Holacanthella* Börner, 1906 (Neanuridae: Uchidanurinae). *N. Z. J. Zool.* **2007**, *34*, 63–78. [CrossRef]
16. Wilmshurst, J.M.; Anderson, A.J.; Higham, T.F.; Worthy, T.H. Dating the late prehistoric dispersal of Polynesians to New Zealand using the commensal Pacific rat. *Proc. Natl. Acad. Sci. USA* **2008**, *105*, 7676–7680. [CrossRef]
17. Ewers, R.M.; Kliskey, A.D.; Walker, S.; Rutledge, D.; Harding, J.S.; Didham, R.K. Past and future trajectories of forest loss in New Zealand. *Biol. Conserv.* **2006**, *133*, 312–325. [CrossRef]
18. Clarkson, B.D.; Wehi, P.M.; Brabyn, L.K. A spatial analysis of indigenous cover patterns and implications for ecological restoration in urban centres, New Zealand. *Urban Ecosyst.* **2007**, *10*, 441–457. [CrossRef]
19. Clarkson, B.D.; Kirby, C.L. Ecological restoration in urban environments in New Zealand. *Ecol. Manag. Restor.* **2016**, *17*, 180–190. [CrossRef]
20. Nomura, M.; Ohlemüller, R.; Lee, W.G.; Lloyd, K.M.; Anderson, B.J. Can we predict which species win when new habitat becomes available? *PLoS ONE.* **2019**, *14*, e0213634. [CrossRef]
21. Statistics New Zealand. 2021 Statistics New Zealand. Functional Urban Areas—Methodology and Classification. Available online: https://www.stats.govt.nz/methods/functional-urban-areas-methodology-and-classification (accessed on 28 February 2021).
22. McDonald, J.D.N. New Zealand land legislation. *Aust. Hist. Stud.* **1952**, *5*, 195–211. [CrossRef]
23. Arnold, R.; Dudding, A. *New Zealand's Burning: The Settler's World in the Mid 1880's*; Victoria University of Wellington Press: Wellington, New Zealand, 1994; pp. 118–120. Available online: http://nzetc.victoria.ac.nz/tm/scholarly/tei-ArnNewZ-b2-2.html (accessed on 28 February 2021).
24. Perry, G.L.; Wilmshurst, J.M.; McGlone, M.S. Ecology and long-term history of fire in New Zealand. *N. Z. J. Ecol.* **2014**, *38*, 157–176.
25. Graham, R.; Kenny, P.; Moohan, S.; Smith, C.; Woolford, N.; Boardman, R.; Whittaker, S. The notable trees of New Zealand. *Arboric. J.* **2012**, *34*, 83–90. [CrossRef]
26. Hall, M. Enrichment Planting of Late-Successional Plant Species within Restored Urban Forests. Master's Thesis, University of Waikato, Hamilton, New Zealand, 2020.
27. Brockerhoff, E.G.; Hoffmann, J.H.; Roques, A. Is biological control an option for the management of wilding pines (*Pinus* spp.) in New Zealand? In *Managing Wilding Conifers in New Zealand: Present and Future, Proceedings of the Annual Conference of the New Zealand Plant Protection Society, Christchurch, New Zealand, 1 August 2003*; Hill, R.L., Zydenbos, S.M., Bezar, C., Eds.; New Zealand Plant Protection Society: Auckland, New Zealand, 2003; pp. 65–78.
28. Sullivan, J.J.; Timmins, S.M.; Williams, P.A. Movement of exotic plants into coastal native forests from gardens in northern New Zealand. *N. Z. J. Ecol.* **2005**, *29*, 1–10.
29. Howell, C.J. *Consolidated list of Environmental Weeds in New Zealand*; Science & Technical Publishing Department of Conservation (DOC): Wellington, New Zealand, 2008; pp. 5–42.
30. Messier, C.; Bauhus, J.; Sousa-Silva, R.; Auge, H.; Baeten, L.; Barsoum, N.; Zemp, D.C. For the sake of resilience and multifunctionality, let's diversify planted forests! *Conserv. Lett.* **2021**, e12829. [CrossRef]
31. Ministry for the Environment. Environment New Zealand 2007. In *Chapter 9: Land: Current State and Trends*; Ministry for the Environment (MfE): Wellington, New Zealand, 2007. Available online: https://www.mfe.govt.nz/publications/environmental-reporting/ (accessed on 28 February 2021).
32. Landcare Research New Zealand. *Carbon Stocks and Changes in New Zealand's Soils and Forests, and Implications of Post-2012 Accounting Options for Land-Based Emissions Offsets and Mitigation Opportunities*; Landcare Research: Lincoln, New Zealand, 2009. Available online: https://www.mpi.govt.nz/dmsdocument/28230/direct (accessed on 28 February 2021).
33. Stewart, G.H.; Ignatieva, M.E.; Meurk, C.D.; Earl, R.D. The re-emergence of indigenous forest in an urban environment, Christchurch, New Zealand. *Urban For. Urban Green.* **2004**, *2*, 149–158. [CrossRef]
34. Stewart, G.H.; Meurk, C.D.; Ignatieva, M.E.; Buckley, H.L.; Magueur, A.; Case, B.S.; Parker, M. URban Biotopes of Aotearoa New Zealand (URBANZ) II: Floristics, biodiversity and conservation values of urban residential and public woodlands, Christchurch. *Urban For. Urban Green.* **2009**, *8*, 149–162. [CrossRef]
35. Rahman, M.A.; Moser, A.; Gold, A.; Rötzer, T.; Pauleit, S. Vertical air temperature gradients under the shade of two contrasting urban tree species during different types of summer days. *Sci. Total Environ.* **2018**, *633*, 100–111. [CrossRef]
36. Werner, P.; Kelcey, J.G. Greening cities. In *Urban Green and Biodiversity*, 1st ed.; Tan, P., Jim, C., Eds.; Springer: Singapore, 2017; pp. 131–154.
37. Sonti, N.F.; Hallett, R.A.; Griffin, K.L.; Trammell, T.L.; Sullivan, J.H. Chlorophyll fluorescence parameters, leaf traits and foliar chemistry of white oak and red maple trees in urban forest patches. *Tree Physiol.* **2021**, *41*, 269–279. [CrossRef]
38. Osone, Y.; Kawarasaki, S.; Ishida, A.; Kikuchi, S.; Shimizu, A.; Yazaki, K.; Matsumoto, G.I. Responses of gas-exchange rates and water relations to annual fluctuations of weather in three species of urban street trees. *Tree Physiol.* **2014**, *34*, 1056–1068. [CrossRef]
39. Rodriguez, M.E.; Power, S.A.; Tjoelker, M.G.; Marchin, R.M.; Rymer, P.D. Contrasting heat tolerance of urban trees to extreme temperatures during heatwaves. *Urban For. Urban Green.* **2021**, *66*, 127387. [CrossRef]

40. Department of Internal Affairs. *Sustainable Urban Development: Setting the Scene*; Department of Internal Affairs (DIA): Wellington, New Zealand, 2016. Available online: https://www.dia.govt.nz/diawebsite.nsf/wpg (accessed on 24 February 2020).
41. Statistics New Zealand. 2020 Statistics New Zealand. National Population Projections: 2020(Base)–2073. 2020. Available online: https://www.stats.govt.nz/information-releases/national-population-projections-2020base (accessed on 28 February 2021).
42. MacDonald, K.; Rudel, T.K. Sprawl and forest cover: What is the relationship? *Appl. Geogr.* **2005**, *25*, 67–79. [CrossRef]
43. Stocchero, A.; Seadon, J.K.; Falshaw, R.; Edwards, M. Urban Equilibrium for sustainable cities and the contribution of timber buildings to balance urban carbon emissions: A New Zealand case study. *J. Clean. Prod.* **2017**, *143*, 1001–1010. [CrossRef]
44. Statistics New Zealand. 2020 National Population Estimates. 2020. Available online: https://www.stats.govt.nz/information-releases/national-population-estimates-at-31-march-2020-infoshare-tables (accessed on 28 February 2021).
45. Statistics New Zealand. 2018 Statistics New Zealand. New Zealand's Environmental Reporting Series: Our air 2018. Available online: https://www.stats.govt.nz/information-releases/new-zealands-environmental-reporting-series-our-air-2018 (accessed on 28 February 2021).
46. Ministry for the Environment. *Climate Change Response (Zero Carbon) Amendment Act*; Ministry for the Environment (MfE): Wellington, New Zealand, 2019. Available online: https://www.mfe.govt.nz/climate-change/zero-carbon-amendment-act (accessed on 28 February 2021).
47. Hasan, M.A.; Frame, D.J.; Chapman, R.; Archie, K.M. Emissions from the road transport sector of New Zealand: Key drivers and challenges. *Environ. Sci. Pollut. Res.* **2019**, *26*, 23937–23957. [CrossRef] [PubMed]
48. Environment Canterbury Regional Council. *Regional Land Transport Plan 2015–2025*; Environment Canterbury (ECan): Christchurch, New Zealand, 2018. Available online: https://www.ecan.govt.nz (accessed on 28 February 2021).
49. Chapman, R. Transitioning to low-carbon urban form and transport in New Zealand. *Polit. Sci.* **2008**, *60*, 89–98. [CrossRef]
50. Ministry of Transport. *Briefing to the Incoming Minister 2020*; Ministry of Transport (MoT): Wellington, New Zealand, 2020. Available online: https://www.beehive.govt.nz/sites/default/files/2020-12/Transport%20-%20Strategic.pdf (accessed on 28 February 2021).
51. Arampamoorthy, H.; Patrick, J.E. *Failure Probability of New Zealand Pavements, NZ Transport Agency Research Report 421*; NZ Transport Agency: Wellington, New Zealand, 2010; pp. 15–47.
52. Statistics New Zealand. 2018 Statistics New Zealand. Comparison of 'Journey to Work' between New Zealand's Six Biggest Cities. Available online: https://www.stats.govt.nz/information-releases/2018-census-population-and-dwelling-counts (accessed on 30 May 2021).
53. Statistics New Zealand. 2018 Census Data. Available online: https://www.stats.govt.nz/tools/2018-census-place-summaries/christchurch-city (accessed on 30 May 2021).
54. Li, Y.; Spearpoint, M. Analysis of vehicle fire statistics in New Zealand parking buildings. *Fire Technol.* **2007**, *43*, 93–106. [CrossRef]
55. Environmental Health Indicators New Zealand. *Average Age of Vehicle Fleet, Environmental Health Indicators New Zealand (EHINZ) Centre for Public Health Research*; Massy University: Wellington, New Zealand, 2017. Available online https://www.ehinz.ac.nz/assets/Factsheets/Released-2017/EHI10-11-AverageAgeOfVehicleFleetInNZ2000-2015-release-201701.pdf (accessed on 30 May 2021).
56. Ministry of Transport. *Annual Fleet Statistics 2015*; Ministry of Transport (MoT): Wellington, New Zealand, 2020. Available online: https://www.transport.govt.nz/statistics-and-insights/fleet-statistics/sheet/vehicle-age (accessed on 30 May 2021).
57. Kjellström, T.; Mercado, S. Towards action on social determinants for health equity in urban settings. *Environ. Urban.* **2008**, *20*, 551–574. [CrossRef]
58. Kjellström, T.; Friel, S.; Dixon, J.; Corvalan, C.; Rehfuess, E.; Campbell-Lendrum, D.; Bartram, J. Urban environmental health hazards and health equity. *J. Urban Health* **2007**, *84*, 86–97. [CrossRef]
59. Longley, I.; Tunno, B.; Somervell, E.; Edwards, S.; Olivares, G.; Gray, S.; Clougherty, J.E. Assessment of spatial variability across multiple pollutants in Auckland, New Zealand. *Int. J. Environ. Res. Public Health* **2019**, *16*, 1567. [CrossRef]
60. Kowarik, I.; Fischer, L.K.; Kendal, D. Biodiversity conservation and sustainable urban development. *Sustainability* **2020**, *12*, 4964. [CrossRef]
61. Meurk, C.D.; Blaschke, P.M.; Simcock, R. *Ecosystem Services in New Zealand Cities. Ecosystem Services in New Zealand: Conditions and Trends*; Manaaki Whenua Press: Lincoln, New Zealand, 2013; pp. 254–273.
62. Côté, I.M.; Darling, E.S. Rethinking ecosystem resilience in the face of climate change. *PLoS Biol.* **2010**, *8*, e1000438. [CrossRef]
63. Sevegnani, L.; Uhlmann, A.; de Gasper, A.L.; Meyer, L.; Vibrans, A.C. Climate affects the structure of mixed rain forest in southern sector of Atlantic domain in Brazil. *Acta Oecol.* **2016**, *77*, 109–117. [CrossRef]
64. Sasaki, T.; Furukawa, T.; Iwasaki, Y.; Seto, M.; Mori, A.S. Perspectives for ecosystem management based on ecosystem resilience and ecological thresholds against multiple and stochastic disturbances. *Ecol. Indic.* **2015**, *57*, 395–408. [CrossRef]
65. Magurran, A.E. *Ecological Diversity and It's Measurement*; Princeton University Press: Princeton, NJ, USA, 1988; p. 179.
66. Downing, A.L.; Leibold, M.A. Species richness facilitates ecosystem resilience in aquatic food webs. *Freshw. Biol.* **2010**, *55*, 2123–2137. [CrossRef]
67. Pedro, M.S.; Rammer, W.; Seidl, R. Tree species diversity mitigates disturbance impacts on the forest carbon cycle. *Oecologia* **2015**, *177*, 619–630. [CrossRef]
68. Kaiser-Bunbury, C.N.; Mougal, J.; Whittington, A.E.; Valentin, T.; Gabriel, R.; Olesen, J.M.; Blüthgen, N. Ecosystem restoration strengthens pollination network resilience and function. *Nature* **2017**, *542*, 223–227. [CrossRef]

69. Daniels, G.D.; Kirkpatrick, J.B. Does variation in garden characteristics influence the conservation of birds in suburbia? *Biol. Conserv.* **2006**, *133*, 326–335. [CrossRef]
70. Sjöman, H.; Morgenroth, J.; Sjöman, J.D.; Sæbø, A.; Kowarik, I. Diversification of the urban forest—can we afford to exclude exotic tree species? *Urban For. Urban Green.* **2016**, *18*, 237–241. [CrossRef]
71. Aimers, J.; Bergin, D.; Horgan, G. *Review of Non-Timber Values in Sustainably Managed Native Forest in New Zealand;* Tāne's Tree Trust Bulletin: Hamilton, New Zealand, 2021; pp. 4–82.
72. Stevens, F.R.; Gaughan, A.E.; Linard, C.; Tatem, A.J. Disaggregating census data for population mapping using random forests with remotely-sensed and ancillary data. *PLoS ONE.* **2015**, *10*, e0107042. [CrossRef]
73. Chan, W.Y.; Chau, K.C. Planter vegetation in urban Hong Kong. *Arboric. J.* **2000**, *24*, 41–58. [CrossRef]
74. Dos Santos, A.R.; da Rocha, C.F.D.; Bergallo, H.G. Native and exotic species in the urban landscape of the city of Rio de Janeiro, Brazil: Density, richness, and arboreal deficit. *Urban Ecosyst.* **2010**, *13*, 209–222. [CrossRef]
75. Odóñez, C.; Duinker, P.N. An analysis of urban forest management plans in Canada: Implications for urban forest management. *Landsc. Urban Plan.* **2013**, *116*, 36–47. [CrossRef]
76. Escobedo, F.J.; Kroeger, T.; Wagner, J.E. Urban forests and pollution mitigation: Analyzing ecosystem services and disservices. *Environ. Pollut.* **2011**, *159*, 2078–2087. [CrossRef]
77. Johnson, K.H. Trophic-dynamic considerations in relating species diversity to ecosystem resilience. *Biol. Rev.* **2000**, *75*, 347–376. [CrossRef]
78. Kraxner, F.; Aoki, K.; Kindermann, G.; Leduc, S.; Albrecht, F.; Liu, J.; Yamagata, Y. Bioenergy and the city–what can urban forests contribute? *Appl. Energ.* **2016**, *165*, 990–1003. [CrossRef]
79. Dickens, S.J.M.; Allen, E.B. Exotic plant invasion alters chaparral ecosystem resistance and resilience pre-and post-wildfire. *Biol. Invasions* **2014**, *16*, 1119–1130. [CrossRef]
80. Belnap, J.; Phillips, S.L.; Sherrod, S.K.; Moldenke, A. Soil biota can change after exotic plant invasion: Does this affect ecosystem processes? *Ecology* **2005**, *86*, 3007–3017. [CrossRef]
81. van Hengstum, T.; Hooftman, D.A.; Oostermeijer, J.G.B.; van Tienderen, P.H. Impact of plant invasions on local arthropod communities: A meta-analysis. *J. Ecol.* **2014**, *102*, 4–11. [CrossRef]
82. Zhang, B.; Cai, Y.; Hu, S.; Chang, S.X. Plant mixture effects on carbon-degrading enzymes promote soil organic carbon accumulation. *Soil Biol. Biochem.* **2021**, *163*, 108457. [CrossRef]
83. Moser-Reischl, A.; Rahman, M.A.; Pauleit, S.; Pretzsch, H.; Rötzer, T. Growth patterns and effects of urban micro-climate on two physiologically contrasting urban tree species. *Landsc. Urban Plan.* **2019**, *183*, 88–99. [CrossRef]
84. Mulder, C.P.H.; Uliassi, D.D.; Doak, D.F. Physical stress and diversity-productivity relationships: The role of positive interactions. *Proc. Natl. Acad. Sci. USA* **2001**, *98*, 6704–6708. [CrossRef] [PubMed]
85. Steiner, C.F.; Long, Z.T.; Krumins, J.A.; Morin, P.J. Population and community resilience in multitrophic communities. *Ecology* **2006**, *87*, 996–1007. [CrossRef]
86. Wardle, D.A.; Bonner, K.I.; Barker, G.M. Stability of ecosystem properties in response to above-ground functional group richness and composition. *Oikos* **2000**, *89*, 11–23. [CrossRef]
87. Griffiths, B.S.; Ritz, K.; Wheatley, R.; Kuan, H.L.; Boag, B.; Christensen, S.; Bloem, J. An examination of the biodiversity–ecosystem function relationship in arable soil microbial communities. *Soil Biol. Biochem.* **2001**, *33*, 1713–1722. [CrossRef]
88. Caldeira, M.C.; Hector, A.; Loreau, M.; Pereira, J.S. Species richness, temporal variability and resistance of biomass production in a Mediterranean grassland. *Oikos* **2005**, *110*, 115–123. [CrossRef]
89. Cumming, A.B.; Galvin, M.F.; Rabaglia, R.J.; Cumming, J.R.; Twardus, D.B. Forest health monitoring protocol applied to roadside trees in Maryland. *J. Arboric.* **2001**, *27*, 126–138.
90. Golubiewski, N.; Lawrence, G.; Zhao, J.; Bishop, C. *Auckland's Urban Forest Canopy Cover: State and Change (2013–2016/2018);* Auckland Council Research and Evaluation Unit, RIMU: Auckland, New Zealand, 2020; pp. 9–31.
91. Sjöman, H.; Östberg, J.; Bühler, O. Diversity and distribution of the urban tree population in ten major Nordic cities. *Urban For. Urban Green.* **2012**, *11*, 31–39. [CrossRef]
92. Huang, S. Auckland Street Tree Ecology: Current Status and Future Potential. Master's Thesis, University of Auckland, Auckland, New Zealand, 2020.
93. Pickett, S.T.; Grove, J.M. Urban ecosystems: What would Tansley do? *Urban Ecosyst.* **2009**, *12*, 1–8. [CrossRef]
94. Auckland Council. *Auckland's Urban Ngahere (Forest) Strategy, Auckland Plan;* Strategy and Research Department, Auckland Council: Auckland, New Zealand, 2019. Available online: https://www.aucklandcouncil.govt.nz/plans-projects-policies-reports-bylaws/our-plans-strategies/topic-based-plans-strategies/environmental-plans-strategies/Pages/urban-ngahere-forest-strategy.aspx (accessed on 4 October 2021).
95. Wallace, K.J.; Clarkson, B.D. Urban forest restoration ecology: A review from Hamilton, New Zealand. *J. R. Soc. N. Z.* **2019**, *49*, 347–369. [CrossRef]
96. Morgenroth, J.A. *Tree Canopy Cover in Christchurch, New Zealand: Prepared for the Christchurch City Council;* University of Canterbury: Christchurch, New Zealand, 2017; pp. 4–14. Available online: https://hdl.handle.net/10092/101285 (accessed on 4 October 2021).
97. Quan, W.; Sullivan, J.J.; Meurk, C.D.; Stewart, G.H. A taxonomically detailed and large-scale view of the factors affecting the distribution and abundance of tree species planted in private gardens of Christchurch city, New Zealand. *PeerJ.* **2021**, *9*, e10588. [CrossRef] [PubMed]

98. Jang, J. A mini review: Trees' morpho-physio-biochemical attributes for urban ecosystem resilience. *J. For. Res.* **2021**. *under review*.
99. Marshall, J.D. Reducing urban sprawl could play an important role in addressing climate change. Energy-efficient urban form. *Environ. Sci. Technol.* **2008**, *42*, 3133–3137. [CrossRef]
100. Poudyal, N.C.; Siry, J.P.; Bowker, J.M. Urban forests' potential to supply marketable carbon emission offsets: A survey of municipal governments in the United States. *For. Policy Econ.* **2010**, *12*, 432–438. [CrossRef]
101. Chen, X.; de Vries, S.; Assmuth, T.; Dick, J.; Hermans, T.; Hertel, O.; Reis, S. Research challenges for cultural ecosystem services and public health in (peri-) urban environments. *Sci. Total Environ.* **2019**, *651*, 2118–2129. [CrossRef] [PubMed]
102. Jeon, J.Y.; Hong, J.Y. Classification of urban park soundscapes through perceptions of the acoustical environments. *Landsc. Urban Plan.* **2015**, *141*, 100–111. [CrossRef]
103. Yao, R.T.; Barry, L.E.; Wakelin, S.J.; Harrison, D.R.; Magnard, L.A.; Payn, T.W. *Planted Forests. Ecosystem Services in New Zealand—Conditions and Trends*; Manaaki Whenua Press: Lincoln, New Zealand, 2013; pp. 62–78.
104. Roeland, S.; Moretti, M.; Amorim, J.H.; Branquinho, C.; Fares, S.; Morelli, F.; Calfapietra, C. Towards an integrative approach to evaluate the environmental ecosystem services provided by urban forest. *J. For. Res.* **2019**, *30*, 1981–1996. [CrossRef]
105. Ministry for the Environment. *New Zealand's Greenhouse Gas Inventory 1990–2017. Fulfilling Reporting Requirements under the United Nations*; Ministry for the Environment (MfE): Wellington, New Zealand, 2019. Available online: https://www.mfe.govt.nz/sites/default/files/media/Climate%20Change/nz-greenhouse-gas-inventory-2019.pdf (accessed on 4 October 2021).
106. Ministry for the Environment. *New Zealand and the United Nations Framework Convention on Climate Change*; Ministry for the Environment (MfE): Wellington, New Zealand, 2018. Available online: https://www.mfe.govt.nz/climate-change/why-climate-change-matters/global-response/new-zealand-and-united-nations-framework (accessed on 4 October 2021).
107. Clarkson, B.D. *Urban Ecological Restoration in Aotearoa, New Zealand*; Society for Ecological Restoration (SER) Webinar: Washington, DC, USA, 13 January 2021. Available online https://www.ser.org/events/EventDetails.aspx?id=1465479 (accessed on 14 January 2021).
108. Wallace, K.J.; Laughlin, D.C.; Clarkson, B.D. Exotic weeds and fluctuating microclimate can constrain native plant regeneration in urban forest restoration. *Ecol. Appl.* **2017**, *27*, 1268–1279. [CrossRef] [PubMed]
109. Ferrini, F. Native or Exotic? A Never Ending Debate in the Urban Landscape. In Proceedings of the International Society of Arboriculture 2020 Annual Conference Webinar, Atlanta, GA, USA, 7 December 2020. Available online https://www.isa-arbor.com/Events/Annual-Conference/2020-Annual-Conference/Agenda (accessed on 10 December 2020).
110. Livesley, S.J.; Escobedo, F.J.; Morgenroth, J. The biodiversity of urban and peri-urban forests and the diverse ecosystem services they provide as socio-ecological systems. *Forests* **2016**, *7*, 291. [CrossRef]
111. Ministry for the Environment. *Towards a Sustainable New Zealand: Carbon Neutral Public Service Offset Portfolio*; Ministry for the Environment (MfE): Wellington, New Zealand, 2020. Available online: http://www.mfe.govt.nz/more/cabinet-papers-andrelated-material-search/cabinet-papers/pol-07-215-towards-sustainable-new (accessed on 4 October 2021).
112. Akbari, H.; Matthews, H.D.; Seto, D. The long-term effect of increasing the albedo of urban areas. *Environ. Res. Lett.* **2012**, *7*, 024004. [CrossRef]
113. Strohbach, M.W.; Arnold, E.; Haase, D. The carbon footprint of Urban Green.n space—A life cycle approach. *Landsc. Urban Plan.* **2012**, *104*, 220–229. [CrossRef]
114. Pan, Y.; Birdsey, R.A.; Fang, J.; Houghton, R.; Kauppi, P.E.; Kurz, W.A.; Hayes, D. A large and persistent carbon sink in the world's forests. *Science* **2011**, *333*, 988–993. [CrossRef]
115. Food and Agriculture Organization of the United Nations. Forest and Transparency under the Paris Agreement, Custom Data Acquired 2020 via Website. Available online: https://collaborative-learning.fao.org/ (accessed on 11 December 2020).
116. Dixon, R.K.; Solomon, A.M.; Brown, S.; Houghton, R.A.; Trexier, M.C.; Wisniewski, J. Carbon pools and flux of global forest ecosystems. *Science* **1994**, *263*, 185–190. [CrossRef]
117. Cox, P.M.; Pearson, D.; Booth, B.B.; Friedlingstein, P.; Huntingford, C.; Jones, C.D.; Luke, C.M. Sensitivity of tropical carbon to climate change constrained by carbon dioxide variability. *Nature* **2013**, *494*, 341–344. [CrossRef] [PubMed]
118. Vieilledent, G.; Gardi, O.; Grinand, C.; Burren, C.; Andriamanjato, M.; Camara, C.; Rakotoarijaona, J.R. Bioclimatic envelope models predict a decrease in tropical forest carbon stocks with climate change in Madagascar. *J. Ecol.* **2016**, *104*, 703–715. [CrossRef]
119. Lim, H.; Oren, R.; Palmroth, S.; Tor-ngern, P.; Mörling, T.; Näsholm, T.; Linder, S. Inter-annual variability of precipitation constrains the production response of boreal *Pinus sylvestris* to nitrogen fertilization. *For. Ecol. Manag.* **2015**, *348*, 31–45. [CrossRef]
120. O'Donnell, C.F.; Hoare, J.M. Quantifying the benefits of long-term integrated pest control for forest bird populations in a New Zealand temperate rainforest. *N. Z. J. Ecol.* **2012**, *36*, 131–140.
121. Nowak, D.J.; Dwyer, J.F. Urban Forest Structure, Benefits, and Value. In *Proceedings of the National Urban Forest Conference, Washington, DC, USA, 5 September 2001*; Kollin, C., Ed.; American Forests: Washington, DC, USA, 2001; pp. 24–26. Available online: https://www.fs.usda.gov/treesearch/pubs/15523 (accessed on 4 October 2021).
122. Konijnendijk, C.C.; Ricard, R.M.; Kenney, A.; Randrup, T.B. Defining urban forestry–A comparative perspective of North America and Europe. *Urban For. Urban Green.* **2006**, *4*, 93–103. [CrossRef]
123. Venter, Z.S.; Barton, D.N.; Gundersen, V.; Figari, H.; Nowell, M. Urban nature in a time of crisis: Recreational use of green space increases during the COVID-19 outbreak in Oslo, Norway. *Environ. Res. Lett.* **2020**, *15*, 104075. [CrossRef]

124. Zhu, J.; Xu, C. Sina microblog sentiment in Beijing city parks as measure of demand for Urban Green.n space during the COVID-19. *Urban For. Urban Green.* **2021**, *58*, 126913. [CrossRef]
125. Ugolini, F.; Massetti, L.; Calaza-Martínez, P.; Cariñanos, P.; Dobbs, C.; Ostoić, S.K.; Sanesi, G. Effects of the COVID-19 pandemic on the use and perceptions of Urban Green.n space: An international exploratory study. *Urban For. Urban Green.* **2020**, *56*, 126888. [CrossRef]
126. Esperón-Rodríguez, M.; Barradas, V.L. Ecophysiological vulnerability to climate change: Water stress responses in four tree species from the central mountain region of Veracruz, Mexico. *Reg. Environ. Change* **2015**, *15*, 93–108. [CrossRef]
127. Crous, K.Y.; Campany, C.; Lopez, R.; Cano, F.J.; Ellsworth, D.S. Canopy position affects photosynthesis and anatomy in mature *Eucalyptus* trees in elevated CO_2. *Tree Physiol.* **2020**, *117*, 206–222. [CrossRef] [PubMed]
128. Beets, P.N. Amount and distribution of dry matter in a mature beech/podocarp community. *N. Z. J. For. Sci.* **1980**, *10*, 395–418.
129. Silvester, W.B.; Orchard, T.A. The biology of kauri (*Agathis australis*) in New Zealand. Production, biomass, carbon storage, and litter fall in four forest remnants. *N. Z. J. Bot.* **1999**, *37*, 553–571. [CrossRef]
130. Marden, M.; Lambie, S.; Phillips, C. Biomass and root attributes of eight of New Zealand's most common indigenous evergreen conifer and broadleaved forest species during the first 5 years of establishment. *N. Z. J. For. Sci.* **2018**, *48*, 1–26. [CrossRef]
131. Dale, J.M. Evaluation of Methods for Quantifying Carbon Storage of Urban Trees in New Zealand. Honours' Thesis, University of Auckland, Auckland, New Zealand, 2013.
132. Findlay, B.A. *Adaptation of i-Tree Eco to New Zealand*; University of Auckland: Auckland, New Zealand, 2013; p. 9. Available online: https://www.itreetools.org/eco/resources/Adaptation_of_iTree_Eco_to_New_Zealand_Findlay2013.pdf (accessed on 4 October 2021).
133. Schwendenmann, L.; Mitchell, N.D. Carbon accumulation by native trees and soils in an urban park, Auckland. *N. Z. J. Ecol.* **2014**, *38*, 213–220.
134. Carswell, F.E.; Burrows, L.E.; Hall, G.M.; Mason, N.W.; Allen, R.B. Carbon and plant diversity gain during 200 years of woody succession in lowland New Zealand. *N. Z. J. Ecol.* **2012**, *36*, 191–202.
135. Wakelin, S.A.; Gomez-Gallego, M.; Jones, E.; Smaill, S.; Lear, G.; Lambie, S. Climate change induced drought impacts on plant diseases in New Zealand. *Australas. Plant Path.* **2018**, *47*, 101–114. [CrossRef]
136. Fowler, A.M.; Boswijk, G.; Lorrey, A.M.; Gergis, J.; Pirie, M.; McCloskey, S.P.; Wunder, J. Multi-centennial tree-ring record of ENSO-related activity in New Zealand. *Nat. Clim. Change* **2012**, *2*, 172–176. [CrossRef]
137. Dai, A. Increasing drought under global warming in observations and models. *Nat. Clim. Change* **2013**, *3*, 52–58. [CrossRef]
138. Ministry for the Environment. *Climate Change Projections per Region*; Ministry for the Environment (MfE): Wellington, New Zealand, 2021. Available online: https://environment.govt.nz/facts-and-science/climate-change/impacts-of-climate-change-per-region/ (accessed on 4 October 2021).
139. Macinnis-Ng, C.; Schwendenmann, L. Litterfall, carbon and nitrogen cycling in a southern hemisphere conifer forest dominated by kauri (*Agathis australis*) during drought. *Plant Ecol.* **2015**, *216*, 247–262. [CrossRef]
140. Bui, T.T.P.; Wilkinson, S.; Domingo, N. Climate change adaptation in New Zealand's building sector. In *Proceedings of the 54th International Conference of The Australian and New Zealand Architectural Science Association (ANZAScA), Auckland, New Zealand, 2020*; Ali, G., Ed.; The Australian and New Zealand Architectural Science Association: Auckland, New Zealand, 2020; pp. 236–245.
141. Lawrence, J.; Blackett, P.; Cradock-Henry, N.A. Cascading climate change impacts and implications. *Clim. Risk Manag.* **2020**, *29*, 100234. [CrossRef]
142. Yan, P.; Yang, J. Species diversity of urban forests in China. *Urban For. Urban Green.* **2017**, *28*, 160–166. [CrossRef]
143. Kendal, D.; Dobbs, C.; Lohr, V.I. Global patterns of diversity in the urban forest: Is there evidence to support the 10/20/30 rule? *Urban For. Urban Green.* **2014**, *13*, 411–417. [CrossRef]
144. Raupp, M.J.; Cumming, A.B.; Raupp, E.C. Street tree diversity in eastern North America and its potential for tree loss to exotic borers. *Arboric. Urban For.* **2006**, *32*, 297–304. [CrossRef]
145. Blood, A.; Starr, G.; Escobedo, F.; Chappelka, A.; Staudhammer, C. How do urban forests compare? Tree diversity in urban and periurban forests of the southeastern US. *Forests* **2016**, *7*, 120. [CrossRef]
146. Dusenge, M.E.; Duarte, A.G.; Way, D.A. Plant carbon metabolism and climate change: Elevated CO_2 and temperature impacts on photosynthesis, photorespiration and respiration. *New Phytol.* **2019**, *221*, 32–49. [CrossRef] [PubMed]
147. New Zealand Plant Conservation Network (NZPCN). Flora Species, Custom Data Acquired via Website. Available online: https://www.nzpcn.org.nz/ (accessed on 22 November 2021).

land

MDPI

Article

Historical Changes and Future Trajectories of Deforestation in the Ituri-Epulu-Aru Landscape (Democratic Republic of the Congo)

Joël Masimo Kabuanga [1,2,*], Onésime Mubenga Kankonda [1,3], Mehdi Saqalli [4], Nicolas Maestripieri [5], Thomas Mumuni Bilintoh [6], Jean-Pierre Mate Mweru [7], Aimé Balimbaki Liama [8], Radar Nishuli [8] and Landing Mané [9]

1 Département d'Aménagement des Écosystèmes, Faculté de Gestion des Ressources Naturelles Renouvelables, Université de Kisangani, Kisangani P.O. Box 2012, Democratic Republic of the Congo; onesime.kankonda@unikis.ac.cd
2 Royal Zoological Society of Antwerp (KMDA) & JURISTRALE, 25, avenue de l'OUA-Ngaliema, P.O. Box 16576 Kin 1, Kinshasa, Democratic Republic of Congo
3 Centre de Surveillance de la Biodiversité, Université de Kisangani, Kisangani P.O. Box 2012, Democratic Republic of the Congo
4 CNRS (UMR 5602 GEODE) Maison de la Recherche de l'Université du Mirail, 5, Allées A. Machado, CEDEX 1, 31058 Toulouse, France; mehdi.saqalli@univ-tlse2.fr
5 TerraNIS, 12 Avenue de l'Europe, 31520 Ramonville, France; nicolas.maestripieri@terranis.fr
6 School of Geography, Clark University, Worcester, MA 01610, USA; tbilintoh@clarku.edu
7 Ecole Post-Régionale d'Aménagement et de Gestion Intégrés des Forêts et des Territoires Tropicaux (ERAIFT), Université de Kinshasa, Kinshasa P.O. Box 15.373, Democratic Republic of the Congo; jp.mate@eraift-rdc.org
8 Institut Congolais pour la Conservation de la Nature (ICCN), 13 Avenue des Cliniques, Kinshasa P.O. Box 868 Kin1, Democratic Republic of the Congo; aime.joseph@yahoo.fr (A.B.L.); radarnishuli3@gmail.com (R.N.)
9 Satellite Observatory of the Forests of Central Africa, 14, Sergent Moke—Q/Socimat, Concession Safricas, Kinshasa, Democratic Republic of the Congo; lmane@osfac.net
* Correspondence: mkabuang@hotmail.com; Tel.: +243-826-369-021

Citation: Kabuanga, J.M.; Kankonda, O.M.; Saqalli, M.; Maestripieri, N.; Bilintoh, T.M., Mweru, J.-P.M.; Liama, A.B.; Nishuli, R.; Mané, L. Historical Changes and Future Trajectories of Deforestation in the Ituri-Epulu-Aru Landscape (Democratic Republic of the Congo). *Land* **2021**, *10*, 1042. https://doi.org/10.3390/land10101042

Academic Editor: Sofia Bajocco

Received: 17 March 2021
Accepted: 25 August 2021
Published: 3 October 2021

Abstract: The Ituri-Epulu-Aru landscape (IEAL) is experiencing deforestation and forest degradation. This deforestation is at the root of many environmental disturbances in a region characterized by endemism in biodiversity. The importance of this article is to provide useful information for those who wish to discuss a model that can be replicated for other territories affected by deforestation and changes in natural and anthropogenic forest structure. This article focuses on the triangulation of spatialized prospective scenarios in order to identify future trajectories based on the knowledge of historical dynamics through the diachronic analysis of three satellite images (2003–2010–2014–2016). The scenarios were designed in a supervised model implemented in the DINAMICA EGO platform. The three scenarios: business as-usual (BAU), rapid economic growth (REG) and sustainable management of the environment (SME), extrapolating current trends, show that by 2061 this landscape will always be dominated forests (+84%). Old-growth forests occupy 74.2% of the landscape area in the BAU scenario, 81.4% in the SEM scenario and 61.2% in the REG scenario. The SEM scenario gives hope that restoration and preservation of biodiversity priority habitats is still possible if policy makers become aware of it.

Keywords: land use change; modeling; scenario; deforestation; DINAMICA EGO; PFBC landscapes; Democratic Republic of the Congo

1. Introduction

Deforestation is one of the main environmental problems in the Democratic Republic of the Congo (DRC) [1]. Studies show that deforestation and forest degradation cause disturbances at several levels, such as biodiversity loss, soil erosion and global warming [2].

Indeed, these two processes lead to the modification of the composition and configuration of forest landscapes [3]. Old-growth forest is considered as the priority habitat for biodiversity because it corresponds with the undisturbed natural ecosystem [4,5]. Its replacement by other land uses is therefore of significant ecological concern [6]. Moreover, deforestation and habitat loss represent complex phenomena linked to several causes, in particular the expansion of agriculture, the extension of infrastructure, logging, economic, demographic, cultural, technological, political factors and institutional establishment [7–9]. However, the influence of these factors depends on their intensity and the duration of their pressure [10].

Although quantitative and qualitative studies on the influence of various causes remain rare, the literature agrees that shifting slash-and-burn agriculture is the main driver of deforestation in DR Congo [11,12]. Knowledge from studies of land use and occupation changes is available at the national level [12,13], but it remains less numerous at the provincial and local level, particularly in landscape conservation [13]. In the Ituri-Epulu-Aru landscape (IEAL), studies on change stop at estimating forest area and deforestation rates [2,14,15]. Moreover, studies on the spatiotemporal modeling of forests have recently been produced however very few have been developed and applied at the scale of a conservation landscape [1].

The Ituri-Epulu-Aru landscape is one of twelve conservation landscapes under the Congo Basin Forest Partnership (CBFP). This landscape is mainly dominated by tropical rainforests [15]. Furthermore, it abounds in an exceptional biodiversity including in particular more than 1192 species of plants, 62 species of large mammals (including the extremely rare okapi, the forest elephant and the chimpanzee) and 312 species of birds [15,16]. Deforestation and forest degradation are the main threats to this biodiversity.

The changes in land cover and use across the Ituri-Epulu-Aru landscape are poorly understood and poorly documented [14]. Yet, it is the sum of local dynamics that determines change at the national, regional and global scale [17]. Consequently, the expansion of deforestation raises a series of questions regarding the evolution of priority habitats for biodiversity, its impact on the composition and configuration of the landscape, the role of the dominant factors in the past dynamics and the possible future devastation of forests in the short, medium and long term.

Remote sensing is useful for monitoring vegetation [18]. However, the mapping of land use by remote sensing remains a methodological challenge in the tropical region, given the heavy cloudiness there. Access to satellite images also remains limited. In the Ituri-Epulu-Aru landscape, many institutions working in the management of natural resources rely on cartographic material from national studies [2] regional or global due to lack of technology or financial constraints [13]. However, the definition of the legend or the observation time may not always meet the expectations of managers.

The interest of this study was to simulate deforestation in the future based on present and past deforestation. In addition, the simulations were analyzed in contrasting scenarios in order to plan future actions to fight against deforestation [7,8,19–21].

2. Materials and Methods

2.1. Study Area

The Ituri-Epulu-Aru landscape (2°37′022″–0°31′030″ N, 27°34′034″–30°00′039″ E, 40,862 km^2) is one of the twelve CBFP landscapes (Figure 1). It is located in the northeastern part of the Democratic Republic of the Congo. Most of the landscape is located in Ituri province (in the administrative territories of Mambasa, Irumu and Djugu). A part of the landscape is included in the province of Haut-Uélé (territories of Wamba and Watsa). Another part also affects the province of North Kivu from where part of the population leaves and affects the landscape.

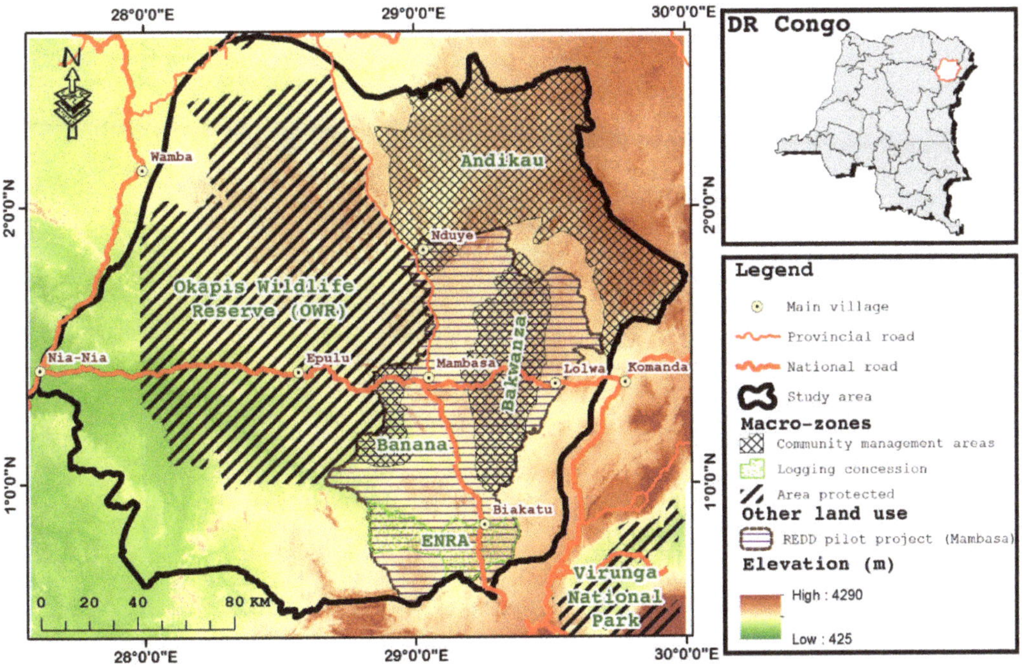

Figure 1. Geographical and topographical context of the study area.

EIAL is characterized by its high biodiversity and number of endemic species. The biophysical occupation of the Ituri-Epulu-Aru landscape is mainly dominated by dense semievergreen dryland to closed canopy forests. These forests include, in particular, the monodominant forests with *Gilbertiodendron dewevrei* and the mixed forests in which no species is predominant. In the extreme northeast of the landscape there is the semideciduous forest, the canopy of which is mainly composed of heliophilous species such as *Entandrophragma* spp. and *Khaya anthotheca*, *Albizia* spp [15,22,23]. Secondary forests and the rural complex are very often along the roads. The region's economic activities are shifting slash-and-burn agriculture, artisanal and semi-industrial mining, artisanal and industrial logging and animal husbandry. The agricultural area is divided into two distinct sectors: the hut gardens and the fields far from the villages. The agriculture practiced by the groups that traditionally live in the forest is based on a rotation of two years of crops and ten years of fallow. The fields are small, generally less than 2 ha, and represent only a small proportion of the agricultural mosaic. Recent immigrants practice more intensive agriculture, with larger fields, shorter fallow periods, and greater clearing of old-growth forest [22,23].

Land use at the landscape scales (Figure 1) includes the Okapis Wildlife Reserve (OWR) (13,720 km^2), the Mai-Tatu Community Reserve (proposed), a logging concession Enzyme Refines Association (ENRA) (520 km^2) and three community management zones: Banana (575 km^2), Andekau (6973 km^2) and Bakwanza (2181 km^2) [22,23].

2.2. Data Used

2.2.1. Satellite Images

Satellite images used for land use dynamics in the Epulu-Ituri-Aru landscape are annual CARPE composites of 0.025 degree resolution. These composites come from Landsat TM, ETM + and OLI images (respectively, Thematic Mapper, Enhanced Thematic Mapper plus and Operational Land Imager). These composites are made up of four spectral bands:

NIR (0.845–0.885 µm), RED (0.63–0.68 µm), SWIR1 (1.56–1.66 µm) and SWIR2 (2.1–2.3 µm). These composites have undergone atmospheric, radiometric and geometric corrections [24].

The Central Africa Regional Program for the Environment (CARPE) composites were chosen because they have no cloud cover and allow the analysis of multi-date changes. They cover all the countries of the Congo Basin and can be downloaded free of charge from the CARPE website (https://carpe.umd.edu/ (accessed on 3 September 2021)) [24]. These images are organized in square tiles of one degree. For this article, 40 tiles were used for the four dates selected (i.e., 10 tiles per date). The CBFP landscapes were created in 2002 and development works started in 2003 in the Epulu-Ituri-Aru landscape. Therefore, the year 2003 was chosen as the reference date. In addition, 2016 was chosen in alignment with a field data collection campaign. And 2010 is the year that roughly halves the observation period (2003 and 2016). The year 2014 was chosen for the validation of the spatialized prospective model. Indeed, 2014 is relatively close to 2010 and 2016 and far enough away from 2003; an ideal time step for validation [25–27].

To ensure multi-temporal comparability, a series of preprocessing were useful. First, the rectified images were projected in the same reference coordinate system: WGS 84, UTM zones 35 North. Then, for each spectral band, a mosaic of tiles was created in the chosen years. Radiometric shifts due to differences in acquisition dates were minimized by doing histogram equalization while taking the sharper tiles as references.

2.2.2. Field Data

Supervised classification generally requires a certain number of training samples and verification samples [26]. Typically, traditional search uses manual visual interpretation to get points. Thus, the sampling consisted of the selection of the objects according to the spectral profiles defined using the GPS field surveys (surveys from 20 December 2016 to 15 January 2017). Then, the training areas that were chosen, on the images after 2016, for each class correspond to areas considered unchanged (built-up areas and inselberg for example) or having signatures close to the profile of 2016. In total, 950 measurement points have been taken (Table 1). This set was split into two groups of data: 665 used for the classification of land use in 2016 (i.e., 70% geographic coordinates) and 285 points used for the validation of the 2016 classification.

Table 1. Description of land use classes.

Land Cover	Code	Number of Points	Description	Sources
Old-growth forest	Pf	257	Woody formation consists of a very dense cover of large trees. Old-growth forest can be semi-deciduous or evergreen, or even swampy. In all cases, the carpet of grasses is absent, and the forest has not undergone significant modification by human activities. The tree layer can reach 50 m in height.	[24,28,29]
Secondary forest	Sf	302	Woody formation corresponding to a stage of reconstitution of forest massifs which have undergone strong anthropogenic interventions, or which have evolved from wastelands. It usually has a strong dominance of moderately fast growing semi-heliophilic species. The tree layer generally reaches 35 m in height	[6,24,29–31]
Non-Forest	NF	315	Non-forest plant formation including wasteland, shrub savannah, land cultivated on an itinerant or intensive basis, as well as recent fallows. This class also includes areas occupied by buildings, dwellings and other high-density constructions as well as areas without vegetation with bare soil, rocky outcrops or even sandy beaches along rivers. This class is represented by the major roads and their right-of-way	[24,28,30–32]
Water	Ww	76	This class includes all bodies of water, including the Ituri River and Epulu	[13,24,28,30]

2.3. Methods

The technical process can be divided into 2 steps:

- Land use and land cover (LULC) classification.
- Modeling of deforestation.

The overall technical process is shown in Figure 2.

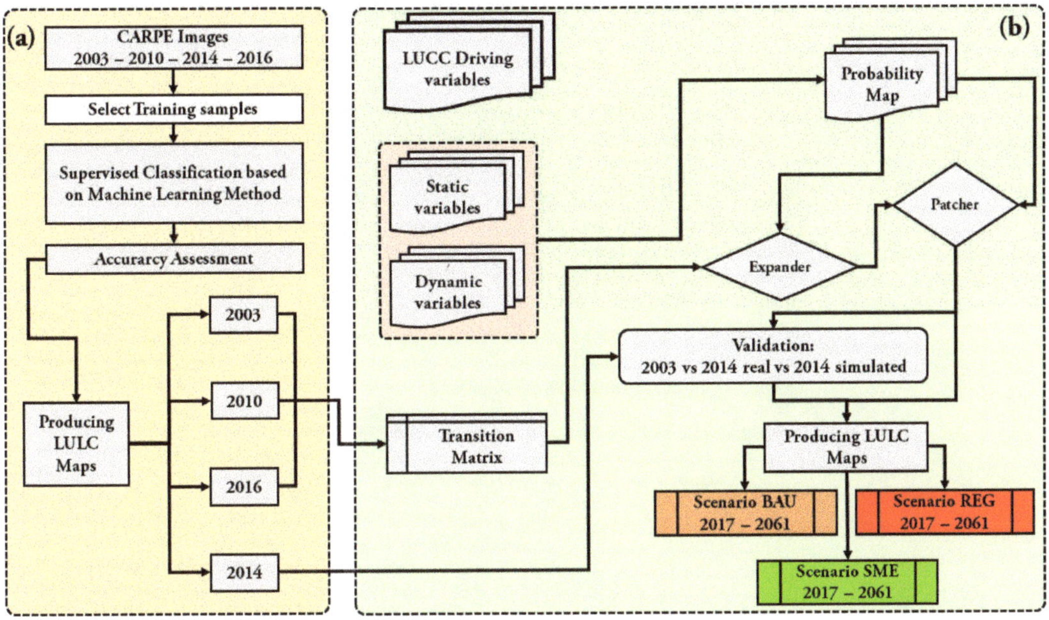

Figure 2. Workflow of the study: (**a**) Land use and land cover (LULC) classification; (**b**) Modeling of deforestation. Abbreviations: CARPE, The Central Africa Regional Program for the Environment; LUCC, Land use and cover changes; LULC, Land use and land cover.

2.3.1. Land Use and Land Cover (LULC) Classification

The Random Forest classification (RFC) was applied to the images of 2003, 2010, 2014 and 2016 via the R software [33] with the R package "RandomForest" [34] to obtain LULC information. RFC [35] is a supervised technique of nonparametric statistical methods [36]. RFC has been used in several studies in the past [24,32,36–43]. In the RFC, when a sample is entered into the model, each decision tree performs a separate evaluation to determine which category the sample should belong to, and the category that is most often selected is ultimately considered the category sample. The RF method can effectively reduce the uncertainty of a particular algorithm and improve the precision of the discriminant classification. The informational dimension of RF processing is larger and more complex than that of other classification algorithms.

In this study, the data entered into the RF model included the full range of raw bands of the annual composites of CARPE (RED, NIR, SWIR1 and SWIR2), Normalized difference vegetation index (NDVI), Normalized difference moisture index (NDMI), Band5/Band 4 ratio (B54R), Normalized brown ratio (NBR), SRTM products. During the study, we found that the classification accuracy of the full band combination was highest when comparing different combinations of bands. Additionally, we have found that SRTM products improve overall accuracy. The number of decision trees was set at 2000 using 70% of all samples.

2.3.2. Modeling of Deforestation

Spatial modeling of deforestation was made on the basis of historical changes in land use assessed between 2003 and 2016. The combination of the transition matrix (2003–2016) adapted to three scenarios: business as usual (BAU), sustainable environmental management (SEM) and rapid economic growth (REC), with maps of transition potential and explanatory factors has enabled regular prospective monitoring up to 2061 to be established using a probabilistic model designed in the DINAMICA EGO platform [10,21,44]. The deforestation simulation included: (i) selection of factors of change, (ii) transitions, (iii) exploratory analysis of deforestation factors, (iv) simulation and (v) validation.

Selection of Variables

Variable selection eliminates overly correlated variables and contributes to the success of the modeling [9,45]. On the basis of the literature, fieldwork and general reflection made it possible to identify the variables (factors) explaining deforestation [11,14,22,23,30,46]. The variables identified were grouped into six categories (Agriculture, Economic factors, Transport, Demographic factors, Sociopolitical factors, Biophysical factors [11]. Only spatially explicit variables were retained for this study. Then, these variables were quantified in a geographic information system (Figure 3). Finally, an exploratory univariate analysis, calculating the correlation between the explanatory variables and deforestation and forest degradation, was carried out to identify the relationships between deforestation and each of the explanatory variables (Table 2).

Table 2. Explanatory variables of deforestation.

Category	Variable Retained	Code	Sources
Agriculture	Distance to agricultural areas	d_agri	Spatial analysis [24]
	Rural complex	Comp	[24]
	Distance to rural complex	d_comp	Spatial analysis [24]
Economic factors	Distances to built-up areas	d_abat	Spatial analysis [24]
	Distances to major center	d_gcent	Spatial analysis [24]
	Forest concessions	Ccf	[47]
	Mining square	Mining	[47]
	Distance to mining squares	d_mining	Spatial analysis [47]
Transport	Distance to national road	d_road1	Spatial analysis [47]
	Distance to provincial road	d_road2	Spatial analysis [47]
	Distance to local road	d_road3	Spatial analysis [47]
Demographic factors	Population density	Dens	[48]
Sociopolitical factors	Protected areas	Ap	[49]
	Agricultural zones delimited	Areaagr	[49]
	community management	Areamngt	[49]
Biophysical factors	Elevation	Dme	[50]
	Slope	Slope	Spatial analysis [50]
	Distance to watercourses	d_w	Spatial analysis [13,24,50]
	Distances to non-forests	d_nf	Spatial analysis [24,30,46]
	Distance to degraded forest	d_fd	Spatial analysis [24]

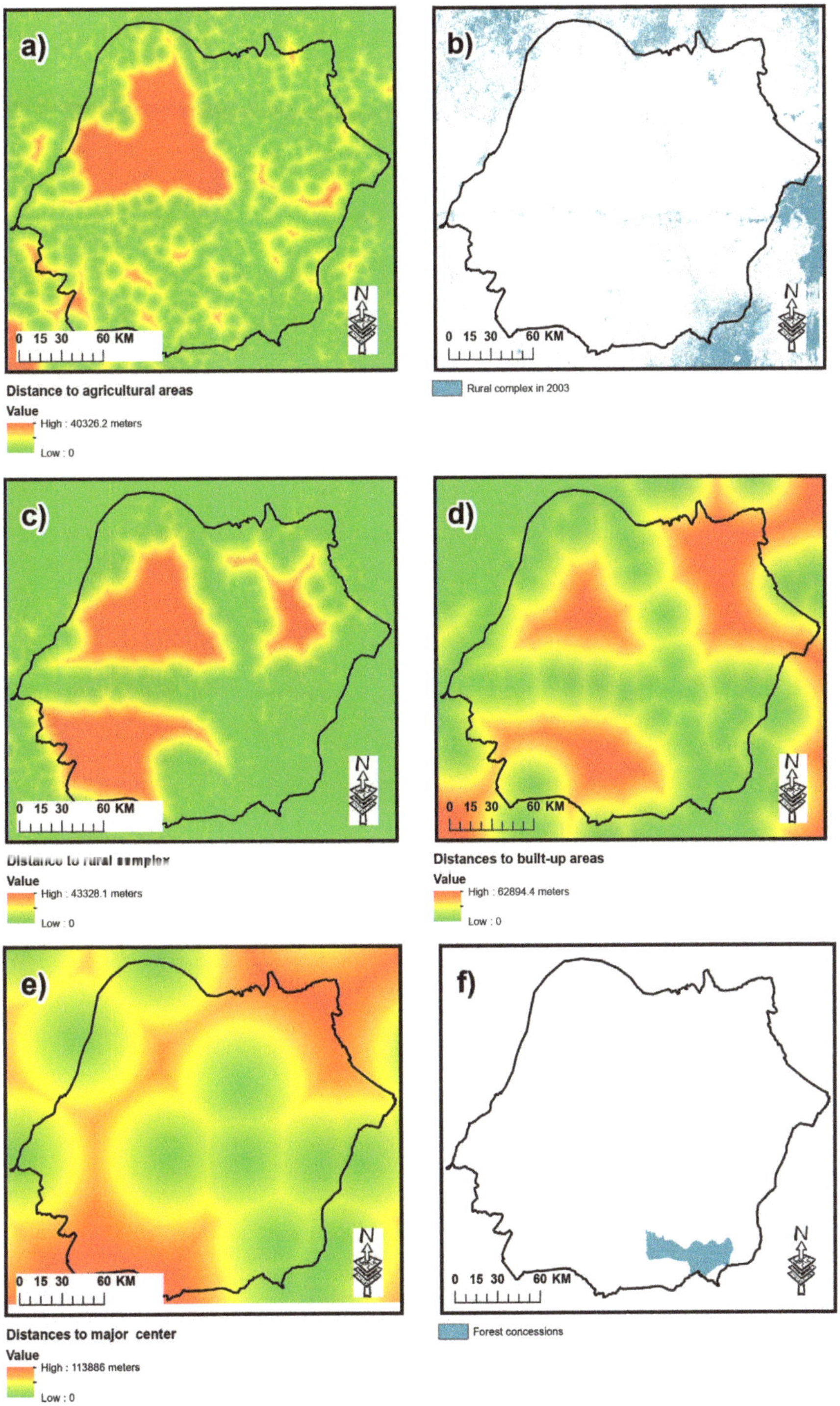

Figure 3. *Cont.*

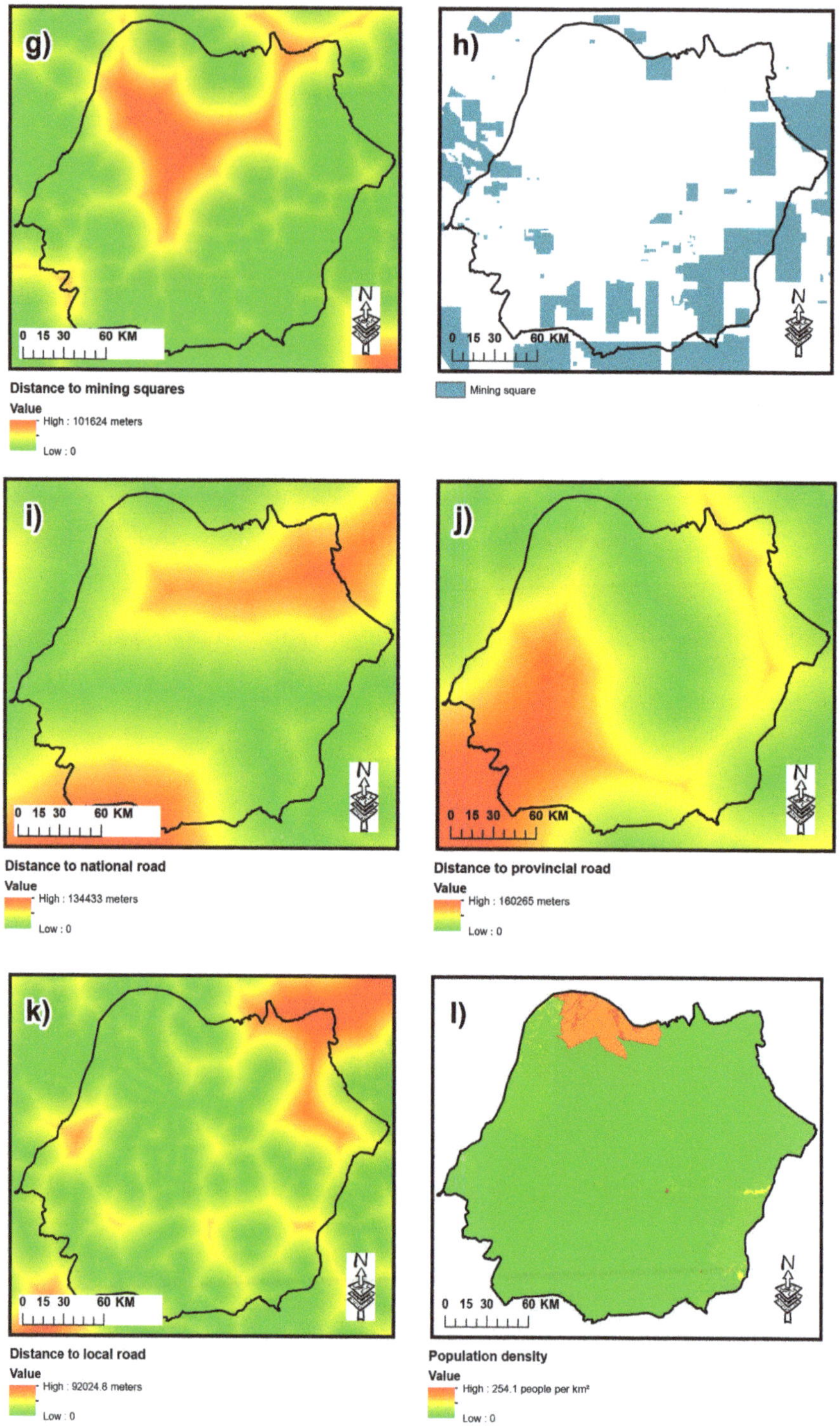

Figure 3. *Cont.*

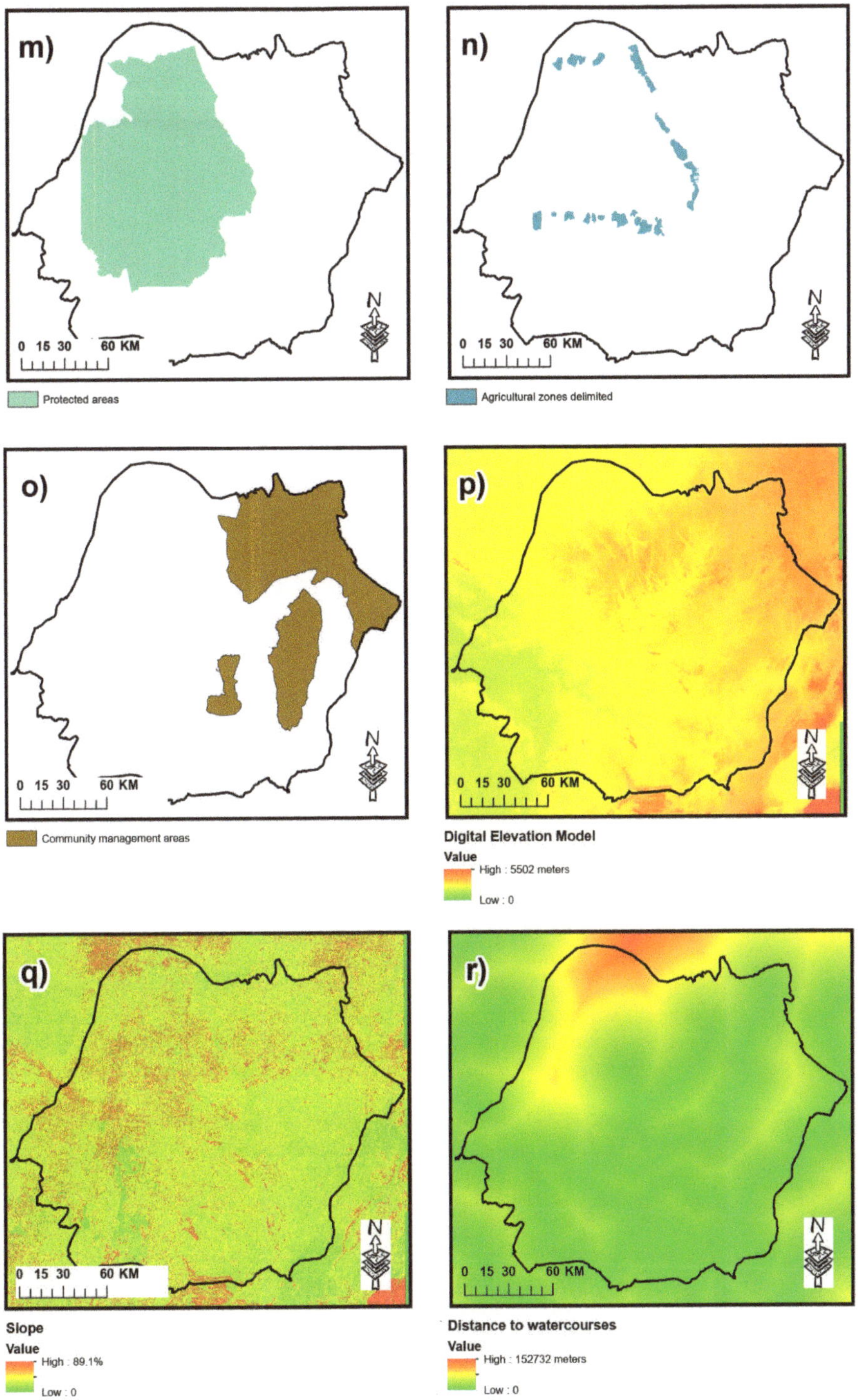

Figure 3. *Cont.*

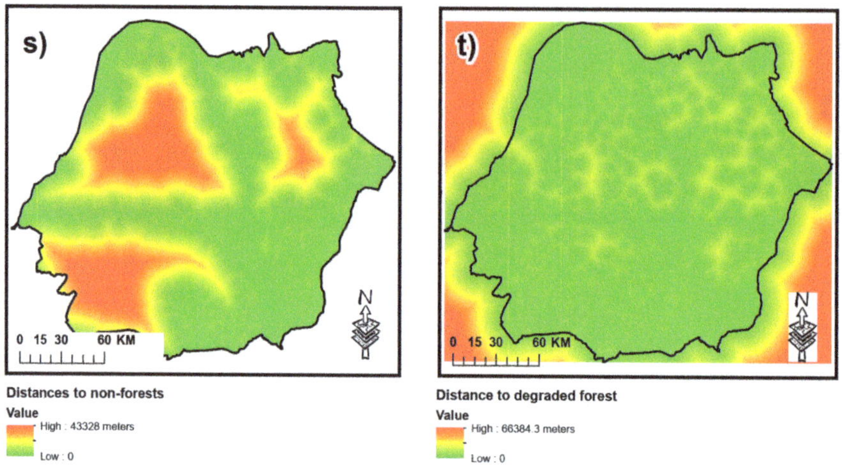

Figure 3. Maps of explanatory variables for deforestation. (**a**) distance to agricultural areas; (**b**) rural complex; (**c**) distance to rural complex; (**d**) distances to built-up areas; (**e**) distances to major center; (**f**) forest concessions; (**g**) distance to mining squares; (**h**) mining square; (**i**) distance to national road; (**j**) distance to provincial road; (**k**) distance to local road; (**l**) population density; (**m**) protected areas; (**n**) agricultural zones delimited; (**o**) community management; (**p**) elevation; (**q**) slope; (**r**) distance to watercourses; (**s**) distances to non-forests; (**t**) distance to degraded forest.

Transitions

The transition corresponds to the total amount of LULCC that occurred during the simulation period. The quantities of change were calculated by the Markovian method. They constitute an essential element in the simulation of changes in land cover and land use because they determine the surface area to be allocated in space according to the change probability maps and the various constraints defined [18]. Transitions modeled in this study include (a) transition from old-growth to secondary forest (degradation), (b) transition from old-growth to non-forest (deforestation), (c) transition from secondary forest to old-growth (maturation), and (d) transition from secondary forest to non-forest (deforestation).

Exploratory Analysis of the Data

When the dynamics of LULCC was modelled, weight of evidence (WoE) was applied to the transition probabilities of the project. The weight of evidence represents the influence of each variable on the spatial probability of an *i-j* transition. A work of adjustment of the weights of evidence was useful because the input data is not always entirely reliable, and some categories resulting from the operations of discretization may be nonexistent. This adjustment, requiring expert knowledge, brought relevant added value. The operation was based on the automatically calculated values and is carried out through a visualization interface made available by DINAMICA EGO. As required by the literature [18,51], no fundamental modifications were applied. The purpose of the adjustment is twofold: to model the most obvious functions (for example in the case of distances or altitude), and to adjust the values deemed unrealistic. Then, pairwise tests were performed for categorical maps to assess the independence hypothesis. The methods used are Chi2, Crammer's V index, contingency, entropy and joint uncertainty information [52]. The purpose of this step was the selection of variables because the study of the past and the present which was not the only way of explaining future deforestation. Its interest was to retain those who best contribute to the establishment of each land use class. Although there is no unanimity on the cut-off that should be used to exclude a variable, a common practice, also adopted in this study, is to choose a cut-off of 0.5 from the Crammer V index. (Measure of the relationship between categorical variables). Above this value, the variables are correlated [21,53].

Simulation of Deforestation

The simulation of land use changes is carried out in order to facilitate decision making. The interest of this simulation lies in its ability to construct the future image of forests according to three contrasting scenarios: a "trend scenario" (business as usual, (BAU)) which starts from the hypothesis of the absence of new economic or environmental policies, a "sustainable environmental management" (SEM) scenario in which legislation and government subsidies encourage the emergence of forestry (multiplication of plantations and agroforestry) and the protection of wood resources, and finally, a "socio-economic" scenario (rapid economic growth, (REG)), i.e., acceleration of the destruction of tree and shrub plant cover and expansion of agricultural land (tendency towards disaster).

Validation

The validation of the simulation model focused on budgeting for errors and correct predictions [54,55]. In practice, this involves comparing three maps: (i) the map of the initial year (2003), (ii) the simulated map in 2014, and (iii) the one produced by satellite image classification in the same year (2014). This three map analysis shows how simulated change compares to baseline change by revealing five components [54,56,57]: (1) the reference change simulated correctly as a change (i.e., hits), (2) reference change simulated incorrectly as persistence (i.e., misses), (3) reference persistence incorrectly simulated as a change (i.e., false alarms), (4) persistence of the correctly simulated reference as persistence (i.e., correct rejections) and (5) reference change simulated incorrectly as a change to the wrong gain category (i.e., false results) (Table 3). Based on these pixels, two types of errors were evaluated in order to judge the accuracy of the overall prediction across the entire landscape. First, the quantity error (Q) was determined by the difference between false alarms and misses (Q = | F − M |). Finally, the allocation error (A) calculated by the difference of the total observed changes (OC = M + H) with the quantity errors [A = (F + M) − Q]. The total observed changes (OC) are given by the sum of misses and hits (OC = M + H). Also, the total predicted changes were determined by the combination of false alarms and hits (PC = F + H).

Table 3. Approach to error budgeting and correct predictions. 1 = Old-growth forest; 2 = Secondary forest; 3 = Non-Forest; 4 = Water.

Comparison of Three Maps				
2003	**2014**	**2014si**	**Components**	
1	1	1	Reference persistence simulated correctly as persistence	Correct rejections
2	2	2		
3	3	3		
4	4	4		
1	2	1	Reference change simulated incorrectly as persistence	Misses
1	3	1		
1	4	1		
2	1	2		
2	3	2		
2	4	2		
3	1	3		
3	2	3		
3	4	3		
4	1	4		
4	2	4		
4	3	4		

Table 3. *Cont.*

Comparison of Three Maps				
2003	**2014**	**2014si**	**Components**	
1	1	2	Reference persistence simulated incorrectly as change	False Alarms
1	1	3		
1	1	4		
2	2	1		
2	2	3		
2	2	4		
3	3	1		
3	3	2		
3	3	4		
4	4	1		
4	4	2		
4	4	3		
1	2	2	Reference change simulated correctly as change	Hits
1	3	3		
1	4	4		
2	1	1		
2	3	3		
2	4	4		
3	1	1		
3	2	2		
3	4	4		
4	1	1		
4	2	2		
4	3	3		
1	2	3	Reference change simulated incorrectly as change to the wrong gaining category	Wrong Hits
1	2	4		
1	3	2		
1	3	4		
1	4	2		
1	4	3		
2	1	3		
2	1	4		
2	3	1		
2	3	4		
2	4	1		
2	4	3		
3	1	2		
3	1	4		
3	2	1		
3	2	4		
3	4	1		
3	4	2		
4	1	2		
4	1	3		
4	2	1		
4	2	3		
4	3	1		
4	3	2		

3. Results

3.1. Assessment of the Quality of Land Use Maps

The overall precision of land cover classifications in the study area from 2003 to 2016 was 0.94 $\pm$ 0.03. In detail, old-growth forest, Non-Forest and Water exhibited a higher classification accuracy than the Secondary Forest class. Their User Precision (UA) and Producer Precision (PA) values were in all cases greater than 0.75 (Table 4). The precision was low for secondary forests, with AU of 0.8 ± 0.02 and PA of 0.78 ± 0.02. In addition, non-forests had intermediate precision values, with AU of 0.82 ± 0.03 and BP of 0.8 ± 0.02. There have been many instances in which secondary forests have been incorrectly classified as old-growth forest and non-Forest.

Table 4. Accuracy assessments of land cover classifications.

Accuracy	Land Use	2003	2010	2014	2016
User	Pf	0.91	0.93	0.89	0.98
	Sf	0.78	0.82	0.79	0.77
	Nf	0.81	0.82	0.79	0.85
	Ww	0.92	0.94	0.91	0.90
Producer	Pf	0.89	0.90	0.86	0.94
	Sf	0.77	0.79	0.76	0.81
	Nf	0.79	0.80	0.78	0.82
	Ww	0.90	0.92	0.89	0.95
Over all		0.91	0.93	0.93	0.97

3.2. Analysis of Historical Changes of Deforestation between 2003 and 2016

Table 5 presents: (i) forest areas (ha), (ii) deforested areas between 2003–2010 and between 2010–2016 (ha) and (iii) observed deforestation rates by axis (%). For all dates, old-growth forest represents more than 3,600,000 ha. The most deforestation event is observed between 2010–2016. It is the deforestation of old-growth forests estimated at more than 108,000 ha. It is worth recalling the dramatic increase in secondary forests. The overall assessment of forest dynamics provides information on increasing deforestation over the two periods. Indeed, 14,983 ha of forests were deforested between 2003–2010 and over 37,000 ha between 2010–2016. As a result, the annual rates of deforestation almost tripled between 2003–2016. They went from 0.05% to 0.14% between 2003 and 2016.

Table 5. Areas and annual rates of deforestation between 2003–2010 and 2010–2016.

Forest Type	Forest Areas						Deforested Areas			
	2003		2010		2016		2003–2010		2010–2016	
	Ha	%	Ha	%	Ha	%	DA	Td	DA	Td
Pf	3,801,767	91.75	3,751,719	91.73	3,643,399	89.28	50,048	0.19	108,319	0.42
Sf	178,472	5.83	213,538	5.28	284,351	6.91	−35,065	−2.56	−70,813	−4.09
Total	3,980,240	97.58	3,965,257	91.73	3,927,751	96.19	14,983	0.05	37,505	0.14

Td = Annual rate of deforestation in percentage; DA = Deforested area in hectares.

3.2.1. Historical Transitions

Table 6 summarizes the transitions observed between 2003 and 2016. From a global point of view, the historical dynamics of the landscape occur to the detriment of old-growth forest s over the entire observation period. Old-growth forests decrease by 2.69% compared to the proportion of 2003. They occupy from 91.75% in 2003. They barely represent 89% in 2016. In addition, secondary forests are experiencing an increase in area. They increased from 5.83% in 2003 to 6.91% of the total landscape area in 2016. This increase in secondary forests results from the conversion of old-growth forest s into secondary forests (2.18%) and non-forests into secondary forests (1.14%). Furthermore, the non-forest class increased by 26%, from 2.12% in 2003 to 3.50% of the total area of the landscape in 2016. The proportion of the landscape occupied by forests in 2003 and converted to non- forest in 2016 are estimated at 2.52% of the total area of the landscape. Indeed, secondary forests are the most affected by the changes. In terms of stability, the old-growth forest class shows great stability. In addition, non-forests are very fluctuating. Indeed, they show a stability of 46% compared to their proportion of 2003. The comparison of the proportions of land use in

2003 with those of 2016 does not reveal any significant changes in the composition of land use occupation. (X-squared = 0.46; df = 3; p = 0.93).

Table 6. Matrix of transitions between 2003 and 2016.

2003–2016		Land Use in 2016				Total 2003
		Pf	**Sf**	**Nf**	**Ww**	
Land use in 2003	Pf	87.66	2.18	1.90	0.00	91.75
	Sf	1.61	3.59	0.62	0.00	5.83
	Nf	0.00	1.14	0.97	0.00	2.12
	Ww	0.00	0.00	0.00	0.31	0.31
	Total 2016	89.28	6.91	3.50	0.31	100

3.2.2. Deforestation Effort between 2003 and 2016

The smallest deforestation spot is 0.06 ha for all periods and for both classes of forest cover. On the other hand, the largest event of deforestation was estimated at 1007 ha over the period 2010–2016 in old-growth forest s. In addition, between 2003–2010, the biggest spot of deforestation in old-growth forest covers an area of 756 ha. The average area of deforestation plots in old-growth forest is estimated at 1 ha and 1.6 ha, respectively between 2003–2010 and between 2010–2016. Indeed, the area of deforestation spots observed in old-growth forest does not change significantly. depending on the observation period (p-value = 0.39). However, in secondary forest, the average area of deforestation is estimated at 1 ha between 2003–2010 and 0.7 ha between 2010–2016. The largest deforestation spot is estimated at 240 ha between 2003–2010 and at 410 ha between 2010–2016. There is also no significant difference between the areas of deforestation in secondary forest between the two periods (p-value = 0.54). For the entire observation period (2003–2016), the average area of deforestation in old-growth forest is 1.3 and 0.8 ha in secondary forest. Comparison of the deforestation spots in old-growth forest with those observed in secondary forests reveals a significant difference between the areas of deforestation spots observed in these two forest types (p-value = 0.04). Old-growth forest appears to be more vulnerable to deforestation than secondary forest. Taken as a whole, the deforestation spots observed between 2003–2010 seem to be smaller than those observed between 2010 –2016. Their average area is 1 ha between 2003–2010 and 1.2 ha between 2010–2016. Indeed, there is no significant difference between the areas of the deforestation spots over the two periods (p-value = 0.08). Figure 4 shows the variation in the areas of deforestation spots according to the type of forest cover. The diamond represents the mean.

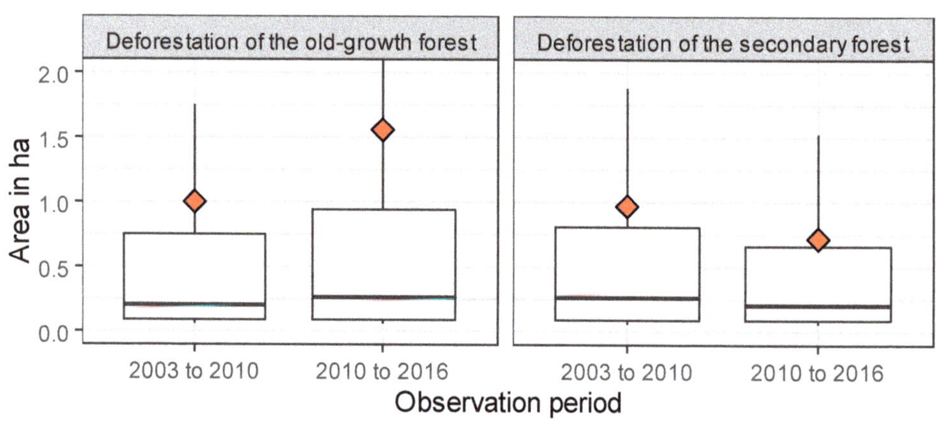

Figure 4. Variation in deforestation areas between 2003 and 2016.

Figure 5 illustrates these changes observed between 2003–2016. Visual analysis reveals that the landscape seems to be more affected by deforestation in the Southeast. Deforestation forms a disturbance gradient linked to the road and to major centers. The area of the wildlife reserve seems to be more stable. Indeed, this variability in forest deforestation seems to be a function of land use (macro-zones).

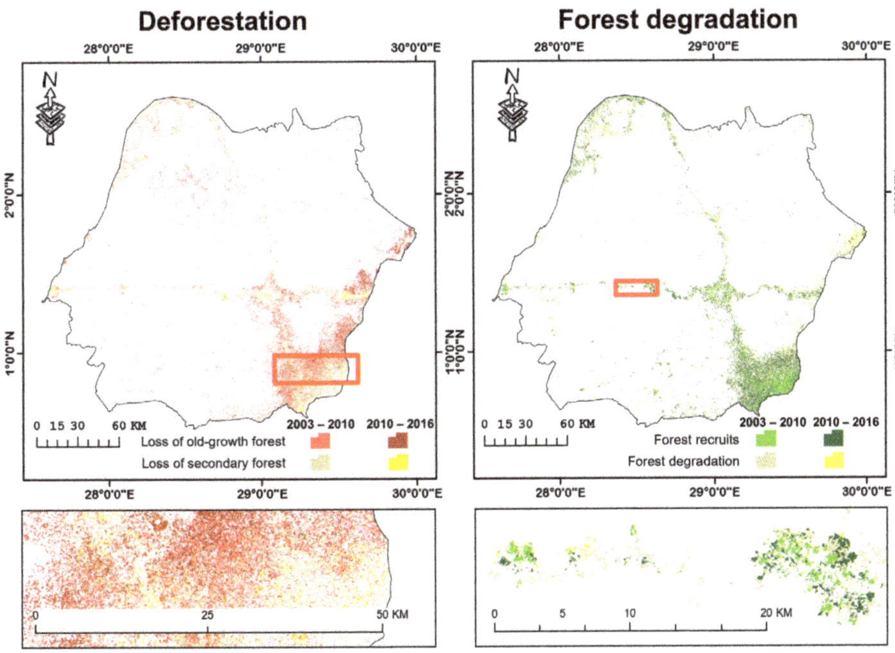

Figure 5. Mapping of forest cover changes: deforestation (**left**) and forest degradation (**right**).

3.3. Future Trajectories of Deforestation

3.3.1. Validation of the Model in 2014

The comparison of the changes observed and predicted between 2003 and 2014 made it possible to validate the simulation model of deforestation. The results of error budgeting and correct prediction reveal that 88.1% of the pixels in the landscape are correct due to observed and predicted consistency (Correct rejections [N]). Additionally, the correct pixels due to an observed and predicted change (Hits [H]) represent 4.25% of the pixels in the landscape. On the other hand, the errors due to a constancy observed but predicted to be changed (False alarms [F]) amount to 1.67% of the pixels in the landscape. The errors due to a change observed but predicted as constant (Misses [M]) are 5.42% (Figure 6). The total observed changes (OC = M + H) are 9.66% while the total predicted changes (PC = F + H) were underestimated with 5.92%.

The accuracy of the global prediction of changes across the entire landscape indicates that the quantity errors ($Q = |F - M|$) are estimated at 3.75% of the landscape pixels while the allocation errors [$A = (F + M) - Q$] represent only 3.34% pixels of the landscape. Therefore, the total error (Q + A) is 7.09% pixels of the landscape.

Figure 7 gives a spatial overview of the distribution of errors and correct predictions.

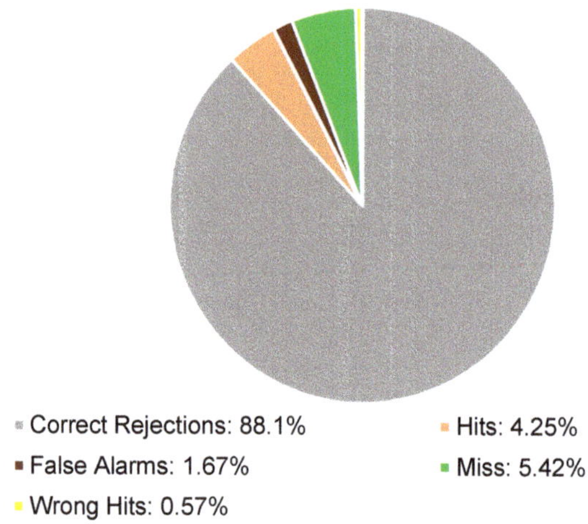

Figure 6. Budgeting for errors and correct predictions.

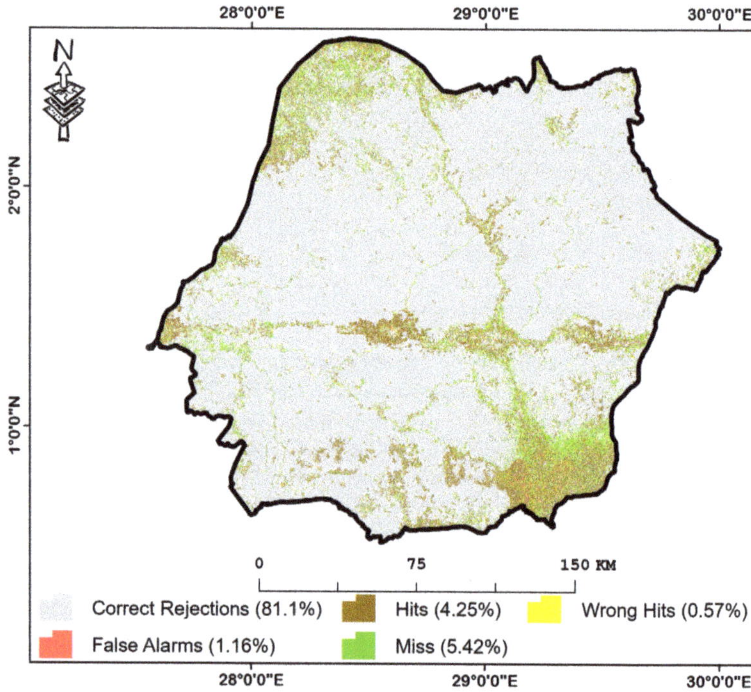

Figure 7. Map of errors and correct predictions.

The landscape observed in 2014 consists of 90.20% of old-growth forests, 6.16% of secondary forests, and 3.36% of non-forests and 0.28% of water. The concordance of simulated and observed land cover is estimated at 84.87% of the landscape for old-growth forests, 3.54% of the landscape for secondary forests and 1.37% of the landscape for non-forests. In addition, the landscape simulated in 2014 consists of 87.42% of old-growth forests, 8.17% of secondary forests, 4.11% of non-forests and 0.28% of water. The model seems to underestimate old-growth forests. It also overestimates secondary forests, non-forests and water (Table 7).

Table 7. Comparison between observed and simulated land use.

Observed–Simulated		Simulated Land Use in 2014				Total Observed
		Pf	Sf	Nf	Ww	
Observed land use in 2014	Pf	84.87	3.64	1.68	0.00	90.20
	Sf	1.55	3.54	1.06	0.00	6.16
	Nf	0.99	0.99	1.37	0.00	3.36
	Ww	0.00	0.00	0.00	0.29	0.28
	Total simulated	87.42	8.17	4.11	0.29	100

3.3.2. Future Trajectories of Deforestation

The combination of the transition matrix adapted to the different BAU, SME and REG scenarios with the transition potential maps and explanatory factors has made it possible to establish regular prospective monitoring until 2061 and the evolving statistics of land use areas (Figures 8–10). In the BUA scenario, the dynamic future of the landscape will come at the expense of old-growth forests (Figure 8).

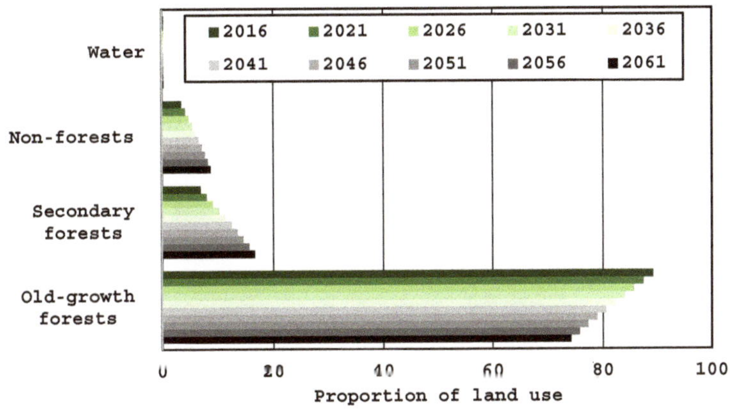

Figure 8. Future evolution of the composition of the occupation according to the trend scenario.

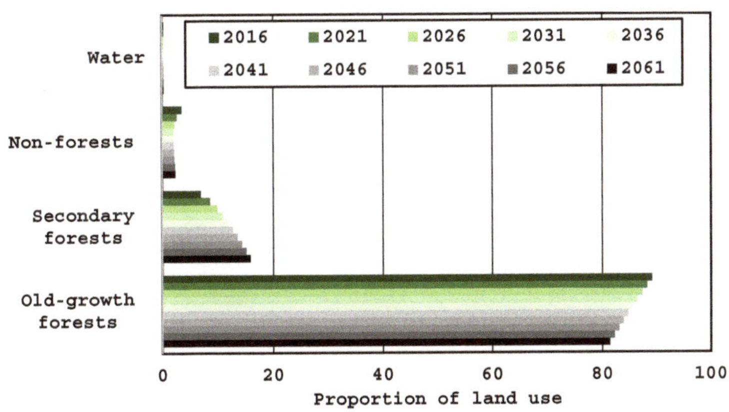

Figure 9. Future evolution of the composition of the occupation according to the scenario of SEM.

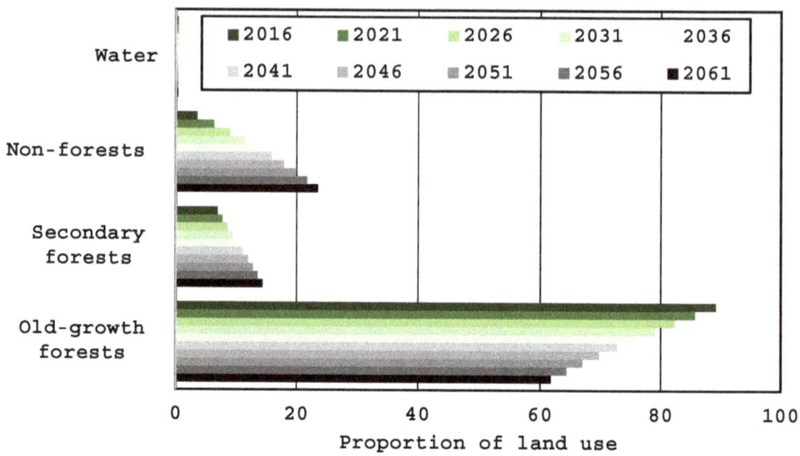

Figure 10. Future evolution of the composition of the occupation according to the REG scenario.

In the SEM scenario, the dynamics of land use will benefit forests (Figure 9). Considering the changes to be observed between 2016 and 2061, 8.84% of the proportion of the landscape occupied by old-growth forests will come from secondary forests (7.86%) and non-forests to secondary forests (2.73%). Furthermore, the non-forest class will experience a decrease in future years. They represent 3.50% of the proportion of the landscape in 2016. They will cover only 2.71% of the landscape by 2061. The proportion of the landscape that will remain unchanged is estimated at 87.03% of the total area of the landscape (respectively, 80.11% covered by old-growth forests, 5.95% by secondary forests, 0.66% non-forest and 0.31% water). Table 8 shows the transitions obtained in the SEM scenario between 2016 and 2061.

Table 8. Transition matrix in the SEM scenario between 2016 and 2061.

2016–2061		**Land Use in 2061**				**Total 2016**
		Pf	**Sf**	**Nf**	**Ww**	
Land use in 2016	Pf	80.11	7.86	1.30	0.00	89.28
	Sf	0.21	5.95	0.75	0.00	6.91
	Nf	0.12	2.73	0.66	0.00	3.50
	Ww	0.00	0.00	0.00	0.31	0.31
	Total 2061	80.44	16.54	2.71	0.31	100

The change in the composition of land cover in the SEM scenario between 2016 and 2061 is illustrated in Figure 9.

In the rapid economic growth scenario, the future dynamics of the landscape will come at the expense of forests (Figure 10).

Between 2016 and 2061, 28.84% of the proportion of the landscape occupied by old-growth forests will be converted to secondary forests (12.62%) and non-forests (17.52%). The non-forest class will experience a dramatic increase in the years to come. It represents 3.51% of the proportion of the landscape in 2016. It will cover more than 25.25% of the landscape by 2061. The proportion of the landscape that will remain unchanged is estimated at 64.22% of the total area landscape (respectively, 59.13% covered by old-growth forests, 1.92% by secondary forests, 2.86% non-forest and 0.31% water). Table 9 shows the transitions obtained in the REG scenario between 2016 and 2061.

Table 9. Transition matrix in the REC scenario between 2016 and 2061.

2016–2061		Land Use in 2061				Total 2016
		Pf	**Sf**	**Nf**	**Ww**	
Land use in 2016	Pf	59.13	12.62	17.52	0.00	89.28
	Sf	0.11	1.92	4.87	0.00	6.91
	Nf	0.07	0.58	2.86	0.00	3.50
	Ww	0.00	0.00	0.00	0.31	0.31
	Total 2061	59.31	15.13	25.25	0.31	100

The comparison of the proportions of land use simulated in 2061 with those observed in 2016 shows significant differences in two different scenarios: BAU (X-squared = 16.46; df = 3; p = 0.03) and REG (X-squared = 17.25; df = 3; p = 0.001). Moreover, the simulated occupation of the SEM is not statistically different from that observed in 2016 (X-squared = 20.71; df = 3; p = 0.06).

4. Discussion

4.1. Historical and Future Trajectories of Deforestation

The dynamics of land use in the study area are characterized by deforestation and forest degradation. Deforestation observed in the Ituri-Epulu-Aru landscape shows significant differences between periods, forest types and macro-zones (protected area, sustainable management zone for natural resources and extraction zone). Indeed, before 2010, the annual rate of deforestation was relatively low (0.05%) and the average area of deforestation spots was 1 ha. It more than doubled between 2010–2016 reaching 0.14% per year and the average area of deforestation spots increased by 1.2 ha. The significant decrease over time in forest area confirms the hypothesis of continual anthropization of Ituri's forests. However, comparing deforestation rates by period does not reveal any significant difference. Likewise, in all cases, the average area of deforestation spots is not significantly different over the two periods, which shows that there is no "period" effect on deforestation rates.

Considering land use, the differences in annual deforestation rates are very large ranging from 0.02 to 3.05% for the period 2003–2010 and from 0.1 to 3.20% for the second observation period. At the landscape level, these rates remain below the national average of 0.22% per year [1]. Moreover, except in the OWR, these rates are above this average, particularly in the second period. Several authors share the same opinion that deforestation is increasing in the majority of forests [2,29].

Comparison of key deforestation figures obtained in this study with those of other similar studies should be done with caution since the methodologies and data used are not always compatible. FACET [2] estimates the area of old-growth forests in 2010 at 3,843,218.88 ha. This area is slightly less than that obtained in the present study. Some scenes used may be different. Statistics from FACET [2] reveal increasing rates of deforestation, a trend shared by our results. Furthermore, Lusana et al. [14] estimate this area at 4,049,204 ha in 2003 and 3,997,690 ha in 2010, i.e., a loss of 51,514 ha between 2003 and 2010. Note that this latest study is based on the mapping materials of Hansen et al. [13] who overestimate the forest area [25]. The main reason given by these researchers is that the scale of analysis does not allow a good definition of the forest. Thus, it is possible that certain wastelands are confused with forests. This explains why the estimates of Lusana et al. [14] seem to exceed those carried out in this study.

Deforestation rates observed in the Ituri-Epulu-Aru landscape remain relatively low compared to other regions of the country, such as in the Bombo-Lumene reserve located not far from Kinshasa (0.46% per year between 2000 and 2015), the Yangambi Biosphere Reserve (4.5% between 2003 and 2016) [29] and very low compared to tropical America (0.51%) or Tropical Asia (0.58%) [58].

4.2. Simulation of Deforestation

This article provides useful information for those who wish to discuss a model that can be replicated for other territories affected by deforestation and changes in natural and anthropogenic forest structure. Fieldwork identified agriculture, forestry, infrastructure, demographic factors, socio-political factors, economic factors and biophysical factors. Among the variables retained, the distance from rural complexes, distance from national roads, artisanal mining and distance from major centers seems to play an important role in view of the main changes observed between 2003 and 2016. These results are similar to those of several authors [11,59].

The development of images from trendy and contrasting prospective scenarios will promote the identification of areas with socio-environmental issues concerning on the one hand the living environment of Pygmy communities and on the other hand the preservation of old-growth forests. For decades, primary and secondary forests have given way to crops in agricultural areas [59]. The use of these deforested areas makes it possible to benefit from new fertile land and therefore to increase agricultural production.

In summary, the trend of regression of the forest landscape in favor of culture and urban spaces has been observed for several decades [2]. This is then done at the expense of urban and village centers but also along the main communication axes (road network, network of tracks) [3,11]. Moreover, this degradation mainly impacts old-growth forests [28]. Suddenly, deforestation leads to a loss of biodiversity due to the destruction of many natural habitats [60]. The different prospective scenarios designed here take into account the different socio-economic activities developed in the study environment.

In general, forest dynamics are regressive although secondary forests are increasing. The trend scenario (BAU) suggests an alarming deforestation in the next four decades, which makes it possible to verify the fourth hypothesis. In this BAU scenario, both non-forests and secondary forests have increased. Indeed, this increase could be explained by the increase in population and therefore the need for food and housing. In compensation for this strong demand for land, there is a reduction in the area of old-growth forests. These results are corroborated by those of Samie et al. [61] obtained in Punjab (Pakistan).

The catastrophic scenario (REG) predicts that natural plant formations will regress in favor of anthropogenic ones. The sustainable environmental management (SEM), which combines both the preservation of plant cover with agricultural activities, empowers the state in its role of controlling deforestation and subsidizing domestic gas to replace fuelwood. This is similar to the densification scenario developed by Lajoie and Hagen-Zanker [62] which encourages the preservation of forests and limits urban sprawl in Reunion Island by 2031.

The validation of the model constitutes a first step in the prospective modeling of forests by 2061. The prospective model designed presents conclusive results and seems to be able to better take into account the evolution trends, the latter, by its unsupervised character.

The prediction model developed in this study to estimate the quantities of land cover changes produced values close to reality. In fact, it confirms, on the whole, the trends in land use. Nevertheless, it presented difficulties in predicting the changes that took place between 2003 and 2014. This is linked to the high observed and simulated constancy which was 88% at the landscape scale. This means that our analysis, both at the landscape level and at the level of land cover classes, highlights interesting results but which should be qualified. Moreover, there are fewer false alarms than failures, indicating that the simulated change is less than the baseline change. The quantity component does not indicate whether the false alarms are less than the misses or vice versa. The quantity component is about the same size as the allocation component. If our false alarms are less than your errors, it may be because the rate of change during the calibration time interval is slower than the rate of change during the validation time interval.

Overall, the observation of errors reveals that they are localized near the non-forests observed. The limitation of the model lies mainly in the fact that there are other variables that may explain the changes in land cover and use. These are, for example, political and institutional factors such as poverty, unemployment, conflicts and the forest code, demographic factors such as migration and population distribution, cultural factors (household consumption) and economic factors (cost of labor and capital). The addition of these additional variables was limited by their non-quantifiable nature and their unavailability in digital format [19,27,55,63]. In addition to the variables used.

5. Conclusions

This article aimed to analyze deforestation in the Ituri-Epulu-Aru landscape. This article provides valuable information on deforestation and forest degradation patterns. The results obtained confirm the trend towards deforestation. Although the landscape has seen a slight increase in the area of secondary forests, that of old-growth forests has declined significantly. Taken as a whole, forests are shrinking as a result of the unsustainable land use pattern characterized by shifting slash-and-burn agriculture with increasingly shorter fallows. The great concern lies in maintaining priority habitats for the biodiversity of this landscape. Our model predicts an increase in secondary forests over the entire landscape studied. This increase is not good news; indeed, it is indicative of a strong deterioration due, in particular, to subsistence activities.

Taking into account the results obtained, we propose that the landscape management consortium initiates local environmental and social management plans around hot zones of deforestation, in particular around Mambasa and Walese-Vonkutu. Land use should favor the restoration of severely degraded landscapes while highlighting sustainable development approaches (agro-ecology, renewable energies, etc.) and biodiversity conservation (particularly in sensitive areas). Raising awareness and improving the agrarian system through agroforestry techniques (particularly agro-forests, which are particularly suited to the region) must be at the center of strategies for the creation and development of village secondary forests, a pledge of the sustainable management of natural resources from which the populations will be able to obtain forest products for their usual needs.

It is also recommended that future analyzes assess the influence of this deforestation on the climate by quantifying the associated emissions. It would be interesting to say the impact of this loss of forests on the well-being of neighboring populations in order to further inform policy choices.

Author Contributions: Conceptualization, J.M.K., O.M.K., R.N. and L.M.; methodology, J.M.K., O.M.K., T.M.B. and L.M.; software, J.M.K.; validation, J.M.K., O.M.K., T.M.B., R.N. and L.M.; formal analysis, J.M.K.; investigation, J.M.K. and A.B.L.; resources, J.M.K., M.S., N.M., T.M.B. and L.M., data curation, J.M.K.; writing—original draft preparation, J.M.K.; writing—review and editing, J.M.K., O.M.K., M.S., N.M., T.M.B., R.N.; visualization, J.M.K.; supervision, O.M.K., R.N., J.-P.M.M. and L.M.; project administration, O.M.K. and L.M. All authors have read and agreed to the published version of the manuscript.

Funding: This research received no external funding.

Institutional Review Board Statement: Not applicable.

Informed Consent Statement: Not applicable.

Data Availability Statement: Data presented in this study are available on request from the corresponding author. The data are not publicly available because they are part of ongoing research.

Acknowledgments: We warmly thank Victor Kadiata for having the Satellite Observatory of the Forests of Central Africa (OSFAC) for ensuring the download of satellite images. We sincerely thank Alastair McNeilage and Toussaint Molenge (CARPE) for recommending us to SOFCA. Big thanks to Salomon Dyoma Ezadri, Alimengo Paul, Atimo Michel, Kimate Henry, Landasi Tsavurume, M'vuge Bijoux, Mamimolo Jean-Marie, Mumbere Jackson, Ngenda Elvis, Mundele Oscar, Yambwa Gloire for assistance in the field. We also thank the former Chief Curator/Site Manager of the Okapis Wildlife Reserve and Directors of Parks and Reserves in DR Congo Paulin TSHIKAYA and the Curator/Head of Research and Biomonitoring Program at the Okapis Wildlife Reserve Balimbaki Aimé for access to the study site. We thank Rosemarie Ruf for its support. We are very grateful to Robert Gilmore

Pontius and Thomas Bilintoh for their valuable comments on the methodology and particularly on the validation of the simulation model.

Conflicts of Interest: The authors declare no conflict of interest.

References

1. European Commission. *Joint Research Centre. Les Forêts du Bassin du Congo: État des Forêts 2010*; Publications Office: Luxembourg, 2012.
2. Potapov, P.V.; Turubanova, S.A.; Hansen, M.C.; Adusei, B.; Broich, M.; Altstatt, A.; Mane, L.; Justice, C.O. Quantifying forest cover loss in Democratic Republic of the Congo, 2000–2010, with Landsat ETM+ data. *Remote Sens. Environ.* **2012**, *122*, 106–116. [CrossRef]
3. Bamba, I.; Barima, Y.S.S.; Bogaert, J. Influence de la densité de la population sur la structure spatiale d'un paysage forestier dans le bassin du Congo en R.D. Congo. *Trop. Conserv. Sci.* **2010**, *3*, 31–44. [CrossRef]
4. Mongo, L.I.W.; Visser, M.; De Cannière, C.; Verheyen, E.; Akaibe, B.D.; Ali-Patho, J.U.; Bogaert, E.J. Anthropisation et effets de lisière: Impacts sur la diversité des rongeurs dans la réserve forestière de Masako (Kisangani, R.D. Congo). *Trop. Conserv. Sci.* **2012**, *5*, 270–283. [CrossRef]
5. Kankonda, O.M.; Akaibe, B.D.; Sylvain, N.M.; Le Ru, B.-P. Response of maize stemborers and associated parasitoids to the spread of grasses in the rainforest zone of Kisangani, DR Congo: Effect on stemborers biological control: Response of stemborers and parasitoids to grasses spreading. *Agric. For. Entomol.* **2018**, *20*, 150–161. [CrossRef]
6. Turubanova, S.; Potapov, P.V.; Tyukavina, A.; Hansen, M.C. Ongoing primary forest loss in Brazil, Democratic Republic of the Congo, and Indonesia. *Environ. Res. Lett.* **2018**, *13*, 074028. [CrossRef]
7. Mas, J.-F.; Kolb, M.; Paegelow, M.; Camacho Olmedo, M.T.; Houet, T. Inductive pattern-based land use/cover change models: A comparison of four software packages. *Environ. Model. Softw.* **2014**, *51*, 94–111. [CrossRef]
8. Mas, J.; Puig, P.; Palacio, J.L.; Sosa-López, A. Modelling deforestation using GIS and artificial neural networks. *Environ. Model. Softw.* **2004**, *19*, 461–471. [CrossRef]
9. Pérez-Vega, A.; Mas, J.-F.; Ligmann-Zielinska, A. Comparing two approaches to land use/cover change modeling and their implications for the assessment of biodiversity loss in a deciduous tropical forest. *Environ. Model. Softw.* **2012**, *29*, 11–23. [CrossRef]
10. Thierry, A.; Martin, P.; Ismaïla, T.I.; Brice, T. Modelisation des changements d'occupation des terres en region soudanienne au nord-ouest du Benin. *ESJ* **2018**, *14*, 248. [CrossRef]
11. Defourny, P.; Delhage, C.; Kibambe, J.-P. *Analyse Quantitative des Causes de la Deforestation et de la Dégradation des Forêts en République Démocratique du Congo*; ULC/ELI Géomatique, Université Catholique de Louvain: Louvain-la-Neuve, Belgium, 2011; p. 104.
12. Tyukavina, A.; Hansen, M.C.; Potapov, P.; Parker, D.; Okpa, C.; Stehman, S.V.; Kommareddy, I.; Turubanova, S. Congo basin forest loss dominated by increasing smallholder clearing. *Sci. Adv.* **2018**, *4*, eaat2993. [CrossRef]
13. Hansen, M.C.; Potapov, P.V.; Moore, R.; Hancher, M.; Turubanova, S.A.; Tyukavina, A.; Thau, D.; Stehman, S.V.; Goetz, S.J.; Loveland, T.R.; et al. High-resolution global maps of 21st-century forest cover change. *Science* **2013**, *342*, 850–853. [CrossRef] [PubMed]
14. Lusuna, M.; Kibambe, J.-P.; Makana, J.-R.; Mwinyihali, R. *Evaluation par Télédétection de l'Etendue de la Couverture Forestière dans le Paysage Ituri-Epulu-Aru (2000–2014)*; WCS: Kinshasa, Democratic Republic of the Congo, 2014.
15. PFBC. Les Forêts du Bassin du Congo: Évaluation Préliminaire; PFBC. 2005. Available online: https://carpe.umd.edu/sites/default/files/focb_aprelimassess_fr.pdf (accessed on 3 March 2021).
16. UICN-PACO. *Parcs et Réserves de la République Démocratique du Congo: Évaluation de l'Efficacité de la Gestion des Aires Protégées*; IUCN: Gland, Switzerland, 2010.
17. Forest Resources Assessment Programme. *Global Forest Resources Assessment 2015*; Food and Agriculture Organization of the United Nations: Rome, Italy, 2015.
18. Soares-Filho, B.; Rodrigues, H.; Follador, M. A Hybrid analytical-heuristic method for calibrating land-use change models. *Environ. Model. Softw.* **2013**, *43*, 80–87. [CrossRef]
19. Pontius, R.G.; Castella, J.-C.; de Nijs, T.; Duan, Z.; Fotsing, E.; Goldstein, N.; Kok, K.; Koomen, E.; Lippitt, C.D.; McConnell, W.; et al. Lessons and challenges in land change modeling derived from synthesis of cross-case comparisons. In *Trends in Spatial Analysis and Modelling*; Behnisch, M., Meinel, G., Eds.; Geotechnologies and the Environment; Springer International Publishing: Cham, Switzerland, 2018; Volume 19, pp. 143–164. ISBN 9783319525204.
20. Verburg, P.H.; Schot, P.P.; Dijst, M.J.; Veldkamp, A. Land use change modelling: Current practice and research priorities. *GeoJournal* **2004**, *61*, 309–324. [CrossRef]
21. Soares-Filho, B.S.; Coutinho Cerqueira, G.; Lopes Pennachin, C. Dinamica—A stochastic cellular automata model designed to simulate the landscape dynamics in an amazonian colonization frontier. *Ecol. Model.* **2002**, *154*, 217–235. [CrossRef]
22. Brown, E.L. Ituri-Epulu-Aru. In *Les Forêts du Basin du Congo—Etat des Forêts 2008*; Publications Office: Luxembourg, 2010.
23. Brown, E.L. Okapi faunal reserve, Ituri-Epulu-Aru landscape, Democratic Republic of the Congo. In *Landscape-scale conservation in the Congo Basin: Lessons learned from the Central African Regional Program for the Environment (CARPE)*; UICN: Gland, Switzerland, 2010; ISBN 9782831712888.
24. Masimo Kabuanga, J.; Mubenga Kankonda, O.; Horning, N.; Saqalli, M.; Maestripieri, N.; Makanzu Imwangana, F.; Mané, L. Random forest algorithm for mapping deforestation in the Ituri-Epulu-Aru landscape (Democratic Republic of the Congo). *Preprints* **2020**, 2020120752. [CrossRef]

25. Grinand, C.; Rakotomalala, F.; Gond, V.; Vaudry, R.; Bernoux, M.; Vieilledent, G. Estimating deforestation in tropical humid and dry forests in Madagascar from 2000 to 2010 using multi-date landsat satellite images and the random forests classifier. *Remote Sens. Environ.* **2013**, *139*, 68–80. [CrossRef]
26. Mikwa, J.-F.; Kabuanga, J.M.; Anitambua, S.; Kahindo, M.; Nshimba, S. Analyse prospective de la déforestation estimée par télédétection dans la réserve de biosphère de Yangambi. *Int. J. Innov. Sci. Res.* **2016**, *24*, 236–256.
27. Varga, O.G.; Pontius, R.G.; Singh, S.K.; Szabó, S. Intensity analysis and the figure of Merit's components for assessment of a cellular automata—Markov simulation model. *Ecol. Indic.* **2019**, *101*, 933–942. [CrossRef]
28. Pontius, R.G.; Millones, M. Death to kappa: Birth of quantity disagreement and allocation disagreement for accuracy assessment. *Int. J. Remote Sens.* **2011**, *32*, 4407–4429. [CrossRef]
29. Kyale Koy, J.; Wardell, D.A.; Mikwa, J.-F.; Kabuanga, J.M.; Monga Ngonga, A.M.; Oszwald, J.; Doumenge, C. Dynamique de la déforestation dans la réserve de biosphère de Yangambi (République Démocratique Du Congo): Variabilité spatiale et temporelle au cours des 30 dernières années. *Bois For. Trop.* **2019**, *341*, 15. [CrossRef]
30. Kabuanga, J.M. *Analyse Diachronique Dynamique de l'Occupation du Sol à Partir de l'Imagerie Landsat TM, ETM+ et OLI*; Gilman International Conservation: Epulu, Democratic Republic of the Congo, 2016; p. 36.
31. Kabuanga, J.M.; Adipalina Guguya, B.; Ngenda Okito, E.; Maestripieri, N.; Saqalli, M.; Rossi, V.; Iyongo Waya Mongo, L. Suivi de l'anthropisation du paysage dans la région forestière de Babagulu, République Démocratique du Congo. *Vertigo Rev. Électronique Sci. Environ.* **2020**. [CrossRef]
32. Rodriguez-Galiano, V.F.; Chica-Olmo, M.; Abarca-Hernandez, F.; Atkinson, P.M.; Jeganathan, C. Random forest classification of mediterranean land cover using multi-seasonal imagery and multi-seasonal texture. *Remote Sens. Environ.* **2012**, *121*, 93–107. [CrossRef]
33. *R Core Team R: A Language and Environment for Statistical Computing*; R Foundation for Statistical Computing: Vienna, Australia, 2018.
34. Liaw, A.; Wiener, M. Classification and regression by RandomForest | BibSonomy. *R News* **2002**, *2*, 18–22.
35. Breiman, L. Random forest. *Mach. Learn.* **2001**, *45*, 5–32. [CrossRef]
36. Pal, M. Random forest classifier for remote sensing classification. *Int. J. Remote Sens.* **2005**, *26*, 217–222. [CrossRef]
37. Gislason, P.O.; Benediktsson, J.A.; Sveinsson, J.R. Random forests for land cover classification. *Pattern Recognit. Lett.* **2006**, *27*, 294–300. [CrossRef]
38. Puissant, A.; Rougier, S.; Stumpf, A. Object-oriented mapping of urban trees using random forest classifiers. *Int. J. Appl. Earth Obs. Geoinf.* **2014**, *26*, 235–245. [CrossRef]
39. Strobl, C.; Boulesteix, A.-L.; Kneib, T.; Augustin, T.; Zeileis, A. Conditional variable importance for random forests. *BMC Bioinform.* **2008**, *9*, 307. [CrossRef]
40. Van Beijma, S.; Comber, A.; Lamb, A. Random forest classification of salt marsh vegetation habitats using quad-polarimetric airborne SAR, elevation and optical RS data. *Remote Sens. Environ.* **2014**, *149*, 118–129. [CrossRef]
41. Akar, Ö.; Güngör, O. Classification of multispectral images using random forest algorithm. *J. Geod. Geoinf.* **2012**, *1*, 105–112. [CrossRef]
42. Immitzer, M.; Atzberger, C.; Koukal, T. Tree Species classification with random forest using very high spatial resolution 8-Band WorldView-2 satellite data. *Remote Sens.* **2012**, *4*, 2661–2693. [CrossRef]
43. Teluguntla, P.; Thenkabail, P.S.; Oliphant, A.; Xiong, J.; Gumma, M.K.; Congalton, R.G.; Yadav, K.; Huete, A. A 30-m Landsat-derived cropland extent product of Australia and China using random forest machine learning algorithm on Google Earth engine cloud computing platform. *ISPRS J. Photogramm. Remote Sens.* **2018**, *144*, 325–340. [CrossRef]
44. Rodrigues, H.; Soares-Filho, B. A short presentation of Dinamica EGO. In *Geomatic Approaches for Modeling Land Change Scenarios*; Camacho Olmedo, M.T., Paegelow, M., Mas, J.-F., Escobar, F., Eds.; Lecture Notes in Geoinformation and Cartography; Springer International Publishing: Cham, Switzerland, 2018; pp. 493–498. ISBN 9783319608013.
45. Miranda-Aragón, L.; Treviño-Garza, E.J.; Jiménez-Pérez, J.; Aguirre-Calderón, O.A.; González-Tagle, M.A.; Pompa-García, M.; Aguirre-Salado, C.A. Modeling susceptibility to deforestation of remaining ecosystems in north central Mexico with logistic regression. *J. For. Res.* **2012**, *23*, 345–354. [CrossRef]
46. Kabuanga, J.M. *Cartographie Participative et Zonage dans la Chefferie de Mambasa*; Gilman International Conservation: Epulu, Congo, 2017; p. 25.
47. Institute, W.R. Atlas Forestier de la République Démocratique du Congo—Accueil. Available online: https://cod.forest-atlas.org/ (accessed on 24 June 2021).
48. Linard, C.; Gilbert, M.; Snow, R.W.; Noor, A.M.; Tatem, A.J. Population distribution, settlement patterns and accessibility across Africa in 2010. *PLoS ONE* **2012**, *7*, e31743. [CrossRef]
49. CARPE. CARPE Open Data Portal. Available online: https://carpe-worldresources.opendata.arcgis.com/ (accessed on 24 June 2021).
50. Earth Resources Observation and Science (EROS). Center Shuttle Radar Topography Mission (SRTM) 1 Arc-Second Global. 2017. Available online: https://www.usgs.gov/centers/eros/science/usgs-eros-archive-digital-elevation-shuttle-radar-topography-mission-srtm-1-arc?qt-science_center_objects=0#qt-science_center_objects (accessed on 28 August 2021).
51. Almeida, C.M.D.; Monteiro, A.M.V.; Câmara, G.; Soares-Filho, B.S.; Cerqueira, G.C.; Pennachin, C.L.; Batty, M. GIS and remote sensing as tools for the simulation of urban land-use change. *Int. J. Remote Sens.* **2005**, *26*, 759–774. [CrossRef]
52. Bonham-Carter, G.F.; Merriam, D.F. *Geographic Information Systems for Geoscientists: Modelling with GIS.*; Elsevier Science: Amsterdam, The Netherlands, 2013; ISBN 9780080571805.

53. Kamusoko, C.; Wada, Y.; Furuya, T.; Tomimura, S.; Nasu, M.; Homsysavath, K. Simulating future forest cover changes in Pakxeng district, Lao People's Democratic Republic (PDR): Implications for sustainable forest management. *Land* **2013**, *2*, 1–19. [CrossRef]
54. Paegelow, M.; Camacho Olmedo, M.T.; Mas, J.F. Techniques for the validation of LUCC modeling outputs. In *Geomatic Approaches for Modeling Land Change Scenarios*; Camacho Olmedo, M.T., Paegelow, M., Mas, J.-F., Escobar, F., Eds.; Lecture Notes in Geoinformation and Cartography; Springer International Publishing: Cham, Switzerland, 2018; pp. 53–80. ISBN 9783319608013.
55. Maestripieri, N.; Paegelow, M. Validation spatiale de deux modèles de simulation: L'exemple des plantations industrielles au Chili. *Cybergeo* **2013**. [CrossRef]
56. Pontius, R.G.; Peethambaram, S.; Castella, J.-C. Comparison of three maps at multiple resolutions: A case study of land change simulation in Cho Don District, Vietnam. *Ann. Assoc. Am. Geogr.* **2011**, *101*, 45–62. [CrossRef]
57. Brown, D.; Band, L.E.; Green, K.O.; Irwin, E.G.; Jain, A.; Lambin, E.F.; Pontius, R.G.; Seto, K.C.; Turner, B.L., II; Verburg, P.H. *Advancing Land Change Modeling: Opportunities and Research Requirements*; National Research Council (U.S.), Ed.; National Academies Press: Washington, DC, USA, 2014; ISBN 9780309288330. [CrossRef]
58. Gillet, P.; Vermeulen, C.; Feintrenie, L.; Dessard, H.; Garcia, C. Quelles sont les causes de la déforestation dans le bassin du Congo? Synthèse bibliographique et études de cas. *Biotechnol. Agron. Société Environ.* **2016**, *20*, 183–194.
59. Megevand, C. *Dynamiques de Déforestation dans le Basin du Congo: Réconcilier la Croissance Économique et la Protection de la Forêt*; The World Bank: Washington, DC, USA, 2013; ISBN 9780821398272.
60. Duguma, L.A.; Atela, J.; Minang, P.A.; Ayana, A.N.; Gizachew, B.; Nzyoka, J.M.; Bernard, F. Deforestation and Forest Degradation as an Environmental Behavior: Unpacking Realities Shaping Community Actions. *Land* **2019**, *8*, 26. [CrossRef]
61. Samie, A.; Deng, X.; Jia, S.; Chen, D. Scenario-based simulation on dynamics of land-use-land-cover change in Punjab Province, Pakistan. *Sustainability* **2017**, *9*, 1285. [CrossRef]
62. Lajoie, G.; Hagen-Zanker, A. La Simulation de l'étalement urbain à la Réunion: Apport de l'automate Cellulaire Metronamica® pour la prospective territoriale. *Cybergeo* **2007**. [CrossRef]
63. Pontius, R.G. Component intensities to relate difference by category with difference overall. *Int. J. Appl. Earth Obs. Geoinf.* **2019**, *77*, 94–99. [CrossRef]

land

MDPI

Review

A Synthesis of Land Use/Land Cover Studies: Definitions, Classification Systems, Meta-Studies, Challenges and Knowledge Gaps on a Global Landscape

Ryan Nedd [1], Katie Light [2], Marcia Owens [3], Neil James [4], Elijah Johnson [1] and Aavudai Anandhi [2,*]

[1] School of Environment, Florida A&M University, FSH Science Research Center, 1515 S. MLK Jr. BLVD, Tallahassee, FL 32307, USA; ryan1.nedd@famu.edu (R.N.); elijah.johnson@famu.edu (E.J.)
[2] Biological Systems Engineering Program, College of Agriculture and Food Science, Florida A&M University, Benjamin Bannekar Tech C, 1409 Wahnish Way, Tallahassee, FL 32307, USA; katie1.light@famu.edu
[3] School of Environment, Florida A&M University, 308L Humphries Science Research Center, Tallahassee, FL 32307, USA; marcia.owens@famu.edu
[4] College of Agriculture and Food Sciences, Florida A&M University, 204 South Perry Paige, Tallahassee, FL 32307, USA; neil.james@famu.edu
* Correspondence: anandhi@famu.edu

Abstract: Land is a natural resource that humans have utilized for life and various activities. Land use/land cover change (LULCC) has been of great concern to many countries over the years. Some of the main reasons behind LULCC are rapid population growth, migration, and the conversion of rural to urban areas. LULC has a considerable impact on the land-atmosphere/climate interactions. Over the past two decades, numerous studies conducted in LULC have investigated various areas of the field of LULC. However, the assemblage of information is missing for some aspects. Therefore, to provide coherent guidance, a literature review to scrutinize and evaluate many studies in particular topical areas is employed. This research study collected approximately four hundred research articles and investigated five (5) areas of interest, including (1) LULC definitions; (2) classification systems used to classify LULC globally; (3) direct and indirect changes of meta-studies associated with LULC; (4) challenges associated with LULC; and (5) LULC knowledge gaps. The synthesis revealed that LULC definitions carried vital terms, and classification systems for LULC are at the national, regional, and global scales. Most meta-studies for LULC were in the categories of direct and indirect land changes. Additionally, the analysis showed significant areas of LULC challenges were data consistency and quality. The knowledge gaps highlighted a fall in the categories of ecosystem services, forestry, and data/image modeling in LULC. Core findings exhibit common patterns, discrepancies, and relationships from the multiple studies. While literature review as a tool showed similarities among various research studies, our results recommend researchers endeavor to perform further synthesis in the field of LULC to promote our overall understanding, since research investigations will continue in LULC.

Keywords: synthesis of land use/land cover definitions; meta-analysis studies in land use/land cover; challenges and knowledge gaps in land use/land cover assessments; literature review

Citation: Nedd, R.; Light, K.; Owens, M.; James, N.; Johnson, E.; Anandhi, A. A Synthesis of Land Use/Land Cover Studies: Definitions, Classification Systems, Meta-Studies, Challenges and Knowledge Gaps on a Global Landscape. *Land* **2021**, *10*, 994. https://doi.org/10.3390/land10090994

Academic Editor: Giuseppe Pulighe

Received: 13 August 2021
Accepted: 14 September 2021
Published: 21 September 2021

1. Introduction

The land is the earth's terrestrial surface (immediately above or below the surface) that is delineable and with attributes [1] and is considered a nexus for environmental challenges [2]. Its characterized by land objects (distinguishing properties) and land key elements [1]. Anandhi et al. (2020) provided a narrow and broad definition of land resources more recently [3]. They broadly defined a land resource to include multiple components such as ecological resources of climate, water, soil, landforms, flora, and fauna, and all the socio-economic systems that interact with agriculture, forestry, and other land uses within some system boundary. Knowledge of land use and land cover is essential for

(1) understanding land development, loss and degradation [4]; (2) food and energy security for the growing population [5]; (3) simulating water and carbon cycles, ecosystem dynamics, and climate change inland surface models [6]; (4) equalizing tax assessment in many states [4]; (5) assessing associated land use related environmental effects and impact on provisioning of ecosystem services (e.g., eutrophication, pollution, biodiversity loss or climate effects) [5]; (6) land management consideration, which account for land cover modifications that influence approximately 71–76% of free range land under land cover conversions [7,8]; (7) change detection analysis (e.g., location where the change occurs, the type of change, and how the change is) [9]; (8) understanding and assessing the effects of landscape changes on the atmosphere, climate and sea level [10,11]; (9) considering the changes of land dynamics, how habitats and biodiversity are impacted [12,13]; (10) use of monitoring tools in policy change, landscape monitoring, and natural resource management within the environment [14]. These contributed to the observation, researching, planning, and implementation of policies that will strike a balance between managing resources on the land, such as agriculture, forestry, and building construction that alters the land surfaces while protecting the environment (ecosystems and wildlife habitat) [15]. (11) Administration of a variety of land conservation programs. The USDA 2019 Economic Research Service report identified the principal need for a better understanding of the drivers that will help with strategic planning and program design. This would considerably improve land conservation programs resulting in billions of dollar savings. Additionally, the United Nations Convention to combat desertification targets land degradation neutrality (LDN), addressing sustainable development goals to strengthen national capacity and quantitatively assess land degradation [16].

Several literature reviews have published investigations in the field of LULC. This study has summarized the review studies since the year 2000. Over the decades, the diverse research literature has defined "Land-use" and "Land-cover" in various ways. Depending on the specific area of interest, they have further broken down each region separately to clarify meaning. Land-use and land-cover can carry separate definitions, where land-use relates to what purpose the land is utilized, e.g., agricultural or recreational use. In contrast, land-cover states specific landscape patterns and characteristics [17]. While the terminologies for LULC may be used interchangeably [18,19], the concept remains the same for any particular region. It focuses on man's utilization in time and space of the various physical, chemical, and cultural factors of the land [20]. A synthesis of LULC definitions fills an apparent gap in the existing literature.

Land-use and land-cover are key physical elements that observe the Earth's surface and answer basic questions: What is this (land-cover)? What is it for (land-use)? [21]. classification systems are required to differentiate between land-use and land-cover. These systems provide the essential functions of tool structuring for classification, naming, and identifying objects on the earth [21]. Classification systems have incorporated mapping and spatial data as an essential function for analysis, and challenges have arrived with these classification system's assessment of land observations. "Continuity in observation" for both fine and coarse resolution satellite data along with in-situ information is an essential issue to addressed [22]. Challenges in LULC have given rise to associated knowledge gaps in data collection, image sampling, naming rules, overlap, and the inclusion of new objects [21]. Finally, a chart summarizes the knowledge gained in this study that will be useful to potential stakeholders working in the field of land use and land cover. Land use and land cover are important aspects of land resources. The general goal of this study is to review land use and land cover literature. Specific objectives are to summarize current knowledge in their definitions, classification systems, meta-studies, challenges, and knowledge gaps while building on past reviews in the field. Finally, this study seeks to systematically interpret and summarize that knowledge for stakeholders who work in land use and land cover.

A literature review was the methodology followed out in this study. It is a procedure used by investigators or researchers to compare the results from various studies. It finds common patterns, accomplishes synthesis, finds discrepancies or relationships using multiple studies [23,24]. This methodology will focus on five areas of interest related to LULC from various articles to give a clear and concise understanding of LULC. The five (5) areas of research interest for this paper are as follows:

1. Land use/land cover definitions and how the various authors define them.
2. The use of land use/land cover classification systems and how they are used by countries, regionally and on a global scale.
3. Land use/land cover meta-analysis studies in the areas of direct and indirect LULC, as well as data and methods of change.
4. Associated challenges correlated to land use/land cover changes.
5. Knowledge gaps and needs associated with land use/land cover change.

2. Methodology

This research project downloaded three hundred and eighty-nine (389) research articles, and one hundred and forty-six (146) were used to present research findings in this study. This research used a systematic meta-analysis framework adopted from Mengist et al. (2020) [25]. This framework takes into consideration protocol (P), search (S), appraisal (A), synthesis (S), analysis (A), and report (R). The methodology used for this research is the PSALSAR framework. Research articles for this study were taken from the year 2000 to 2019 and obtained through search engines accessed during the period 10 January 2020–30 April 2021: Science Direct (https://www.sciencedirect.com/) and Google Scholar (https://scholar.google.com/). There was a specific protocol used for the collection of published articles for the research objectives. According to the aim of this study, articles were downloaded and compiled on the keywords for each research objective. Keywords used in the "search" aspect of the framework are as follows: (1). Land; (2) Land-use/Land-cover; (3) Land classification systems; (4) Land-use/land-cover challenges; (5) Knowledge gaps and needs associated with land-use/land-cover.

These terms were used in the extensive search, appraisal, synthesis, and analysis based on the research questions: (1) How do various authors define land-use and land-cover?; (2) Land use/land cover classification systems used by countries, regionally, and on a global scale; (3) Meta-analysis studies of various LULC investigations for direct and indirect impacts on land use; (4) The associated challenges with LULC; (5) Identification of different LULC knowledge gap areas. The methodology presents a step-by-step process used in the synthesizing of LULC: (1) Definitions, (2) Classification systems used worldwide; (3) Meta-studies of LULC; (4) Challenges related to LULC; (5) Knowledge gaps associated with LULC. This systemic literature review summarizes information, ideas, explanations, and various methods from secondary data (published research articles). This protocol in Table 1 describes the process of acquiring papers to report actual findings. The methodological steps are outlined accordingly and broken down for each research objective.

Table 1. A systematic step by step process to acquire research articles for each objective.

Research Source	Step 1	Step 2	Step 3	Step 4	Step 5	Step 6
1. Google Scholar 2. Science Direct	Research articles were downloaded based on scope and keywords.	The articles were selected for specific objectives.	The articles were appraised for a specific objective.	Specific information was used from the appraised articles.	Tables and figures represent research findings.	Result findings are reported and discussed.

2.1. *The Steps Used According to the PSALSAR Framework for Collecting and Synthesizing Land Use/Land Cover Definitions for This Literature Review*

1. The defined scope and terms "land use and land cover definitions" was used as the keywords as part of the search strategy.
2. A total of thirty-five (35) articles on "land use and land cover definitions" were downloaded based on the terms.

3. "Backward and forward snowball" sampling for a further thirteen (13) research articles based on the terms "land use/land cover definitions".
4. Thirty (30) articles for "land use and land cover definitions" were selected and appraised for information.
5. The definitions were synthesized and placed into a table template, separated into two categories: land use and land cover.
6. The definitions were divided by regions where the specific authors did research.

2.2. The Steps Used According to the PSALSAR Framework for Collecting and Synthesizing Land Use/Land Cover Classification System Used for This Literature Review Worldwide

1. The defined scope and terms "land use and land cover classification systems" was used as the keywords as part of the search strategy.
2. A total of two hundred and thirty-three (233) articles on "land use/land cover classification systems" were downloaded for information.
3. One hundred and seventy-one (171) articles on "land use/land cover classification systems" were selected based on their relevance.
4. Sixty-two (62) articles were reviewed and appraised for "land use/land cover classification systems" information based on classification systems.
5. Twelve (12) articles were synthesized and placed into a table template for "land use/land cover classification systems".
6. The table contains three categories (national, regional, and global) and shows classification systems used by various countries.

2.3. The Steps Used According to the PSALSAR Framework for Collecting and Synthesizing Land Use/Land Cover Meta-Analysis Studies for (1) Direct Changes in LULC; (2) Indirect Changes in LULC and; (3) Meta-Studies of Data/Methods

1. The defined scope and terms "land use and land cover meta-analysis" was used as the keywords for the search strategy in the search engines.
2. A total of fifty-five (55) articles were downloaded for information on "land use/land cover meta-analysis".
3. To conduct the "land use/land cover meta-analysis", forty-eight (48) articles were selected and appraised.
4. Three categories: Direct, indirect, and data/methods associated with "land use/land cover meta-analysis" were used to present the papers in figures.

2.4. The Steps Used According to the PSALSAR Framework in Synthesizing Associated Challenges Correlated to Land Use/Land Cover Changes

1. The defined scope and terms "land use and land cover challenges" was used as the keywords as part of the search strategy.
2. A total of thirty-five (35) articles were downloaded on "land use/land cover challenges" for information.
3. Twenty-nine (29) papers were selected and appraised for information related to "land use/land cover challenges".
4. Two categories: Data quality and data consistency on "land use/land cover challenges", were used for the paper and presented in a table.

2.5. The Steps Used According to the PSALSAR Framework in Synthesizing Knowledge Gaps and Needs Associated with Land Use/Land Cover Change

1. The defined scope and terms "land use/land cover knowledge gaps and needs" were used for land use/land cover,
2. A total of thirty-one (31) articles were downloaded on "land use/land cover knowledge gaps and needs" for information,
3. Twenty-seven (27) articles were selected and appraised for information "land use/land cover knowledge gaps and needs",

4. "Land use/land cover knowledge gaps and needs" were identified and placed into four categories related to LULC, namely ecosystem services, forestry, data, and modeling.

3. Results

3.1. *Land Use/Land Cover Definitions*

The terms "land use" and "land cover" are widely used. Therefore, several definitions are utilized by authors to describe them. This study attempts to analyze the purpose of land-use and land-cover by highlighting the similarities and differences in the diverse descriptions used to explain the terms. The literature review presented various definitions for the terms "land use" and "land cover" from appraised research articles (Table 2). Finally, the "definitions" are illustrated and interpreted in Figure 1 with a paragraph description for potential stakeholders (Figure 1).

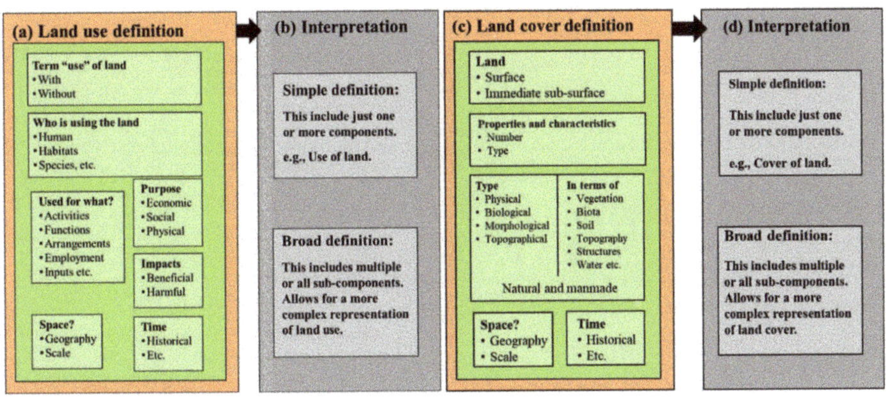

Figure 1. Showing simple and broad definitions of LULC developed and interpreted in this study.

Studies assume these concepts (land use and land cover) to be similar and interchangeable in the literature. Moreover, other concepts regarding land characteristics are defined based on species land class [10]. Other researchers consider them as different concepts [8,26]. The review of thirty (30) definitions revealed variations in the definition/description of land use and land cover (Figure 1). These definitions from different regions worldwide (e.g., North America, European nations, Africa, and Asian countries) showed how the meaning of LULC has similarities based on various authors in different regions worldwide. While exploring the concept of land use, the simplest definition observed coincides with its semantic meaning "What this land is used for?". However, a more complex description can be regarded with the following components:

1. Most of the definitions have the word "use" while describing the term. However, Anderson et al. (1976) defined land use as "Man's activities on land which are directly related to the land" [4], while Sreedhar et al. (2016) describe it as "Human activity or economic functions associated with a specific geography" [20]. These definitions do not have the term "use".
2. Many interpretations focus on who is using the land. The majority of the studies have a human component. Man, human, anthropogenic, land managers are used to describe this component [27]. Other definitions that highlight habitats or species using the land have been described [28,29].
3. Definitions often include the activities or functions related to the use of land. Additionally, terms such as arrangements and inputs have been used in describing this component. Further, the employment of land is used in descriptions by social

scientists [10]. In some cases, this component precisely defines the purpose of these activities, such as economic, social, and physical reasons.

4. Definitions can describe the effect of these activities. Additionally, beneficial or harmful impacts of the changes in the land are included in these interpretations.
5. A few of the definitions have a space component in them. Geographic scales are the terms associated with this component [20,30].
6. A few of the definitions have a time component in them. Terms such as historical are associated with this component [30].

Table 2. Definitions for *"Land use"* and *"Land cover"* from various articles.

S.N.	Land Use	Land Cover	Country	Citation
1	Man's activities on land that are directly related to the ground.	The vegetational and artificial constructions covering the land surface.	USA	[4]
2	Land use denotes the human employment of the land and is primarily studied by social scientists.	Land cover denotes the physical and biotic character of the land surface and is studied mainly by natural scientists.	USA	[10]
3		Land cover, which we define as 'the observed biophysical cover of the earth's surface' is an expression of human activities and, as such, changes with changes in land use and management.	USA	[11]
4	Land use refers to the purposes for which humans exploit the land cover.	The term land cover refers to the attributes of a part of the Earth's land surface and immediate subsurface, including biota, soil, topography, surface and groundwater, and human structures.	Belgium	[12]
5	Land-use areas refer to what this land is used for, such as commercial areas, industrial areas, or residential areas.	Land-cover materials refer to what is actually on the land, such as grass, asphalt, or soil.	USA	[31]
6	Land use is characterized by the arrangements, activities, and inputs people undertake in a certain land cover type to produce, change or maintain it.	Land cover is the observed (bio) physical cover on the earth's surface.	Rome	[32]
7	Land use deals with the socio-economic inputs to land and, thus, describes an activity with an input, a process, and an output.	Land cover is the observed (bio) physical cover on the Earth's surface.	Scotland	[22]
8	Natural scientists define land use in terms of syndromes of human activities such as agriculture, forestry and building construction that alter land surface processes including biogeochemistry, hydrology, and biodiversity.	Land cover refers to the physical and biological cover over the land's surface, including water, vegetation, bare soil, and artificial structures.	Brazil	[15]
9	Land use is related to important changes in species composition on and around the used area.		United Kingdom	[29]
10	Land use is referred to as man's activities and the various uses which are carried on Land.	Land cover is referred to as natural vegetation, water bodies, rock/soil, artificial cover, and others resulting due to land transformation.	India	[30]
11	Land use is the manner in which human beings employ the land and its resources.	Land cover describes the physical state of the land surface.	Malaysia	[26]
12	Land use is defined as the way or manner in which the land is used or occupied by humans.	Land cover refers to the observed biotic and abiotic assemblage of the earth's surface and immediate subsurface (Meyer and Turner, 1992). *	USA	[33]
13	Land use includes the human activities and management practices for which land is used.	Land cover includes the status of vegetation, bare soil, developed structures (for example, building, roads, and other infrastructure), and water bodies, including wetlands.	Kenya	[34]
14	Land use, in contrast, refers to the purposes for which humans exploit the land cover.	Land cover addresses the layer of soils and biomass, including natural vegetation, crops, and human structures that cover the land surface.	Netherlands	[13]
15	Land use corresponds to the description of the former areas in terms of their socio-economic purpose (the function they serve): areas used for residential, industrial, or commercial purposes, for farming or forestry, for recreational or conservation purposes, etc.	Land cover corresponds to a physical description of Earth, leading to a simple definition: the observed physical cover of Earth's surface.	USA	[21]
16	Land use is characterized by anthropogenic activities to modify, manage and use certain types of land cover.	Land cover describes the physical cover of the Earth's surface, including vegetation, non-vegetation, and man-made features.	Germany	[35]
17	Forest land use is a function of the social and economic purposes for which land is managed.	Forest land cover is a human definition of the biological cover observed on the land (Watson et al., 2000). *	USA	[36]

Table 2. *Cont.*

S.N.	Land Use	Land Cover	Country	Citation
18	Land use normally refers to the arrangements, activities, and inputs people engage in a certain land cover type to produce, change or maintain it (Liang, 2008). *	Land cover is defined as the observed biophysical state of the earth's surface and is largely described by the presence or absence of various vegetation types (Anderson, 2005). *	Germany	[37]
19	Land use is determined by environmental factors such as soil characteristics, climate, topography, vegetation, basic human forces that motivate production, and its responses to environmental changes. (Dinakar S., 2005; Dinakar and Basavarajappa., 2005). *		India	[38]
20	Land use denotes the approach in which land has been used by humans for economic activities. (Mengistu and Salami, 2007; Reis, 2008; Forkuo and Frimpong, 2012; Olokeoguna et al., 2014). *	"In common, land cover is defined as the perceived (bio)-physical cover on the Earth's surface which may include vegetation, man-made features, bare rock, bare soil, and inland water surfaces, etc."	India	[39]
21	"In general, the term "land use" refers to the human activity or economical functions associated with a specific geography."	Land cover as a type of natural features present on the surface of the earth. (Lillesand and Kiefer, 2000). *	India	[20]
22	Land use is more complex. On the one hand, it can be equally approached by natural scientists by analysing the "syndromes of human activities" in the context of biodiversity, hydrology, or biochemistry (Ellis, 2013). *	Land cover describes the directly observable bio-/physical overlay of the Earth's surface (Fisher et al., 2005; Verheye, 2009). *	Germany	[1]
23	Land-use refers to the way in which humans and their habitat have used land, usually with accent on the functional role of land for economic activities (Kumar et al., 2013). *	Land cover refers to the physical characteristics of earth's surface, captured in the distribution of vegetation, water, soil and other physical features of the land, including those created solely by human activities, e.g., settlements (Kumar et al., 2013). *	India	[28]
24	Land use can be broadly defined as the manner in which the observed biophysical cover is actually used by humans (Cihlar and Jansen, 2001). *	Land cover can be broadly defined as the manner in which the observed biophysical cover is actually used by humans (Di Gregorio, 2005). *	China	[40]
25	Land use is commonly defined as a series of operations on land, carried out by humans, with the intention to obtain products and benefits through using land resources.	Land cover is commonly defined as the vegetation (natural or planted) or man-made constructions (buildings, etc.) which occur on the earth's surface. Water, ice, bare rock, sand, and similar surfaces also count as land cover.	Ethiopia	[8]
26	Land use describes the social, economic, and cultural utility of the land (Turner 1997) and is known to alter how ecosystems function (DeFries, Foley, and Asner 2004). *	Land cover informs the functional relationship between terrain, climate, and soils, providing biophysical insights into the environment and drivers of change.	Canada	[41]
27	land use refers to the conversion or transformation of the land cover into the desired human purposes which are associated with that cover, e.g., cropping, conservation, or settlement.	The formation of a given land cover results complex processes and can be considered as the biophysical state of the earth's surface and immediate subsurface.	Ethiopia	[42]
28		Land cover is a biophysical indicator that refers to both the observed biotic and abiotic assemblage of Earth's surface, including the vegetation and anthropogenic structures covering the land (Hansen and Loveland 2012, Meyer and Turner 1992). *		[43]
29	Land use documents how people are using the land for development, conservation, or mixed uses (NOAA, 2015). *	land cover refers to the physical land type, such as how much of a region is covered by forests, impervious surfaces, agricultural lands, wetlands, and open water (NOAA, 2015). *	Bangladesh	[44]
30	The events that take place in the land represent the current use of the properties such as built-up institutions, shopping centers, parks, and reservoirs are described as land use categories (Fonji and Taff 2014). *	Natural and biological landscapes such as forests, marshlands, grasslands, water lands, and urbanized and built areas denote the land cover.	Germany	[45]

All definitions are direct quotes from research articles. * Multiple authors have similar definitions.

These components in the complex definition establish direct links between land cover and the people's action in their environment on a space and time scale.

While exploring the land cover definitions, most descriptions describe the land surface and immediate sub-surface properties and characteristics. The definitions vary with the number and type of properties and characteristics used in the interpretations. They include the physical, biological, morphological, and topographical cover of land in terms of vegetation, biota, soil, topography, water, structures, etc., which can be anthropogenic or natural. A simple definition of land cover can be its semantic meaning "What's this land cover". A more complex description can have more specific components.

3.2. *Land Use/Land Cover Classification System Used Worldwide*

3.2.1. Classification Systems

Over the years, the need for structuring information systems on LULC has developed into "classification systems" that are abstract representations of the situation in the field using well-defined diagnostic criteria. Earlier stages it was defined by Sokal in 1984 as: "the ordering or arrangement of objects into groups or sets based on their relationships [46]". Jansen and Gregorio (2002) also described this coordination of objects as: "the systematic framework with the names of the classes and the criteria used to distinguish them, and the relation between classes" [9]. Understanding LULC classification systems have caused researchers to investigate further how they work, and their definitions have broadened. According to Duhamel (2012), classification systems have three main functions of structuring information, facilitating communication and exchange among users of these systems [21]. These are (1) classification (assignment of all objects in a hierarchical series); (2) nomenclature (naming and describing the groups of objects); and (3) identification (allowing to assigning the membership status of individual objects in the classification) [21]. For classification systems to work efficiently, there is a need for land cover maps. Land cover maps are the foundation for accurate extraction of information from land covers to remote sensing modeling [6]. A literature review has shown various studies that provided the accuracies and a comparison of different LULC classification systems [47]. There are several classification systems examined in this study, and they are categorized based on their use as national, regional, and global systems (Table 3).

Table 3. Classification Systems used at National, Regional and Global Scales.

Category	Classification System	Year	Scale	Location	Citation
National	1. National Land Cover Data Classification System	1992; 2006; 2011	1:5000–1:10,000	U.S.A	[48,49]
	2. US National Vegetation Classification Standard	1997			
	National Forest Inventory Land Cover Classification Scheme	1999	1:5000–1:10,000	Canada	[41]
	National Institute of Statistics, Geography and Informatics	1993; 2000	1:25,000	Mexico	[50]
	National Land Use Database (NLUD)	2001	1:100,000	United Kingdom	[51]
	Sistema de Información de Ocupación del Suelo en España (SIOSE)	2000	1:25 000	Spain	[51,52]
	National Land Survey Classification System	1984; 2007	1:100,000–1:125,000	China	[53]
	NRSA LULC Classification System	2007	1:250,000	India	[30]
	South African Standard Land Cover Classification System	1996	1:100,000	South Africa	[49]
	The MapBiomas LULC Classification Scheme	2020	1:125,000	Brazil	[54]
	ALUM Classification System	2005	1:100,000–1:125,000	Australia	[51]
	New Zealand Land use Class.	1984	1:100,000–1:125,000	New Zealand	[51]
Regional	CORINE/Land Cover2006	1985–2018	1:100,000–1:125,000	Europe	[55]
	AFRICOVER Land Cover Classification System	1995–2002	1:100,000–1:125,000	Africa	[11]
	AARS Land Cover Classification	1999	1:100,000–1:125,000	Asia	[49]
	North American Land Change Monitoring System	2005	1:100,000–1:125,000	North America	[49]
Global	Land Cover Classification System (FAO)	1996	1:100,000–1:125,000	FAO	[11]
	USGS Land Use/Land Cover Classification Systems (National)	1972/1976	1:100,000–1:125,000	USGS	[4,49]
	International Geosphere-Biosphere Programme-Data and Information System	1996	1:100,000–1:125,000	IGBP	[49]

3.2.2. National Classification System

Classification systems at the national level assist policymakers, leading to the sustainable development of land resources [56]. National classification systems use land cover maps from remote sensing data to model, monitor, and understand landscape changes [6]. Table 3 shows how some countries have specific national classification systems for LULC at lower spatial scales. These classification systems describe the structure and relationship of various land objects [57]. To understand this process, land cover products that aggregate maps and multiple datasets are used for different land cover projects. For example, technological advances have used remote sensing. Wulder et al. (2018) showed "Land Cover 2.0" to be a great tool that emerged in that it enabled free and open access to data; it is high performing, accurate, and rapidly developed data processing and analysis capabilities [41]. This system is user-friendly and efficient in generating land object data for classification systems, and it's used widely for various LULC model projects.

On the national scale, many countries have developed classification systems. Table 3 shows countries with LULC classification systems for the identification of land cover objects. Notable examples of the national classification system are found in the North American region, which has employed consecutive land cover classification data for the past decade. The National Land Cover Database (NLCD) contains classifications for the United States for 1992, 2001, 2006, 2011, and 2016 [48,58]. There are many different sources of information on existing land use and land cover and the various changes occurring over time in the United States of America, including the U.S. Geological Survey (USGS) Land Cover Trends (LCT), National Land Cover Database (NLCD), North American Forest Dynamics (NAFD), Monitoring Trends in Burn Severity (MTBS), Protected Areas Database (PADUS), and North American Forest Age [59].

While Canada has its land cover classification system (National Forest Inventory Land Cover Classification Scheme), most land classifications use the NLCD. While countries like Mexico have the Instituto Nacional de Estadística y Geografía (INEGI) Uso del Suelo y Vegetación land cover product (INEGI, 2014) contains consistent classifications over Mexico for years 1985, 1993, 2002, 2007, and 2011 [59].

Table 3 shows national classification systems used by countries with significant land development and changing landscapes pushed by population density. The synthesis indicates that various literature resources have identified these different classification systems at the national level, their scale of use, and year of development. Due to the small scale ranges, these systems update readily with changing land covers by human activities and are utilized locally. The use of accurate land cover maps identifies land classes and objects, classifying at the sub-national scales, division, district, sub-district, county, city, and municipality levels [60]. While classification at the local level is for the observation of land objects, the result is for advancement and documentation of changing land covers. The literature review has shown that many researchers use remotely-sensed information for land cover classification. The typical practice involves using raw numerical data or calibrated reflectance from other land cover studies [61]. This information from national land cover classification (detected land classes and objects) is adapted for regional and global classifications.

3.2.3. Regional Classification Systems

Regional classification systems are sometimes known as continental classification systems, and they are used at large scales (from 1:250,000 to 1:100,000) compared to those used at national levels [62]. As shown in Table 3, they would periodically involve multiple countries or overlapping landscapes, depending on project type. Large projects employ continental classification systems. For example, while the services of other LULC classification systems are used locally, the CORINE (Coordination of Information on the Environment) classification system is used as a system to classify significant areas in Europe [49]. Eurostat was developed at the early stages in the European Union as a LULC statistical system [63]. There has been significant land development through decision-making for most European countries. The AFRICOVER Land Cover Classification System (LCCS) developed by the Food and Agricultural Organization (FAO), classified twenty-one African countries [49]. There is also the AARS land cover classification that develops land products for Asia [64]. Similarly, the North American Land Change Monitoring System covers the North American region, including Canada, the United States, and Mexico. This system provided continental information and met the need for country-specific monitoring programs with improved land cover maps for accurate databases [65].

Regional land classification systems are employed for continental landscape monitoring. Regions, such as North and South America, Asia, Africa, and Europe, did initial work for general land classification. However, with multiple accurate land products, land maps, and numerous land features, objects, and legends, more countries have developed a localized classification system for land objects for decision-making related to land management, monitoring, and ecosystem preservation. The key to many regional classification systems is land-cover products. While many land products may not accurately update changing land covers [66], these land products are used at the local and regional levels for classifying and decision making. Numerous LULC methods have been used at the local and regional levels to develop and amend various landscape policies, safeguarding ecosystems and biodiversity [67]. These regional classification systems update land characteristics and are also pivotal in regional decision-making from these landscape surveys.

3.2.4. Global Classification System

Global LULC classification systems have been around since the 1980s. As shown in Table 3, international organizations developed such systems for use worldwide. These include The Food and Agricultural Organization (FAO), which created the Land Cover Classification System (LCCS); The CORINE Land Cover (CLC) of the European Union and classification system designed by the International Geosphere-Biosphere Programme (IGBP) [1]. The MODIS Land Cover, GlobCover, or Global Land Cover [1], are examples. Others include UMD land-cover product, Globeland-30, Corine-2012, GlobeCover-2009, and Global Historical Land-Cover Change [62,68]. Once these land-cover products are validated, they provide accurate information and datasets related to land cover classes, objects, and features for global use in LULC classification systems. Classified land products lead to enhanced decision-making related to landscape management, monitoring, and change.

The FAO LCCS, a land-cover classification system, is used regionally and globally [32]. According to Keil (2016), the LCCS is a standardized multi-purpose system usable for any land cover condition independent of collection method and hierarchy [1]. The LCCS is a hierarchical classification scheme. Its classification focuses on specific classes, containing twenty-three (23) exclusive categories, divided into three layers with dimensions for land use, land cover, and surface hydrology [69]. The LCCS system is widely used as a global classification system and using various datasets from various sources. However, the Global Land Cover Network (GLCN) (http://www.fao.org/geonetwork/srv/en/main.home (accessed on 17 September 2021)) is one of the primary dataset sources used by LCCS created by the FAO [70]. There are other global land-cover dataset maps used as part of land classification, and these are as follows: (1) International Geosphere-Biosphere Project (IGBP), http://www.igbp.net/ (accessed on 17 September 2021); (2) University of Maryland (UMD) https://geog.umd.edu/research/landingtopic/land-cover-land-use-change (accessed on 17 September 2021) [71]; (3) Global Land Cover 2000 (GLC2000) https://ec.europa.eu/jrc/en/scientific-tool/global-land-cover (accessed on 17 September 2021), and (4) Moderate Resolution Imaging Spectroradiometer (MODIS) https://modis.gsfc.nasa.gov/data/ (accessed on 17 September 2021) [72].

3.3. *Synthesis of Meta-Analysis Studies in LULC*

Land use and land cover changes have been investigated extensively over the years. These research studies have been linked to the landscape's various changes from human modification to the earth's surface. Meta-analysis is a valuable technique that employs a combination of peer-reviewed studies to determine relationships [73]. Research studies have used meta-analysis to synthesize studies that show direct impacts on the changes in LULC. The focus of the analysis is on specific sites and landscapes [47]. Most peer reviews concentrated on deforestation and reforestation studies [74–76], while some meta-studies have focused on the indirect effects of LULCC [77–79]. The analysis of such studies addressed categories of direct and indirect LULCC, then further subcategorized into specific areas of interest (Figure 2). A further category was added to show meta-studies of data and method changes in LULCC. The information for these studies was also placed into a diagram (Figure 3) between the years 2000 and 2020, showing how the meta-analysis studies have developed over the years. More peer reviews in the area of LULC analyzed by this method provided qualitative and quantitative results from published research studies.

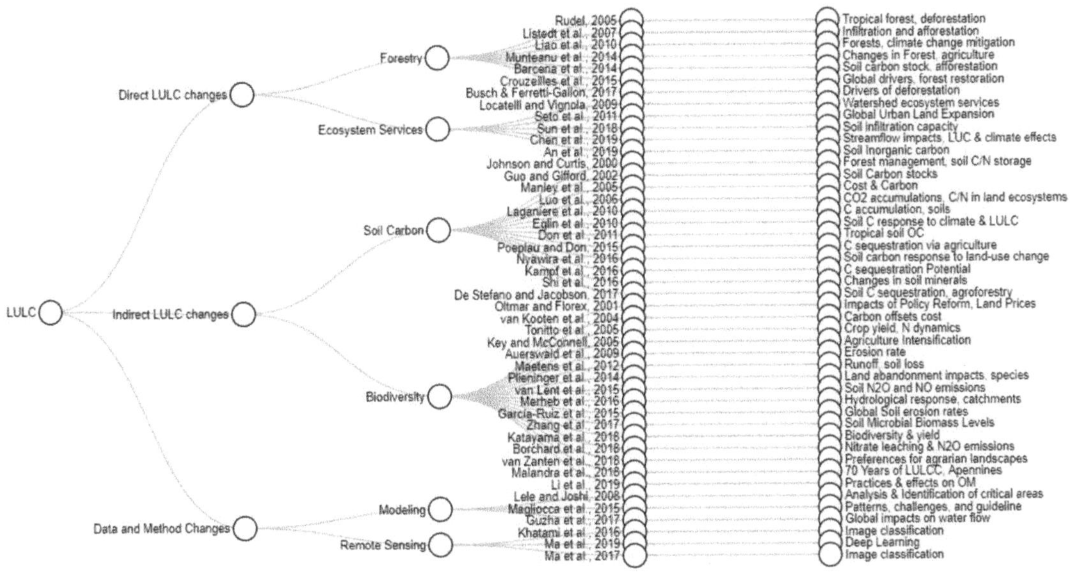

Figure 2. Diagram showing meta-studies divided based on categories.

3.3.1. Meta-Analysis Related to Direct Changes in LULC (Forest)

Land use has essential impacts on forests, the use of forest products, and changes in forest cover. These changes appear primarily in the United States but also occur globally [34]. In the context of LULC forestry and forestry cover, meta-studies look at the public response concerned with deforestation and the implementation of policies that have slowed down deforestation of some protected areas [76].

The meta-analysis allows a researcher to test specific hypotheses about the effect of a treatment by considering research information from various studies in the past in a particular topical area. This approach and technique are utilized for medical research [80,81]. LULC in forestry is unique and appropriate for meta-analysis studies. While other studies focus on agents of deforestation, for example, growing population, new settlements, roads, even topography, and other general factors that affect forest loss [82]. A meta-analysis is a valuable tool for understanding how LULC changes in forestry are changing [83]. Some case studies looked at the forest and agricultural land change [74], while others have looked at forest restoration, attributing it to the enhancement of 15–84% of biodiversity and 36–77% of vegetation structure compared to other ecosystems that are showing signs of degradation [75]. It is also helpful to identify the various services the forest may provide, specifically the storage of carbon, biodiversity for habitat, disease suppression, water filtration, storm mitigation, food, medicines, recreation, timber, and non-timber products [75].

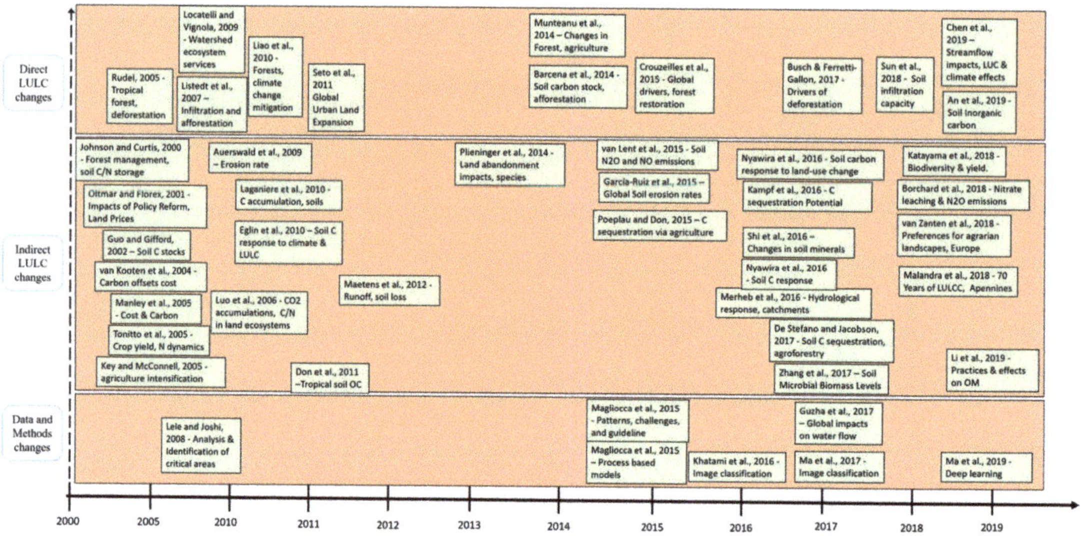

Figure 3. Diagram showing meta-studies timeline of articles.

The literature review showed meta-analysis studies focused exclusively on specific services that afforestation has benefited soil [84]. Various studies, including their respective experiments and statistical software, have shown that water infiltration increased due to afforestation [81]. Our synthesis revealed similar meta-analysis work by another researcher, offering a different meta-analytical approach to determine land use and climate effects on streamflow and infiltration [84]. Some studies focused on land use capacity and its impacts on soil infiltration [85]. Some others investigated the potential of soil carbon stock, the effects after reforestation, and the abilities to perform carbon sequestration [86]. Additional studies showed differing results between the carbon sequestration in a natural forest against a plantation-type forest. The meta-analysis technique showed the differences based on stand age, stand types, tree species, and origin of the plantation [87]. Some studies look at various factors that make up a forest, including investigations within watersheds, focusing on tropical forest services, and comparing hydrological flows between plantation and natural forests [88].

3.3.2. Meta-Analysis Related to Indirect Changes in LULC (Climate Change)

Various landscape projects utilize land use/land cover meta-analysis, significant catalysts of soil-carbon changes in recent decades center around land-use changes. A notable example would be replacing natural forest and vegetation for cropland and urban areas, leading to soil carbon loss. However, the reverse can lead to soil carbon being replenished [89,90]. Older meta-analyses focus on forest soil being a sink for Carbon (C) and Nitrogen (N). Johnson and Curtis (2001) conducted a meta-analysis on various management techniques and determined the mean response of forest soil C and N [91]. This method is adaptable for quantitative analysis of studies with multiple experiments for soil investigations [91]. In contrast, Luo et al. (2006) showed patterns among various studies due to C and N processes within the soil interaction with plants in response to high carbon dioxide [92]. Soil N_2O and NO emissions and their effects on the level of greenhouse gases in the atmosphere have been reviewed and synthesized from multiple studies from land-use changes in tropical and sub-tropical areas [93]. Meta-analysis identified interactions at the soil level to lower some of these emissions into the atmosphere [79]. Synthesizing is the appropriate approach for studies that seek an understanding of forestry-based projects and how researchers can select specific topics for a particular research area. Researchers would have compiled studies on the costs of sequestering carbon in terrestrial ecosystems by activities within the forests [94]. While another study focused on the forests' ability to accumulate carbon, their analysis examines the various experimental techniques used to determine carbon content among studies, thus drawing rational conclusions [95].

While many land-use change projects are formed after deforestation or in cooperation with reforestation practices, some land use characteristics such as agriculture have been responsible for carbon loss from the soil by converting forest and grasslands to arable lands. This depleted soil carbon and biomass have potentially seen carbon emissions into the atmosphere from the biosphere due to intensified agricultural activities [96]. Meta-analyses studies focus on the microbial biomass levels at various land-use sites and their effects on the ecosystems [97]. Soil provides significant storage for carbon. As such, depleted soil from agricultural enterprises can remove soil carbon by destroying primary forests, releasing soil carbon that results in higher CO_2 in the atmosphere [98]. These depleted soil can take atmospheric carbon to replenish what they would have lost [99,100]. There are numerous strategies to sequester atmospheric carbon into the soil through agricultural measures. These include reduced tillage intensity, increasing residue inputs from higher yields, eliminating summer fallows, nutrient and manure management, and restoring permanent grasslands or forests [100,101].

While replanting forest has been a sure way of carbon sequestration, the meta-analysis of various studies has shown numerous ways of soil loss. The investigation of soil loss shows a positive correlation with annual rainfall, plot runoff, and annual runoff coefficient [102]. This used investigative approaches of the effect at the subcontinental scale based on various environmental conditions. Land-use changes have significant impacts on soil degradation. While studies have shown the effects of land-use change through research, a meta-analysis methodology shows studies on specific areas. Shi et al. (2016) conducted a global meta-analysis across broad climatic zones using one hundred and thirty-nine (139) papers that investigated the changes in soil carbon, nitrogen, phosphorus, sulfur, and their stoichiometry in the soils of planted forest investigation [103]. Other meta-analyses on carbon stocks and sequestration investigated soil organic carbon in cultivated soils using cover crops [104]. They quantified soil organic carbon changes accumulation as a response to the use of cover crops. In comparison, researchers use meta-analysis to investigate the effects of agroforestry systems on carbon stocks [24]. Another meta-analysis may look at carbon stock based on land-use changes in a general sense [89]. Meta-analysis can analyze case studies on historical and future global soil carbon response to land-use change [105].

Land-use change meta-analyses have documented research in the area of soil erosion studies globally. Some studies meta-analytically investigated the rate of sheet and rill erosion in Germany [106]. Comparatively, others used meta-analysis to investigate soil infiltration rate effects in China from comparisons of various studies from LULC changes [85]. The use of soil erosion studies to understand the rates of erosion from global sites showed the estimation method for erosion rates and the rationale for improving the practices and theory on soil erosion studies [107]. Land use meta-analysis studies have been centered around agriculture, from investigations into an intensification of agriculture and global changes [78]. Meta-analysis investigations on crop yield and nitrogen dynamics, using cover crops in fertilizer-intensive cropping systems [108]. This methodology has also been applied in no-till cultivation operations, focusing on cost and carbon benefits [109]. The impact of agricultural policy reform on land prices shows quantitative analysis as a focal area [77] and their effects on plant density on a global scale [110], as well as their concentration of dissolved organic matter impacts as a result of land management [111]. This method has contributed to land-use science and its effects on a global scale. Using case studies and other literature as exploratory focal points for identifying impacts of land use/land cover change on climate change, scientific references of anthropogenic effects on the biosphere have shifted to a new geological epoch [112].

Meta-analytical approaches have informed other researchers' use of multiple studies to observe land-use change, anthropogenic activities, and the impacts on climate change within a particular region. Zanten et al. (2014) conducted a meta-analysis study to examine generic preferences of particular landscape attribute across Europe [113]. The assessment conveyed responses from the general public on their willingness to pay for goods and services provided by the environment. In contrast, another study used meta-analysis to investigated land abandonment effects on plants species richness and animal abundance. The results showed the impacts on the biodiversity of the Mediterranean Basin, specifically on arable land, pastures, agroforestry systems, and other permanent crops [114]. Another looked at biodiversity measures under various land covers, assessing different land-use types [115].

Furthermore, this methodology identified objectives and models adopted in the hydrological response attribute of Mediterranean catchment areas [116]. The use of meta-analysis to show the various impacts of land-use changes on the environment compares and shows similarities in specific areas, but it can be used as a guide for multiple researchers, whether regionally or globally, to synthesize, contrast, and show correlations in specific topical areas, perform quality checks, combine and aggregate information from various studies, and then provide estimates of the average magnitudes on particular topics [90].

3.3.3. Meta-Analysis Related to Data and Methods

According to our research findings, meta-analysis is considered a method of analysis in land-use change [117]. Various research can provide different explanatory interpretations and for land-use change. Analytical differences analysis is dependent on data use. For example, one can use spatial units such as pixel images and political units or inform individual decision-making [118,119]. Remote imaging and maps of land-use changes have equipped researchers to identify, summarize, develop, and document the use of various factors of object-based land-cover image classification with the help of meta-analysis [73]. Their results provided instructions on the use of these classifiers for land cover mapping, whereas another researcher used meta-analysis to provide systematic guidance classification process performance, using research literature to inform supervised per-pixel classification over fifteen years [47]. Meta-analysis has been used to quantify how researchers used various studies, showing LULCC impacts through hydrological influxes on discharge, surface runoff, and low flow in the East African Region [120]. Other researchers have used the meta-analytical approach to research forest cover maps from various data sets over a long period throughout various studies, observation deforestation rate and forest cover change through comparison of land cover images [121]. Conducting meta-analysis at the remote sensing level has been observed over several years to develop land cover models, to analyze unrealized synergies between land change meta-studies and the evaluation, framework, and designs of land change models [117]. Additionally, they navigated data types and relevant research questions for land-use change data, using the typological synthesis approach.

3.4. *Land Use/Land Cover Associated Challenges*

Land use and land cover change remain an urgent environmental challenge related to sustainable management of the earth's surface [120,122]. Anthropogenic activities on a global, regional, and local scale have seen significant landscape changes, land degradation, ecosystem changes, and a shift in the biodiversity of numerous areas that were once forest. These economic and environmental changes to the landscape have characterized LULC changes that provide livelihood (e.g., urban settlements and agriculture enterprises) for persons occupying these land spaces. The last two decades have seen significant encounters of humans and changing land surfaces, and these changes accelerated due to socioeconomic and biophysical drivers from anthropogenic activities [123,124]. Generalized knowledge on the impacts of local, regional, and global land change continues to be an essential challenge of land use science [121,125,126]. Some of the leading documented challenges identified in LULCC are logging, fires, drainage, forest cover change, and other changes to wetlands that degrade soils in cropland areas. Additionally, alterations in these land's volume and beneficial capacity could result in these changes [60,125]. This research investigated LULC change for two areas using meta-analysis, there are "data quality and data consistency". These categories were chosen since they were so familiar in relation to LULC remote sensing and modeling. Significant challenges determined from research in Table 4 have been identified based on specific research articles in the literature review. A word cloud is used in Figure 4 to identify the crucial challenges determined in the research papers, and they are associated with the challenges in landscape monitoring.

Table 4. Showing significant challenges and recommendations.

Category	Major Challenges Highlighted	Recommendation	Citation
Data Quality	• Incomplete data coverage, • Various changes in definitions of categories, • Different methods used by source agencies, • Various data age, • Incompatible classification systems.	The correct classification and standardization of land objects and features.	[4]
	• The difference in datasets for biogeography of contrasting regions.	Measure and determine the impacts of land-use changes on land quality and biogeography.	[29,126]
	• No concept application at the landscape level, • Inappropriate data for quantification.	Engaging frameworks and methods with the use of classification systems to track ecosystem goods and services	[127]
	• Unrecorded and undocumented information.	The classification, quantification, and validation of ecosystem services for past land-use data.	[128,129]
	• Incorrect data, • Use of wrong classification system, • Use of the insufficient resolution.	The use of land cover polygons and valuations in dollars/hectare/year to show the total value of ecosystem services.	[130]
	• The lack of reliable or comprehensive data, • There are different levels of resolution and quality of datasets.	Good data pools are needed to analyzes dynamics between ecosystem services.	[131]
	• Few data sets are designed to provide very similar atmospheres over crops and forests.	Test how models capture LULCC impacts on weather.	[132]
	• Absent of comprehensive knowledge base for datasets associated with remote-sensing.	Correct data must be used at the right time for policy change and decision making (e.g., climate change)	[133]
	• Difficulty in mapping global land -use, • The need for local-based data from local-based study areas are required for producing, • No accurate LULC datasets.	There are databases worldwide that offer free access to current and past information on LULC changes globally.	[40]
	Very High Resolution (VHR) images to develop national, regional, or global maps have proven to be challenging: • The high cost associated with VHR imagery, • Their low spatial extent (a few hundreds of km^2) (Gibbs et al., 2007), • Low availability due to their low temporal resolution and lack of global coverage (Pangra et al., 2015), • The variation of radiometric properties among sensors, • The influence of acquisition conditions (i.e., Sun-scene-sensor angles) (Anser et al., 2003; Barbier et al., 2011; Bastin et al., 2014; Ryan et al., 2016), • Classic atmospheric perturbations (e.g., cloud, fires) (Pangra et al., 2015).	"Collect Earth" as a free search engine for past and present LULC change information can be used for many investigations.It's readily updated and accurate with data at multiple scales.	[134]
	• Time series data (various image composite for land cover mapping); • Movement of algorithms from a research to operational phase (such as data handling and processing).	Progressive work over the years in technology and data availability has seen advance/updated algorithms used for time series data.	[135]

Table 4. *Cont.*

Category	Major Challenges Highlighted	Recommendation	Citation
Data Consistency	• Low accuracy and quality of assessment, • Challenges obtaining national land cover maps for distinctive timestamps.	Accurately mapping global land cover maps.	[136]
	• The challenge of representing decision-making mechanisms in models to show land change.	Model coupling—focused on representing human decision-making, the coupling between human and environmental systems.	[137]
	• Inconsistent global maps.	Mapping projects use accuracy assessment as a tool to accept land cover components.	[36]
	• Discrepancies between maps, • Different features and maps, • Comparison between maps from the same source, • Difference between product features in the same year.	Data consistency is essential when given the capacity to produce maps that have acquired data from a single sensor.	[138]
	• Comparison of different legend information from various classification schemes, • Inconsistencies for land class definitions, • Errors arising from different methods of data collection, • Incomplete compilation of remotely-sensed datasets.	Validation efforts are needed to assess precise accuracy at the regional and global scales for LULC classification.	[133]
	• Large data volume, • Unavailable data in certain seasons, • Challenges obtaining cloud-free images, • Technical difficulties in Landsat satellites, • Revisiting the cycle of the Landsat model, making it more difficult to trust.	The Landsat model needs a continued upgrade.	[139]

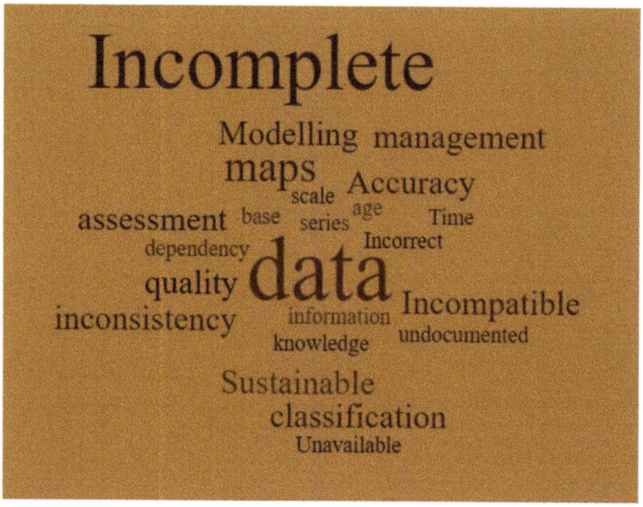

Figure 4. Showing significant challenges associated with Land-use and Land-cover.

3.4.1. Data Quality

The diverse challenges associated with LULCC impact assessment are centered around data quality and consistency. Some conference presentations have shown a need for land classification standardization of land feathers [4]. However, some of the challenges are incomplete data coverage, change in definitions of categories, changes in methods used by source agencies, varying age of data, and incompatible classification systems by agencies. Researchers need to measure and determine the impacts of land-use changes on land quality and biogeography. The life cycle impact assessment (LCIA) is a method developed to undertake such a task. As such, it has provided accurate data regarding land quality changes [29]. However, challenges are present in extricating changes and the quality of biogeography of contrasting regions. Therefore, additional maps and images should distinguish other land feathers, such as varying vegetation, soil types, and climates [29,126]. Various studies have shown engaging frameworks and methods using classification systems to track ecosystem goods and services [43]. At the same time, there is an urgency to centralize and determine land, ecosystem, biodiversity quality, and climate changes under one approach, a significant restriction to ecosystem goods and services would be the application of concepts at the landscape level as a result of inappropriate data for quantification [127]. Quantifying ecosystem services for past land use, one of the severe challenges is understanding what occurred in the past related to land use for a particular area. The information might be known but not documented, characteristics such as land cover practices or specific resources from the land. The information may be passed on orally by indigenous dwellers from generation to generation [128,129].

The use of land cover polygons and valuations in dollars/hectare/year shows the total value of ecosystem services given the land cover type. However, the challenge is that the data for the particular area must be correct, and using the right classification system at a suitable resolution is essential [130]. A central challenge in the earth systems and resources is sustainable land management. There's a high demand for the supply of food and other resources for a growing human population. However, land management has potential negative impacts on the environment and ecosystem globally. Effects are observed in global climate change, the loss of biodiversity, pollution of soil, water, and the atmosphere [15]. The management of ecosystem services in agricultural landscapes is challenging, specifically in agro-ecological conditions and topographic areas; thus, ecosystem services must be assessed by various approaches that provide multi-temporal information at a national level [131]. The only regions globally that are not directly affected or have evaded large-scale land alteration with limited LULC changes are Antarctica, boreal/tundra areas in Siberia, parts of the Amazon, and parts of the Congo [132]. Other regions have been affected globally. Therefore, LULC change impacts must be made locally, with a literature review to understand the effects on a regional and global scale [132]. However, there are some challenges to using Geographic Information System (GIS) data. Researchers must have a comprehensive knowledge base of the datasets for remote sensing. Correct data must be used at the right time for policy change and decision-making, especially in climate change, monitoring deforestation, urban landscape planning, and policy changes by governmental offices [133].

Various research models have been developed to study LULCC, specifically in urban land use and ecosystem services. The model InVEST Framework observes the sensitivity of ecosystem services through spatial resolution from input data [140]. The data models conceptual approaches of the scale dependence of different ecosystem services, the model compares spatial patterns and measured future ecosystem services in a particular area [140]. In contrast, other research findings used meta-analysis to quantify LULC change impacts through hydrological influxes on discharge, surface runoff, and low flow in the East African Region [120]. While global land use may be difficult to map, data from local-based study areas are required to produce accurate LULC datasets. There is a need for datasets to provide information on LULC changes and the human impact [40]. Technological advances for LULC mapping and remote sensing have seen some of the previous challenges annulled. There are databases worldwide that offer free access to current and past information on LULC changes globally. "Collect Earth" has been identified as a free search engine for past and present LULC change information, developed by the Food and Agriculture Organization (FAO) [134]. Collect Earth provides satellite imagery of high spatial and temporal resolution (e.g., Google Earth) and uses archived images with multiple resolutions to enable land monitoring suitability. The need for accurate and up-to-date data is essential to understand changes in LULC monitoring. This will alleviate most of the challenges of understanding continuous differences and help researchers and decision-makers monitor change, observe trends, and efficiently manage land resources.

3.4.2. Data Consistency

Progressive work over the years in technology and data availability has seen its challenges related to data quality. The challenges related to time series data (various image composite for land cover mapping) and challenges involved in the movement of algorithms from research to operational phase (such as data handling and processing) are mostly user-based [135]. There have been various efforts to map global land cover maps [141]. There are still existing challenges regarding the accuracy and quality of assessment, and developing countries face challenges in obtaining national land cover maps for unique timestamps [136].

A major challenge identified in the literature review was representing decision-making in these models as a mechanism by which land changes. Research has focused on human decision-making, namely the coupling between human and environmental systems while answering questions associated with ecological sustainability challenges through model coupling [137]. Mapping projects use accuracy assessment as a tool to accept land cover components. However, the challenges associated with global maps related to data quality are imperative for successful projects [36]. Data consistency is essential when given the capacity to produce maps that have acquired data from a single sensor [138]. While this is an advantage, it poses various challenges (e.g., in previous studies, Bontemps and others compared the GlobCover 2005 and 2009 maps features discrepancies between products).

Various studies have identified land cover classifications with multiple challenges related to mapping, inconsistent class definitions, descriptions in data collection for land characteristics, legends, objects, and other errors with various systems [133]. The use of Landsat data poses some significant challenges, including (1) large data volume in comparison to coarse spatial resolution data, (2) data inconsistency among changing seasons, (3) land images are challenging to obtain as a result of the sixteen-day cycle of Landsat satellites, and (4) there's a great concern for the Landsat satellites that need maintenance. Such technical issues can delay land cover data if not addressed within a timely manner [139]. Another challenge associated with LULC classification systems data is their ability not to be user-friendly. Some databases have unvalidated data or confusion among datasets; in other occurrences, the data are not broken down by specific classes or objects, making it difficult for researchers to understand the best fit for their projects [11]. The challenges described and summarized critical areas of importance in LULCC. While this meta-analysis focused on secondary data from other research articles, it shows essential areas of existing challenges, observed gaps, and even trends among similar research topics, implying that such recognized challenges must be addressed for LULCC.

3.5. Land Use/Land Cover Knowledge Gaps

Land use and land cover have encountered various challenges based on numerous studies. A literature review can determine many knowledge gaps associated with LULCC. Multiple studies have identified specific needs for particular research areas within this discipline to work efficiently. According to Mengist et al. (2020), an accurate meta-analysis with minimal errors can contribute reliable conclusions for a particular area of interest, which leads to the decision-making process [25]. The knowledge gaps for this section separated specific needs into four categories (ecosystem services, forestry, data modeling, and hydrology). Knowledge gaps identified from the literature review were presented in a word cloud showed in Figure 5. The more prominent words indicate how often particular knowledge gaps were highlighted from the synthesis and their importance to land use and land cover.

3.5.1. Ecosystem Services

Initial examination of LULC impacts on ecosystem services has given rise to the need for assessments on severe LULCC within various ecological systems. A valuable tool of assessment identifies the spatiotemporal approach [131]. Ecosystem services would need internationally accepted land cover classification systems with consistent time-series maps, considering their validation and accuracy to identify land characteristics (e.g., distinguish between cropland, fallow, barren, and wasteland) [142]. While there are various areas of ecosystem services, researchers recommend intermittent surveying and mapping of these services provided to monitor their quality, which helps with the overall management and control of ecosystems [143].

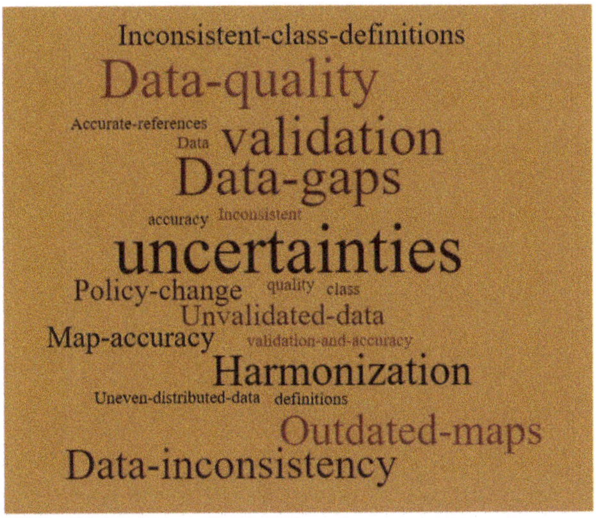

Figure 5. Showing knowledge gaps associated with Land-use and Land-cover.

3.5.2. Forestry

The need in any forestry system fall into either land degradation, deforestation, or restoration of forest lands. While many forested areas have converted to urban and agricultural areas, the meta-analysis for this study has focused on areas within a forest system with identified knowledge gaps in various literature. There is a need for land use planning and reforestation of barren regions, such as degraded lands, hillsides, and the expansion of cultivated land for sustainable resource management [42]. Another knowledge gap has been understanding environmental changes in the forest and predicting climate-induced changes in mature trees [144]. Some authors have identified specific areas of interest. For example, Kayet et al. (2016) undertook research on land surface temperature (LST) and described knowledge gaps related to LST being affected on hilltops, highlighting the evaluation of impacts along with policy changes [145]. Other researchers have focused on analyzing LULC classifications systems, specifically on land products such as map accuracy for these systems to function [36].

3.5.3. Data/Images/Modeling

The meta-analysis also focused on studies in the area of data, image, and modeling. Our analysis identified studies that stressed the need for research in land cover change, with a focus on modeling future spatial patterns [146]. Meanwhile, another focused on spatial data modeling of classification systems, emphasizing areas of need such as accuracy, changing land monitoring, and current [134]. In contrast, another researcher focused on validating of data and models for unexplained points in data [147]. While most of these knowledge gaps are addressed with the multiple studies on LULC data, images, and modeling, research studies have described the need for user-friendly images with high resolution for LULC classification. This contributes to decision-making for urban planners [148]. Another research has focused on land surface climate-change models and their simulation at different scales [149] and provides information for changing landscapes and human impacts [40]. While technological advances have seen data, images, and modeling improvements, their practical use is encouraged for further advancements. This helps to optimize model ensembles, compare images of a given area and identify erroneous output data; validation and accuracy are critical factors for transparent LULC classification of objects and classes [150]. Satellite images need to be preprocessed and accurate, which helps with information advancement, accuracy, reliability, and appropriate estimates of LULCC at the global level [139].

The synthesis identified the following needs related to data, images, and modeling:

1. The general need for:
    - Accurate statistical testing;
    - Identical land-cover configurations;
    - Reduction of model uncertainty;
    - Clear experimental protocols [151].
2. Systematic monitoring and management of land use systems [38].
3. Automating image classification processes for accessible data and processing results in a shorter time [35].
4. The assessment of the performance and sensitivities in LULC classification algorithms [111].
5. Improve accuracy, eliminate uncertainties and discrepancies in the spatio-temporal changes [60].
6. Consistency and comparability of different land cover maps, understanding their suitability and limitations for specific applications [152].
7. Detailed datasets for environmental change studies, resource management, climate modeling, and sustainable development of terrestrial land cover are needed [62].
8. Available data for modeling the advancement and collection of new datasets are needed [8].

3.5.4. Hydrology

The meta-analysis focused on the effect of LULCC on hydrology, with a specific interest in groundwater flow and management. The need for further studies of LULCC assessment impacts on groundwater fluxes is paramount since water management is essential [120]. Researchers are concerned with accurate references of existing information related to landscape changes. There is a need for the assessment and harmonization of information [153]. The measurement of land quality impact indicators by various units to measure pathways affected is also essential for water assessment [29].

4. Discussion

4.1. Land Use and Land Cover Definition

The extent to which "Land-use and Land-cover" definitions differ among research articles has changed over the decades. The direct descriptions for each have been used directly by authors to show differences [8]. There are some uncertainties related to the understanding of each term for definitive use [51]. Researchers have stated that definitions can be contingent on the sector based on their description [154]. This research investigation shows the author's intention as being simple, clear, and concise, where "Land use" is defined base on the activities done by human intervention and "Land cover" corresponds to the physical structures that may occupy the land. This research focused on the synthesis of research journal articles that defined "Land-use and Land-cover". Showing similarities based on keywords used (land use/land cover) and how each research journal interpreted these terms while synthesizing these definitions from the meta-analysis resulted in a general description used for each term. The definitions were placed into a table format to show from researchers (Table 2) and interpreted into a form that researchers can use to define "land use" and "land cover" based on the attributes each term represents (Figure 1).

This synthesis found various similarities as it relates to multiple articles using "land use" as part of the definition to define "Land use" and "cover of land" for the description of "Land cover". Limitations related to this literature review identified research articles collectively defining these terms together, conferring broad base definitions to explain the terms holistically or in contrast to each other [26,41]. Other reports have shown definitions to overlap [155], even established that the purpose for each term is not absolute [51]. This was identified from research findings related to definitions by particular region and research year. The use of a standard and straightforward explanation for "Land-use and Land-cover" should always be clear for readers to understand the dynamics of each description. Hence, the findings from this research propose that definitions for each term should be direct, suitable, and relevant for clear understanding, separated to show a disparity between each term and transparent as it relates to their concepts. Inferences such as "use of land" and "cover of land" will continue to be part of the definitions for both "Land use" and "Land cover". Further, researchers will curtail these definitions to suit specific land management projects types or for decision making.

4.2. Land Use and Land Cover Classification System

The use of "Land-use and Land-cover" classification systems has contributed to the awareness of land-use/land-cover objects by country, continent, and global scale. Countless research articles have shown remote sensing of land images and objects (e.g., forest, cropland, urban areas), land cover mapping databases, and empirical datasets to develop classifications for particular regions. "land use and land cover" classification systems give a spatial analysis of objects for specific landscapes [156], the monitoring of land change from temporal and spatial scales [157], and clear insight on land cover dynamics related to land resources and ecosystem services [158]. This research synthesized articles representing classification systems using the key terms "Land-use and Land-cover" and "classification system and scheme". These classification systems were recognized based on their use by country, regionally, and globally. Yang et al. (2017) have identified a solid base for our research findings on classification systems used globally [49]. Remote sensing technology for LULC mapping across a range of spatial scales has guided many nations to establish land cover mapping and monitoring programs that use moderate resolution satellite data [43]. These LULC information sources have been vital for various disciplines (e.g., urban planning, forestry, etc.) and regularly updated for land monitoring. Remote sensing technological advances have led to a more accurate assessment from LULC classification systems and model simulations. This aids in improved monitoring, observation, and analysis of land used and alteration by anthropogenic activities [34].

The literature review identified, collected, evaluated, and review various research articles that give specific information on each classification system related to (1) the year established; (2) the country, region, or global use; and (3) the scale of use. The synthesis provided information on classification systems, with a direct approach to LULC changes, whether hierarchical or used based on their ease of access, spatial scale, and user-friendly qualities. It identified the ability for classification systems suited for a country as against continental use. The synthesis identified key challenges and limitations related to gathering information for each classification system. Contingent on the region of use, classifications are either "scheme" or "system". This is highly dependent on their temporal scale of use, type of projects, meeting requirements, and bypass limitations [11]. This synthesis showed similarities of particular classification systems among research articles used in this study. While many classification systems are associated with their country of origin, the literature review shows that researchers favor particular ones. Primarily, they are chosen based on their constant upgrades, accuracy, validation, and fitness of use.

4.3. Land Use and Land Cover Meta-Analysis

This research took an analysis of various "Land-use and Land-cover" meta-studies as a means to investigate (1) direct changes, (2) indirect changes, (3) data and method changes of LULC. These meta-studies were identified based on each category and placed into a tree diagram (Figure 2), further breaking down each to show how each category's meta-studies were synthesized. Meta-studies used meta-analysis to congregate various data sources, methods, and ideas to advance a missing knowledge base for a particular study area [119]. This study used a literature review to show the various meta-analysis studies for LULC. A total of fifty-five (55) articles were downloaded and sorted, forty-eight (48) were appraised and used to build the analysis tree (Figure 2). A timestamp for the appraised articles presented in Figure 3 is for the years 2000 to 2019. They were compiled on their synergistic efforts in LULC studies, their effects on direct and indirect landscape changes and their affiliation to land use/land cover remote sensing and modeling.

This research showed how various meta-analyses for each category used different protocols for information collection and synthesis. This research identifies LULC meta-studies in areas of direct land influence (e.g., forest related and for ecosystem services) and indirect land influences (e.g., biodiversity and carbon sequestration). While researchers focused on LULC modeling and dataset imagery [119], this research used narrow criteria to identify ideal meta-studies for this category. Therefore, there were few meta-studies. However, it would be advisable to adopt different measures, focusing on future research objectives related to modeling and remote sensing. The benefits of this analysis will help in understanding the various complexities for synthesizing meta-studies in LULCC. A literature review of meta-studies can provide inferences for researchers determining research journals as "best use" for particular interest categories. This analysis shows how studies in LULC were more fitted to direct and indirect land use and land cover. Most studies complied information for decision-making for land management, ecosystem services, and biodiversity. The next step relates to meta-studies used as a premise to understand land monitoring for various ecosystems. The information is regularly updated with more research conducted in these areas internationally and at the local level. Further synthesis revealed meta-studies that identified necessary action for specific land projects, leading to proper decision-making and, in some cases, policy modification or changes.

4.4. Land Use and Land Cover Challenges

Research findings have identified several challenges related to "Land-use and Land-cover", some were minor, while others were crucial. The literature review identified significant articles that comprise LULC challenges; the PSALSAR framework characterized the challenges into two categories for LULC, namely (1) data quality and (2) data consistency (Table 4). The focus of this synthesis considered remote sensing and imaging in LULC. Our investigation identified some keywords associated with data quality and consistency to develop a word cloud (Figure 4), where significant challenges for each category were highlighted and deemed necessary. There were limitations since our synthesis was category-specific and focused on challenges related to data quality and consistency. However, these categories identified are significant in the field of LULC [119].

The importance of data value and verification for GIS has increased with the development of satellite data for land-cover mapping [159]. As a result, large amounts of data are required for global land-cover mapping [160]. This research highlighted significant challenges related to data accuracy and consistency. The analysis, classification, and description of valid data are essential. Therefore, the focus on data processing, validation, verification, access, and accuracy are vital to address major challenges in LULC changes. This information can determine how LULC projects form conclusions and make decisions. While some of the challenges identified in this study are continuously being addressed, research synthesis describes and determines which challenges are still primary concerns. The idea is to highlight significant problems and encourage researchers to approach continued technological advancement in such areas. Comparable findings reported by Verburg et al. (2011) support that the purpose of LULC challenges is to see them dissolved over time [13].

4.5. Land Use and Land Cover Knowledge Gaps

The combination of Land use/Land cover challenges and knowledge gaps correlates to each other. While many knowledge gaps associated with remote sensing studies are related to LULC [161,162], further LULC investigations have various knowledge gaps. Related to the importance of our studies, we identified significant articles in four categories and created a word cloud (Figure 5) based on the critical knowledge gaps for "land Use and land Cover". Our synthesis found most knowledge gaps corresponded to the category of remote sensing, specifically for data, imaging, and modeling. The information identified specific knowledge gaps that were time-sensitive and relevant. Our finding highlights many knowledge gaps to be data-related: Verification/validation of data, data quality, data harmonization, data gaps, data inconsistency, and the uncertainties associated with data were the major knowledge gaps from our analysis. Our findings revealed that there was no single approach to addressing the knowledge gaps identified. There need for comparison and assessment of updated data to improve data quality is paramount to enhance information for LULC projects [13].

Each category gives information on knowledge gaps and the importance of them being addressed critically. The results indicate that some knowledge gaps will be addressed over time. However, this accomplishment will occur with ongoing investigations in LULC [62]. Research findings did not specify a single path to address the knowledge gaps. The consensus was clear that if researchers did not focus on them, they would continue to cause significant consequences in LULC investigations for global change [144]. Consequently, the significant gaps identified have provided areas of analysis that represent the challenges in this field. Data quality, availability, and harmonization emerged from our synthesis as direct reasons for the LULC challenges. This literature review presented the necessary information that requires focus and assessment for LULC research.

5. Conclusions

The primary objective of the present work was to present five critical areas of LULC in the form of a literature review. Enormous work on LULC has been executed over the decades in numerous topical areas by multiple researchers over time. The PSALSAR framework for this systemic review will support scientific and straightforward synthesis. This synthesis has proven to be a method to synthesize, organize, and present LULC information and make inferences related to specific areas of interest. Our study showed a systemic approach to how secondary information on LULC is presented so researchers can interpret and make assumptions. Figure 6 shows a step-by-step process of our research procedure, as well as how secondary information from research articles was collected, appraised, and presented on the five topical areas of our research interest. This flowchart provides a holistic view of how research articles were used to answer questions in each specific category.

Our research findings recommend understanding various land use/land cover areas. A literature review helps synthesize, compare, and present information among multiple studies. Our bias related to articles selected was based on the keywords for each to keep our research concise. We would recommend using more keywords for LULC studies to get a broader range of articles for analysis. Future research for LULC studies should focus on remote sensing, land-cover monitoring, and their effects on ecosystem services. We recommend a research procedure presented in Figure 6. This will aid the researcher in analyzing the information required for new LULC initiatives accurately. This methodology can find similarities for missing knowledge in LULC studies, giving researchers a base method to apply detailed synthesis. The identification and assortment of summarized knowledge in particular LULC research areas are vital. Thus, using this methodological framework, LULC investigations will report findings accurately and provide inputs for environmental monitoring.

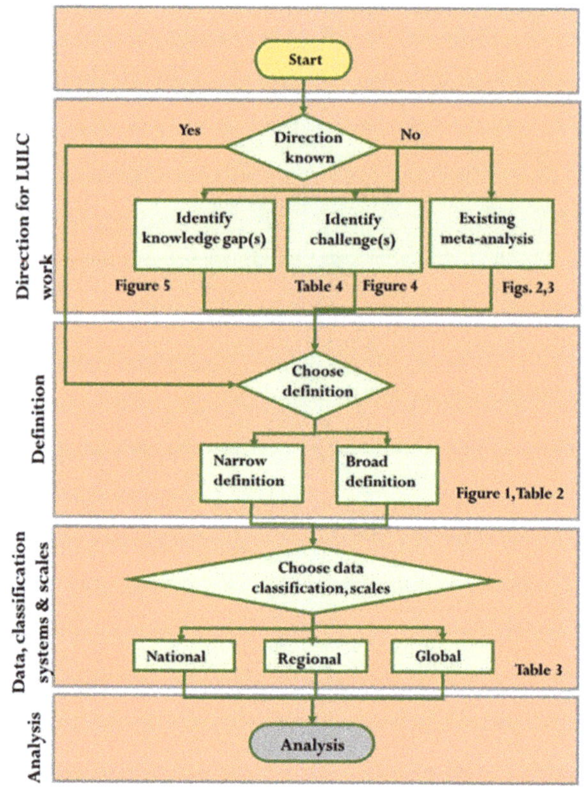

Figure 6. Flowchart of the synthesized research process, the procedure developed for a continuous flow of information from information acquisition to analysis.

Author Contributions: Conceptualization, visualization, methodology, R.N. and A.A.; formal analysis, R.N. and K.L.; resources, supervision, project administration, funding acquisition, A.A.; data curation, writing—original draft preparation, R.N., K.L. and A.A.; writing—review and editing, M.O., N.J. and E.J. All authors have read and agreed to the published version of the manuscript.

Funding: This material is based upon work partially supported by the USDANIFA capacity building grant 2017-38821-26405, USDA-NIFA Evans-Allen Project, Grant 11979180/2016-01711, USDA-NIFA grant No. 2018-68002-27920, and National Science Foundation under Grant No.1735235 awarded as part of the National Science Foundation Research Traineeship.

Institutional Review Board Statement: Not applicable.

Informed Consent Statement: Not applicable.

Data Availability Statement: Data here is the literature and available online.

Acknowledgments: The author express the gratitude to the three anonymous reviewers for their constructive comments and suggestion on the earlier draft of the paper. Sharma's and Deepa's support for scientific & scholarly editing of this draft.

Conflicts of Interest: The authors declare no conflict of interest.

References

1. Keil, A. OpenStreetMap Land Classification System (OSM-LCS). 2016. Available online: https://labor.beuth-hochschule.de/fileadmin/labor/gem/arbeiten/Keil_Master_Arbeit.pdf (accessed on 17 September 2021).

2. Meyfroidt, P.; Roy Chowdhury, R.; de Bremond, A.; Ellis, E.C.; Erb, K.-H.; Filatova, T.; Garrett, R.D.; Grove, J.M.; Heinimann, A.; Kuemmerle, T.; et al. Middle-Range Theories of Land System Change. *Glob. Environ. Chang.* **2018**, *53*, 52–67. [CrossRef]
3. Anandhi, A.; Douglas-Mankin, K.R.; Srivastava, P.; Aiken, R.M.; Senay, G.; Leung, L.R.; Chaubey, I. DPSIR-ESA Vulnerability Assessment (DEVA) Framework: Synthesis, Foundational Overview, and Expert Case Studies. *Trans. ASABE* **2020**, *63*, 741–752. [CrossRef]
4. Anderson, J.R. *A Land Use and Land Cover Classification System for Use with Remote Sensor Data*; U.S. Government Printing Office: Washington, DC, USA, 1976.
5. Erb, K.-H.; Luyssaert, S.; Meyfroidt, P.; Pongratz, J.; Don, A.; Kloster, S.; Kuemmerle, T.; Fetzel, T.; Fuchs, R.; Herold, M.; et al. Land Management: Data Availability and Process Understanding for Global Change Studies. *Glob. Chang. Biol.* **2017**, *23*, 512–533. [CrossRef] [PubMed]
6. Ran, Y.H.; Li, X.; Lu, L.; Li, Z.Y. Large-Scale Land Cover Mapping with the Integration of Multi-Source Information Based on the Dempster–Shafer Theory. *Int. J. Geogr. Inf. Sci.* **2012**, *26*, 169–191. [CrossRef]
7. Luyssaert, S.; Jammet, M.; Stoy, P.C.; Estel, S.; Pongratz, J.; Ceschia, E.; Churkina, G.; Don, A.; Erb, K.; Ferlicoq, M.; et al. Land Management and Land-Cover Change Have Impacts of Similar Magnitude on Surface Temperature. *Nat. Clim. Chang.* **2014**, *4*, 389–393. [CrossRef]
8. Pongratz, J.; Dolman, H.; Don, A.; Erb, K.-H.; Fuchs, R.; Herold, M.; Jones, C.; Kuemmerle, T.; Luyssaert, S.; Meyfroidt, P.; et al. Models Meet Data: Challenges and Opportunities in Implementing Land Management in Earth System Models. *Glob. Chang. Biol.* **2018**, *24*, 1470–1487. [CrossRef]
9. Jansen, L.J.M.; Gregorio, A.D. Parametric Land Cover and Land-Use Classifications as Tools for Environmental Change Detection. *Agric. Ecosyst. Environ.* **2002**, *91*, 89–100. [CrossRef]
10. Meyer, W.B.; Turner, B.L. Human Population Growth and Global Land-Use/Cover Change. *Annu. Rev. Ecol. Syst.* **1992**, *23*, 39–61. [CrossRef]
11. Gregorio, A.D.; Jansen, L.J.M. A New Concept for a Land Cover Classification System. *Land* **1998**, *2*, 55–65.
12. Lambin, E.F.; Rounsevell, M.D.A.; Geist, H.J. Are Agricultural Land-Use Models Able to Predict Changes in Land-Use Intensity? *Agric. Ecosyst. Environ.* **2000**, *82*, 321–331. [CrossRef]
13. Verburg, P.H.; Neumann, K.; Nol, L. Challenges in Using Land Use and Land Cover Data for Global Change Studies: Land Use and Land Cover Data For Global Change Studies. *Glob. Chang. Biol.* **2011**, *17*, 974–989. [CrossRef]
14. Pradeep Kumar, B.; Raghu Babu, K.; Rajasekhar, M.; Ramachandra, M. Change Detection of Land Use/Land Cover Using Geospatial Techniques: A Case Study of Narpala Mandal, Anantapur District, Andhra Pradesh, IndiâPradeep. *J. Remote Sens. GIS* **2019**, *10*, 6–12.
15. Ellis, E.; Pontius, R. Land-Use and Land-Cover Change—Encyclopedia of Earth. Available online: https://web.archive.org/web/20070503192914/http://www.eoearth.org/article/Land-use_and_land-cover_change (accessed on 12 November 2020).
16. United Nations SDG Indicators. Available online: https://unstats.un.org/sdgs/report/2020/ (accessed on 17 November 2020).
17. US EPA. Land Use. Available online: https://www.epa.gov/report-environment/land-use (accessed on 17 November 2020).
18. Dimyati, M.; Mizuno, K.; Kobayashi, S.; Kitamura, T. An Analysis of Land Use/Cover Change in Indonesia. *Int. J. Remote Sens.* **1996**, *17*, 931–944. [CrossRef]
19. Singh, R.K.; Sinha, V.S.P.; Joshi, P.K.; Kumar, M. A Multinomial Logistic Model-Based Land Use and Land Cover Classification for the South Asian Association for Regional Cooperation Nations Using Moderate Resolution Imaging Spectroradiometer Product. *Environ. Dev. Sustain.* **2020**, *23*, 6106–6127. [CrossRef]
20. Sreedhar, Y.; Nagaraju, A.; Murali Krishna, G. An Appraisal of Land Use/Land Cover Change Scenario of Tummalapalle, Cuddapah Region, India—A Remote Sensing and GIS Perspective. *Adv. Remote Sens.* **2016**, *5*, 232–245. [CrossRef]
21. Duhamel, C. Duhamel Land Use and Land Cover, Including Their Classification. *Encylopedia Life Support Syst.* **2012**, *1*, 9.
22. Herold, M.; Latham, J.S.; Di Gregorio, A.; Schmullius, C.C. Evolving Standards in Land Cover Characterization. *J. Land Use Sci.* **2006**, *1*, 157–168. [CrossRef]
23. Borenstein, M.; Hedges, L.V.; Higgins, J.P.T.; Rothstein, H.R. *Introduction to Meta-Analysis*; John Wiley & Sons: Hoboken, NJ, USA, 2011; ISBN 978-1-119-96437-7.
24. De Stefano, A.; Jacobson, M.G. Soil Carbon Sequestration in Agroforestry Systems: A Meta-Analysis. *Agrofor. Syst.* **2017**, *92*, 285–299. [CrossRef]
25. Mengist, W.; Soromessa, T.; Legese, G. Method for Conducting Systematic Literature Review and Meta-Analysis for Environmental Science Research. *MethodsX* **2020**, *7*, 100777. [CrossRef] [PubMed]
26. Mohd Hasmadi, I.; Pakhriazad, H.Z.; Shahrin, M.F. Evaluating Supervised and Unsupervised Techniques for Land Cover Mapping Using Remote Sensing Data. *Geogr. Malays. J. Soc. Space* **2009**, *5*, 1–10.
27. Briassoulis, H. Land Use, Land Cover and Soil Sciences—Volume I. In *Land Use, Land Cover and Soil Sciences—Volume I: Land Cover, Land Use and the Global Change*; EOLSS Publishers Co.: Abu Dhabi, United Arab Emirates, 2009; ISBN 978-1-84826-235-5.
28. Veeraswamy, G.; Nagaraju, A.; Balaji, E.; Sreedhar, Y. Land Use and Land Cover Analysis Using Remote Sensing and GIS:A Case Study In Gudur Area, Nellore District, Andhra Pradesh, India. *Int. J. Res.* **2017**, *4*, 11.
29. Milài Canals, L.; Bauer, C.; Depestele, J.; Dubreuil, A.; Freiermuth Knuchel, R.; Gaillard, G.; Michelsen, O.; Müller-Wenk, R.; Rydgren, B. Key Elements in a Framework for Land Use Impact Assessment Within LCA (11 pp). *Int. J. Life Cycle Assess.* **2007**, *12*, 5–15. [CrossRef]

30. Roy, P.S.; Giriraj, A. Land Use and Land Cover Analysis in Indian Context. *J. Appl. Sci.* **2008**, *8*, 1346–1353. [CrossRef]
31. Hung, M.-C. Urban Land Cover Analysis from Satellite Images. *Conf. Proc.* **2002**, *6*. Available online: https://www.semanticscholar.org/paper/URBAN-LAND-COVER-ANALYSIS-FROM-SATELLITE-IMAGES-Hung/0459d0478b22ca5b628fba3d71c26ad1adf692c1 (accessed on 17 September 2021).
32. Gregorio, A.D. *Land Cover Classification System: Classification Concepts and User Manual: LCCS*; Food & Agriculture Org.: Yokohama, Japan, 2005; ISBN 978-92-5-105327-0.
33. Longley, P. Global Mapping Of Human Settlement: Experiences, Datasets, and Prospects: Book Reviews. *Photogramm. Rec.* **2010**, *25*, 205–207. [CrossRef]
34. Drummond, M.A.; Loveland, T.R. Land-Use Pressure and a Transition to Forest-Cover Loss in the Eastern United States. *BioScience* **2010**, *60*, 286–298. [CrossRef]
35. Huth, J.; Kuenzer, C.; Wehrmann, T.; Gebhardt, S.; Tuan, V.Q.; Dech, S. Land Cover and Land Use Classification with TWOPAC: Towards Automated Processing for Pixel- and Object-Based Image Classification. *Remote Sens.* **2012**, *4*, 2530–2553. [CrossRef]
36. Congalton, R.; Gu, J.; Yadav, K.; Thenkabail, P.; Ozdogan, M. Global Land Cover Mapping: A Review and Uncertainty Analysis. *Remote Sens.* **2014**, *6*, 12070–12093. [CrossRef]
37. Sun, L.; Schulz, K. The Improvement of Land Cover Classification by Thermal Remote Sensing. *Remote Sens.* **2015**, *7*, 8368–8390. [CrossRef]
38. Manjunatha, M.C.; Narayan, J. Climate change and its impact on groundwater table fluctuation in precambrian terrain of chitradurga district, karnataka, india using geomatics application. *Int. J. Civ. Eng. Technol.* **2015**, *6*, 83–96.
39. Kumar Ranjan, A.K.; Anand, A.; Kumar Singh, R. Centre for Land Resource Management, Central University of Jharkhand, Brambe 835205(JH), India LU/LC Change Detection and Forest Degradation Analysis in Dalma Wildlife Sanctuary Using 3S Technology: A Case Study in Jamshedpur-India. *AIMS Geosci.* **2016**, *2*, 273–285. [CrossRef]
40. Fritz, S.; See, L.; Perger, C.; McCallum, I.; Schill, C.; Schepaschenko, D.; Duerauer, M.; Karner, M.; Dresel, C.; Laso-Bayas, J.-C.; et al. A Global Dataset of Crowdsourced Land Cover and Land Use Reference Data. *Sci. Data* **2017**, *4*, 170075. [CrossRef] [PubMed]
41. Wulder, M.A.; Coops, N.C.; Roy, D.P.; White, J.C.; Hermosilla, T. Land Cover 2.0. *Int. J. Remote Sens.* **2018**, *39*, 4254–4284. [CrossRef]
42. Miheretu, B.A.; Yimer, A.A. Land Use/Land Cover Changes and Their Environmental Implications in the Gelana Sub-Watershed of Northern Highlands of Ethiopia. *Environ. Syst. Res.* **2018**, *6*, 7. [CrossRef]
43. Hermosilla, T.; Wulder, M.A.; White, J.C.; Coops, N.C.; Hobart, G.W. Disturbance-Informed Annual Land Cover Classification Maps of Canada's Forested Ecosystems for a 29-Year Landsat Time Series. *Can. J. Remote Sens.* **2018**, *44*, 67–87. [CrossRef]
44. Chowdhury, M.; Hasan, M.E.; Abdullah-Al-Mamun, M.M. Land Use/Land Cover Change Assessment of Halda Watershed Using Remote Sensing and GIS. *Egypt. J. Remote Sens. Space Sci.* **2020**, *23*, 63–75. [CrossRef]
45. MohanRajan, S.N.; Loganathan, A.; Manoharan, P. Survey on Land Use/Land Cover (LU/LC) Change Analysis in Remote Sensing and GIS Environment: Techniques and Challenges. *Environ. Sci. Pollut. Res.* **2020**, *27*, 29900–29926. [CrossRef]
46. Sokal, R.R. Classification: Purposes, Principles, Progress, Prospects. *Science* **1974**, *185*, 1115–1123. [CrossRef]
47. Khatami, R.; Mountrakis, G.; Stehman, S.V. A Meta-Analysis of Remote Sensing Research on Supervised Pixel-Based Land-Cover Image Classification Processes: General Guidelines for Practitioners and Future Research. *Remote Sens. Environ.* **2016**, *177*, 89–100. [CrossRef]
48. Homer, C.; Dewitz, J.; Yang, L.; Jin, S.; Danielson, P.; Xian, G.; Coulston, J.; Herold, N.; Wickham, J.; Megown, K. Completion of the 2011 National Land Cover Database for the Conterminous United States—Representing a Decade of Land Cover Change Information. *Photogramm. Eng. Remote Sens.* **2015**, *81*, 346–354. [CrossRef]
49. Yang, H.; Li, S.; Chen, J.; Zhang, X.; Xu, S. The Standardization and Harmonization of Land Cover Classification Systems towards Harmonized Datasets: A Review. *ISPRS Int. J. Geo-Inf.* **2017**, *6*, 154. [CrossRef]
50. Velázquez, A.; Mas, J.-F.; Bocco, G.; Palacio-Prieto, J.L. Mapping Land Cover Changes in Mexico, 1976–2000 and Applications for Guiding Environmental Management Policy. *Singap. J. Trop. Geogr.* **2010**, *31*, 152–162. [CrossRef]
51. Rutledge, D.; Price, R.; Briggs, C.; Cowell, S. Geospatial Land-Use Classification for New Zealand: Review and Recommendations. *Off. Stat. Res. Ser.* **2009**, *5*, 9.
52. Arozarena, A.; Villa, G.; Valcárcel, N.; Peces, J.J.; Domenech, E.; Porcuna, A. New concept on land cover/land use information system in Spain. Design and production. *Cent. Remote. Sens. Land Surf.* **2006**, *28*, 30.
53. Lin, G.C.S.; Ho, S.P.S. China's Land Resources and Land-Use Change: Insights from the 1996 Land Survey. *Land Use Policy* **2003**, *20*, 87–107. [CrossRef]
54. Souza, C.M.; Shimbo, J.; Rosa, M.R.; Parente, L.L.; Alencar, A.; Rudorff, B.F.T.; Hasenack, H.; Matsumoto, M.; Ferreira, L.; Souza-Filho, P.W.M.; et al. Reconstructing Three Decades of Land Use and Land Cover Changes in Brazilian Biomes with Landsat Archive and Earth Engine. *Remote Sens.* **2020**, *12*, 2735. [CrossRef]
55. Büttner, G. CORINE Land Cover and Land Cover Change Products. In *Land Use and Land Cover Mapping in Europe: Practices & Trends*; Manakos, I., Braun, M., Eds.; Remote Sensing and Digital Image Processing; Springer: Dordrecht, The Netherlands, 2014; pp. 55–74. ISBN 978-94-007-7969-3.
56. Schößer, B.; Helming, K.; Wiggering, H. Assessing Land Use Change Impacts—A Comparison of the SENSOR Land Use Function Approach with Other Frameworks. *J. Land Use Sci.* **2010**, *5*, 159–178. [CrossRef]

57. Li, C.; Wang, J.; Wang, L.; Hu, L.; Gong, P. Comparison of Classification Algorithms and Training Sample Sizes in Urban Land Classification with Landsat Thematic Mapper Imagery. *Remote Sens.* **2014**, *6*, 964–983. [CrossRef]
58. Bohn, T.J.; Vivoni, E.R. MOD-LSP, MODIS-Based Parameters for Hydrologic Modeling of North American Land Cover Change. *Sci. Data* **2019**, *6*, 144. [CrossRef]
59. Sleeter, B.M.; Liu, J.; Daniel, C.; Rayfield, B.; Sherba, J.; Hawbaker, T.J.; Zhu, Z.; Selmants, P.C.; Loveland, T.R. Effects of Contemporary Land-Use and Land-Cover Change on the Carbon Balance of Terrestrial Ecosystems in the United States. *Environ. Res. Lett.* **2018**, *13*, 045006. [CrossRef]
60. Rai, R.; Zhang, Y.; Paudel, B.; Li, S.; Khanal, N. A Synthesis of Studies on Land Use and Land Cover Dynamics during 1930–2015 in Bangladesh. *Sustainability* **2017**, *9*, 1866. [CrossRef]
61. Stefanov, W.L.; Ramsey, M.S.; Christensen, P.R. Monitoring Urban Land Cover Change: An Expert System Approach to Land Cover Classification of Semiarid to Arid Urban Centers. *Remote Sens. Environ.* **2001**, *13*, 173–185. [CrossRef]
62. Manakos, I.; Tomaszewska, M.; Gkinis, I.; Brovkina, O.; Filchev, L.; Genc, L.; Gitas, I.; Halabuk, A.; Inalpulat, M.; Irimescu, A.; et al. Comparison of Global and Continental Land Cover Products for Selected Study Areas in South Central and Eastern European Region. *Remote Sens.* **2018**, *10*, 1967. [CrossRef]
63. European Commission, Statistical Office of the European Communities. *Land Cover and Land Use Information Systems for European Union Policies Needs*; Theme 5—Agriculture and Fisheries; Office for Official Publications of the European Communities: Luxembourg; Bernan Associates: Lanham, MD, USA, 1999; ISBN 978-92-828-7450-9.
64. Tateishi, R. Contribution to the Development of Global Land Related Dataset from Asia. *Proc. KSRS Conf.* **1998**, 116–121.
65. Latifovic, R.; Homer, C.; Ressl, R.; Pouliot, D.A.; Hossian, S.; Colditz, R.; Olthof, I.; Chandra, G.; Victoria, A. North American Land Change Monitoring System. *Remote Sens. Land Use Land Cover Princ. Appl.* **2012**, 303–324. [CrossRef]
66. Tateishi, R. Urban Mapping from Space. 2018. Available online: https://www.ide.go.jp/library/English/Publish/Reports/Brc/pdf/23_01.pdf (accessed on 18 September 2021).
67. Conrad, E.; Christie, M.; Fazey, I. Is Research Keeping up with Changes in Landscape Policy? A Review of the Literature. *J. Environ. Manag.* **2011**, *92*, 2097–2108. [CrossRef]
68. Mora, B.; Tsendbazar, N.-E.; Herold, M.; Arino, O. Global Land Cover Mapping: Current Status and Future Trends. In *Land Use and Land Cover Mapping in Europe*; Manakos, I., Braun, M., Eds.; Remote Sensing and Digital Image Processing; Springer: Dordrecht, The Netherlands, 2014; Volume 18, pp. 11–30. ISBN 978-94-007-7968-6.
69. Sulla-Menashe, D.; Gray, J.M.; Abercrombie, S.P.; Friedl, M.A. Hierarchical Mapping of Annual Global Land Cover 2001 to Present: The MODIS Collection 6 Land Cover Product. *Remote Sens. Environ.* **2019**, *222*, 183–194. [CrossRef]
70. Pérez-Hoyos, A.; Rembold, F.; Kerdiles, H.; Gallego, J. Comparison of Global Land Cover Datasets for Cropland Monitoring. *Remote Sens.* **2017**, *9*, 1118. [CrossRef]
71. Loveland, T.R.; Reed, B.C.; Brown, J.F.; Ohlen, D.O.; Zhu, Z.; Yang, L.; Merchant, J.W. Development of a Global Land Cover Characteristics Database and IGBP DISCover from 1 Km AVHRR Data. *Int. J. Remote Sens.* **2000**, *21*, 1303–1330. [CrossRef]
72. McCallum, I.; Obersteiner, M.; Nilsson, S.; Shvidenko, A. A Spatial Comparison of Four Satellite Derived 1km Global Land Cover Datasets. *Int. J. Appl. Earth Obs. Geoinf.* **2006**, *8*, 246–255. [CrossRef]
73. Ma, L.; Li, M.; Ma, X.; Cheng, L.; Du, P.; Liu, Y. A Review of Supervised Object-Based Land-Cover Image Classification. *ISPRS J. Photogramm. Remote Sens.* **2017**, *130*, 277–293. [CrossRef]
74. Munteanu, C.; Kuemmerle, T.; Boltiziar, M.; Butsic, V.; Gimmi, U.; Lúboš, H.; Kaim, D.; Király, G.; Konkoly-Gyuró, É.; Kozak, J., et al. Forest and Agricultural Land Change in the Carpathian Region—A Meta-Analysis of Long-Term Patterns and Drivers of Change. *Land Use Policy* **2014**, *38*, 685–697. [CrossRef]
75. Crouzeilles, R.; Curran, M.; Ferreira, M.S.; Lindenmayer, D.B.; Grelle, C.E.V.; Rey Benayas, J.M. A Global Meta-Analysis on the Ecological Drivers of Forest Restoration Success. *Nat. Commun.* **2016**, *7*, 11666. [CrossRef] [PubMed]
76. Busch, J.; Ferretti-Gallon, K. What Drives Deforestation and What Stops It? A Meta-Analysis. *Rev. Environ. Econ. Policy* **2017**, *11*, 3–23. [CrossRef]
77. Oltmer, K.; Florex, R. Impacts of Agricultural Policy Reform on Land Prices: A Quantitative Analysis of the Literature. *Electron. Resour.* **2001**, *28*. Available online: https://ageconsearch.umn.edu/record/20507/ (accessed on 17 September 2021).
78. Keys, E.; McConnell, W.J. Global Change and the Intensification of Agriculture in the Tropics. *Glob. Environ. Chang.* **2005**, *15*, 320–337. [CrossRef]
79. Borchard, N.; Schirrmann, M.; Cayuela, M.L.; Kammann, C.; Wrage-Mönnig, N.; Estavillo, J.M.; Fuertes-Mendizábal, T.; Sigua, G.; Spokas, K.; Ippolito, J.A.; et al. Biochar, Soil and Land-Use Interactions That Reduce Nitrate Leaching and N_2O Emissions: A Meta-Analysis. *Sci. Total Environ.* **2019**, *651*, 2354–2364. [CrossRef]
80. Osenberg, C.W.; Sarnelle, O.; Cooper, S.D.; Holt, R.D. Resolving Ecological Questions Through Meta-Analysis: Goals, Metrics, and Models. *Ecology* **1999**, *80*, 1105–1117. [CrossRef]
81. Ilstedt, U.; Malmer, A.; Verbeeten, E.; Murdiyarso, D. The Effect of Afforestation on Water Infiltration in the Tropics: A Systematic Review and Meta-Analysis. *For. Ecol. Manag.* **2007**, *251*, 45–51. [CrossRef]
82. Seto, K.C.; Fragkias, M.; Güneralp, B.; Reilly, M.K. A Meta-Analysis of Global Urban Land Expansion. *PLoS ONE* **2011**, *6*, e23777. [CrossRef]
83. Rudel, T.K. Changing Agents of Deforestation: From State-Initiated to Enterprise Driven Processes, 1970–2000. *Land Use Policy* **2007**, *24*, 35–41. [CrossRef]

84. Chen, H.; Fleskens, L.; Baartman, J.; Wang, F.; Moolenaar, S.; Ritsema, C. Impacts of Land Use Change and Climatic Effects on Streamflow in the Chinese Loess Plateau: A Meta-Analysis. *Sci. Total Environ.* **2020**, *703*, 134989. [CrossRef]
85. Sun, D.; Yang, H.; Guan, D.; Yang, M.; Wu, J.; Yuan, F.; Jin, C.; Wang, A.; Zhang, Y. The Effects of Land Use Change on Soil Infiltration Capacity in China: A Meta-Analysis. *Sci. Total Environ.* **2018**, *626*, 1394–1401. [CrossRef] [PubMed]
86. Bárcena, T.G.; Kiær, L.P.; Vesterdal, L.; Stefánsdóttir, H.M.; Gundersen, P.; Sigurdsson, B.D. Soil Carbon Stock Change Following Afforestation in Northern Europe: A Meta-Analysis. *Glob. Chang. Biol.* **2014**, *20*, 2393–2405. [CrossRef] [PubMed]
87. Liao, C.; Luo, Y.; Fang, C.; Li, B. Ecosystem Carbon Stock Influenced by Plantation Practice: Implications for Planting Forests as a Measure of Climate Change Mitigation. *PLoS ONE* **2010**, *5*, e10867. [CrossRef] [PubMed]
88. Locatelli, B.; Vignola, R. Managing Watershed Services of Tropical Forests and Plantations: Can Meta-Analyses Help? *For. Ecol. Manag.* **2009**, *258*, 1864–1870. [CrossRef]
89. Guo, L.B.; Gifford, R.M. Soil Carbon Stocks and Land Use Change: A Meta Analysis. *Glob. Chang. Biol.* **2002**, *8*, 345–360. [CrossRef]
90. Nyawira, S.S.; Nabel, J.E.M.S.; Don, A.; Brovkin, V.; Pongratz, J. Soil Carbon Response to Land-Use Change: Evaluation of a Global Vegetation Model Using Observational Meta-Analyses. *Biogeosciences* **2016**, *13*, 5661–5675. [CrossRef]
91. Johnson, D.W.; Curtis, P.S. Effects of Forest Management on Soil C and N Storage: Meta Analysis. *For. Ecol. Manag.* **2001**, *140*, 227–238. [CrossRef]
92. Luo, Y.; Hui, D.; Zhang, D. Elevated CO_2 stimulates net accumulations of carbon and nitrogen in land ecosystems: A meta-analysis. *Ecology* **2006**, *87*, 53–63. [CrossRef]
93. Van Lent, J.; Hergoualc'h, K.; Verchot, L.V. Reviews and Syntheses: Soil N_2O and NO Emissions from Land Use and Land-Use Change in the Tropics and Subtropics: A Meta-Analysis. *Biogeosciences* **2015**, *12*, 7299–7313. [CrossRef]
94. Van Kooten, G.C.; Eagle, A.J.; Manley, J.; Smolak, T. How Costly Are Carbon Offsets? A Meta-Analysis of Carbon Forest Sinks. *Environ. Sci. Policy* **2004**, *7*, 239–251. [CrossRef]
95. Laganiere, J.; Angers, D.A.; Pare, D. Carbon Accumulation in Agricultural Soils after Afforestation: A Meta-Analysis: Soc accumulation following afforestation. *Glob. Chang. Biol.* **2010**, *16*, 439–453. [CrossRef]
96. Huh, J.-H.; Kim, K.-Y. Time-Based Trend of Carbon Emissions in the Composting Process of Swine Manure in the Context of Agriculture 4.0. *Processes* **2018**, *6*, 168. [CrossRef]
97. Zhang, Q.; Yang, J.; Koide, R.T.; Li, T.; Yang, H.; Chu, J. A Meta-Analysis of Soil Microbial Biomass Levels from Established Tree Plantations over Various Land Uses, Climates and Plant Communities. *Catena* **2017**, *150*, 256–260. [CrossRef]
98. Don, A.; Schumacher, J.; Freibauer, A. Impact of Tropical Land-Use Change on Soil Organic Carbon Stocks—A Meta-Analysis: Soil organic carbon and land-use change. *Glob. Chang. Biol.* **2011**, *17*, 1658–1670. [CrossRef]
99. Kurganova, I.; Lopes de Gerenyu, V.; Kuzyakov, Y. Large-Scale Carbon Sequestration in Post-Agrogenic Ecosystems in Russia and Kazakhstan. *Catena* **2015**, *133*, 461–466. [CrossRef]
100. Kämpf, I.; Hölzel, N.; Störrle, M.; Broll, G.; Kiehl, K. Potential of Temperate Agricultural Soils for Carbon Sequestration: A Meta-Analysis of Land-Use Effects. *Sci. Total Environ.* **2016**, *566–567*, 428–435. [CrossRef]
101. Lu, F.; Wang, X.; Han, B.; Ouyang, Z.; Duan, X.; Zheng, H.; Miao, H. Soil Carbon Sequestrations by Nitrogen Fertilizer Application, Straw Return and No-Tillage in China's Cropland. *Glob. Chang. Biol.* **2009**, *15*, 281–305. [CrossRef]
102. Maetens, W.; Vanmaercke, M.; Poesen, J.; Jankauskas, B.; Jankauskiene, G.; Ionita, I. Effects of Land Use on Annual Runoff and Soil Loss in Europe and the Mediterranean: A Meta-Analysis of Plot Data. *Prog. Phys. Geogr. Earth Environ.* **2012**, *36*, 599–653. [CrossRef]
103. Shi, S.; Peng, C.; Wang, M.; Zhu, Q.; Yang, G.; Yang, Y.; Xi, T.; Zhang, T. A Global Meta-Analysis of Changes in Soil Carbon, Nitrogen, Phosphorus and Sulfur, and Stoichiometric Shifts after Forestation. *Plant Soil* **2016**, *407*, 323–340. [CrossRef]
104. Poeplau, C.; Don, A. Carbon Sequestration in Agricultural Soils via Cultivation of Cover Crops—A Meta-Analysis. *Agric. Ecosyst. Environ.* **2015**, *200*, 33–41. [CrossRef]
105. Eglin, T.; Ciais, P.; Piao, S.L.; Barre, P.; Bellassen, V.; Cadule, P.; Chenu, C.; Gasser, T.; Koven, C.; Reichstein, M.; et al. Historical and Future Perspectives of Global Soil Carbon Response to Climate and Land-Use Changes. *Tellus B Chem. Phys. Meteorol.* **2010**, *62*, 700–718. [CrossRef]
106. Auerswald, K.; Fiener, P.; Dikau, R. Rates of Sheet and Rill Erosion in Germany—A Meta-Analysis. *Geomorphology* **2009**, *111*, 182–193. [CrossRef]
107. García-Ruiz, J.M.; Beguería, S.; Nadal-Romero, E.; González-Hidalgo, J.C.; Lana-Renault, N.; Sanjuán, Y. A Meta-Analysis of Soil Erosion Rates across the World. *Geomorphology* **2015**, *239*, 160–173. [CrossRef]
108. Tonitto, C.; David, M.B.; Drinkwater, L.E. Replacing Bare Fallows with Cover Crops in Fertilizer-Intensive Cropping Systems: A Meta-Analysis of Crop Yield and N Dynamics. *Agric. Ecosyst. Environ.* **2006**, *112*, 58–72. [CrossRef]
109. Manley, J.; van Kooten, G.C.; Moeltner, K.; Johnson, D.W. Creating Carbon Offsets in Agriculture through No-Till Cultivation: A Meta-Analysis of Costs and Carbon Benefits. *Clim. Chang.* **2005**, *68*, 41–65. [CrossRef]
110. Gerstner, K.; Dormann, C.F.; Stein, A.; Manceur, A.M.; Seppelt, R. EDITOR'S CHOICE: REVIEW: Effects of Land Use on Plant Diversity—A Global Meta-Analysis. *J. Appl. Ecol.* **2014**, *51*, 1690–1700. [CrossRef]
111. Li, M.; Wang, J.; Guo, D.; Yang, R.; Fu, H. Effect of Land Management Practices on the Concentration of Dissolved Organic Matter in Soil: A Meta-Analysis. *Geoderma* **2019**, *344*, 74–81. [CrossRef]

112. Malandra, F.; Vitali, A.; Urbinati, C.; Garbarino, M. 70 Years of Land Use/Land Cover Changes in the Apennines (Italy): A Meta-Analysis. *Forests* **2018**, *9*, 551. [CrossRef]
113. Van Zanten, B.T.; Verburg, P.H.; Koetse, M.J.; van Beukering, P.J.H. Preferences for European Agrarian Landscapes: A Meta-Analysis of Case Studies. *Landsc. Urban Plan.* **2014**, *132*, 89–101. [CrossRef]
114. Plieninger, T.; Hui, C.; Gaertner, M.; Huntsinger, L. The Impact of Land Abandonment on Species Richness and Abundance in the Mediterranean Basin: A Meta-Analysis. *PLoS ONE* **2014**, *9*, e98355. [CrossRef]
115. Katayama, N.; Bouam, I.; Koshida, C.; Baba, Y.G. Biodiversity and Yield under Different Land-Use Types in Orchard/Vineyard Landscapes: A Meta-Analysis. *Biol. Conserv.* **2019**, *229*, 125–133. [CrossRef]
116. Merheb, M.; Moussa, R.; Abdallah, C.; Colin, F.; Perrin, C.; Baghdadi, N. Hydrological Response Characteristics of Mediterranean Catchments at Different Time Scales: A Meta-Analysis. *Hydrol. Sci. J.* **2016**, *61*, 2520–2539. [CrossRef]
117. Magliocca, N.R.; Rudel, T.K.; Verburg, P.H.; McConnell, W.J.; Mertz, O.; Gerstner, K.; Heinimann, A.; Ellis, E.C. Synthesis in Land Change Science: Methodological Patterns, Challenges, and Guidelines. *Reg. Environ. Change* **2015**, *15*, 211–226. [CrossRef]
118. Overmars, K.P.; Verburg, P.H.; Veldkamp, T. Comparison of a Deductive and an Inductive Approach to Specify Land Suitability in a Spatially Explicit Land Use Model. *Land Use Policy* **2007**, *24*, 584–599. [CrossRef]
119. Magliocca, N.R.; van Vliet, J.; Brown, C.; Evans, T.P.; Houet, T.; Messerli, P.; Messina, J.P.; Nicholas, K.A.; Ornetsmüller, C.; Sagebiel, J.; et al. From Meta-Studies to Modeling: Using Synthesis Knowledge to Build Broadly Applicable Process-Based Land Change Models. *Environ. Model. Softw.* **2015**, *72*, 10–20. [CrossRef]
120. Guzha, A.C.; Rufino, M.C.; Okoth, S.; Jacobs, S.; Nóbrega, R.L.B. Impacts of Land Use and Land Cover Change on Surface Runoff, Discharge and Low Flows: Evidence from East Africa. *J. Hydrol. Reg. Stud.* **2018**, *15*, 49–67. [CrossRef]
121. Lele, N.; Joshi, P.K. Analyzing Deforestation Rates, Spatial Forest Cover Changes and Identifying Critical Areas of Forest Cover Changes in North-East India during 1972–1999. *Environ. Monit. Assess.* **2009**, *156*, 159–170. [CrossRef]
122. Mustard, J.F.; Defries, R.S.; Fisher, T.; Moran, E. Land-Use and Land-Cover Change Pathways and Impacts. In *Land Change Science: Observing, Monitoring and Understanding Trajectories of Change on the Earth's Surface*; Gutman, G., Janetos, A.C., Justice, C.O., Moran, E.F., Mustard, J.F., Rindfuss, R.R., Skole, D., Turner, B.L., Cochrane, M.A., Eds.; Remote Sensing and Digital Image Processing; Springer: Dordrecht, The Netherlands, 2004; pp. 411–429. ISBN 978-1-4020-2562-4.
123. Lambin, E.F.; Turner, B.L.; Geist, H.J.; Agbola, S.B.; Angelsen, A.; Bruce, J.W.; Coomes, O.T.; Dirzo, R.; Fischer, G.; Folke, C.; et al. The Causes of Land-Use and Land-Cover Change: Moving beyond the Myths. *Glob. Environ. Chang.* **2001**, *11*, 261–269. [CrossRef]
124. Agaton, M.; Setiawan, Y.; Effendi, H. Land Use/Land Cover Change Detection in an Urban Watershed: A Case Study of Upper Citarum Watershed, West Java Province, Indonesia. *Procedia Environ. Sci.* **2016**, *33*, 654–660. [CrossRef]
125. Lambin, E.F.; Geist, H.J.; Lepers, E. Dynamics of Land-Use and Land-Cover Change in Tropical Regions. *Annu. Rev. Environ. Resour.* **2003**, *28*, 205–241. [CrossRef]
126. Udo de Haes, H. How to Approach Land Use in LCIA or, How to Avoid the Cinderella Effect? *Int. J. Life Cycle Assess.* **2006**, *11*, 219–221. [CrossRef]
127. Burkhard, B.; Kroll, F.; Müller, F.; Windhorst, W. Landscapes' Capacities to Provide Ecosystem Services—A Concept for Land-Cover Based Assessments. *Landsc. Online* **2009**, *15*, 1–22. [CrossRef]
128. Turner, N.J.; Ignace, M.B.; Ignace, R. Traditional Ecological Knowledge and Wisdom of Aboriginal Peoples in British Columbia. *Ecol. Appl.* **2000**, *10*, 1275–1287. [CrossRef]
129. Bürgi, M.; Östlund, L.; Mladenoff, D.J. Legacy Effects of Human Land Use: Ecosystems as Time-Lagged Systems. *Ecosystems* **2017**, *20*, 94–103. [CrossRef]
130. Troy, A.; Wilson, M.A. Mapping Ecosystem Services: Practical Challenges and Opportunities in Linking GIS and Value Transfer. *Ecol. Econ.* **2006**, *60*, 435–449. [CrossRef]
131. Egarter Vigl, L.; Tasser, E.; Schirpke, U.; Tappeiner, U. Using Land Use/Land Cover Trajectories to Uncover Ecosystem Service Patterns across the Alps. *Reg. Environ. Chang.* **2017**, *17*, 2237–2250. [CrossRef]
132. Pielke, R.A.; Pitman, A.; Niyogi, D.; Mahmood, R.; McAlpine, C.; Hossain, F.; Goldewijk, K.K.; Nair, U.; Betts, R.; Fall, S.; et al. Land Use/Land Cover Changes and Climate: Modeling Analysis and Observational Evidence: Land Use/Land Cover Changes and Climate: Modeling Analysis and Observational Evidence. *Wiley Interdiscip. Rev. Clim. Chang.* **2011**, *2*, 828–850. [CrossRef]
133. Sidhu, N.; Pebesma, E.; Wang, Y.-C. Usability Study to Assess the IGBP Land Cover Classification for Singapore. *Remote Sens.* **2017**, *9*, 1075. [CrossRef]
134. Bey, A.; Sánchez-Paus Díaz, A.; Maniatis, D.; Marchi, G.; Mollicone, D.; Ricci, S.; Bastin, J.-F.; Moore, R.; Federici, S.; Rezende, M.; et al. Collect Earth: Land Use and Land Cover Assessment through Augmented Visual Interpretation. *Remote Sens.* **2016**, *8*, 807. [CrossRef]
135. Gómez, C.; White, J.C.; Wulder, M.A. Optical Remotely Sensed Time Series Data for Land Cover Classification: A Review. *ISPRS J. Photogramm. Remote Sens.* **2016**, *116*, 55–72. [CrossRef]
136. Jokar Arsanjani, J.; Tayyebi, A.; Vaz, E. GlobeLand30 as an Alternative Fine-Scale Global Land Cover Map: Challenges, Possibilities, and Implications for Developing Countries. *Habitat Int.* **2016**, *55*, 25–31. [CrossRef]
137. Brown, D.G.; Verburg, P.H.; Pontius, R.G.; Lange, M.D. Opportunities to Improve Impact, Integration, and Evaluation of Land Change Models. *Curr. Opin. Environ. Sustain.* **2013**, *5*, 452–457. [CrossRef]
138. Bontemps, S.; Herold, M.; Kooistra, L.; van Groenestijn, A.; Hartley, A.; Arino, O.; Moreau, I.; Defourny, P. Revisiting Land Cover Observation to Address the Needs of the Climate Modeling Community. *Biogeosciences* **2012**, *9*, 2145–2157. [CrossRef]

139. Giri, C.; Pengra, B.; Long, J.; Loveland, T.R. Next Generation of Global Land Cover Characterization, Mapping, and Monitoring. *Int. J. Appl. Earth Obs. Geoinformation* **2013**, *25*, 30–37. [CrossRef]
140. Grafius, D.R.; Corstanje, R.; Warren, P.H.; Evans, K.L.; Hancock, S.; Harris, J.A. The Impact of Land Use/Land Cover Scale on Modelling Urban Ecosystem Services. *Landsc. Ecol.* **2016**, *31*, 1509–1522. [CrossRef]
141. Tayyebi, A.; Pijanowski, B.C. Modeling Multiple Land Use Changes Using ANN, CART and MARS: Comparing Tradeoffs in Goodness of Fit and Explanatory Power of Data Mining Tools. *Int. J. Appl. Earth Obs. Geoinformation* **2014**, *28*, 102–116. [CrossRef]
142. Roy, P.S.; Roy, A.; Joshi, P.K.; Kale, M.P.; Srivastava, V.K.; Srivastava, S.K.; Dwevidi, R.S.; Joshi, C.; Behera, M.D.; Meiyappan, P.; et al. Development of Decadal (1985–1995–2005) Land Use and Land Cover Database for India. *Remote Sens.* **2015**, *7*, 2401–2430. [CrossRef]
143. Pullanikkatil, D.; Palamuleni, L.G.; Ruhiiga, T.M. Land Use/Land Cover Change and Implications for Ecosystems Services in the Likangala River Catchment, Malawi. *Phys. Chem. Earth Parts A/B/C* **2016**, *93*, 96–103. [CrossRef]
144. Zuidema, P.A.; Baker, P.J.; Groenendijk, P.; Schippers, P.; van der Sleen, P.; Vlam, M.; Sterck, F. Tropical Forests and Global Change: Filling Knowledge Gaps. *Trends Plant Sci.* **2013**, *18*, 413–419. [CrossRef] [PubMed]
145. Kayet, N.; Pathak, K.; Chakrabarty, A.; Sahoo, S. Spatial Impact of Land Use/Land Cover Change on Surface Temperature Distribution in Saranda Forest, Jharkhand. *Model. Earth Syst. Environ.* **2016**, *2*, 127. [CrossRef]
146. Lambin, E.F.; Geist, H.J. *Land-Use and Land-Cover Change: Local Processes and Global Impacts*; Springer: Berlin, Germany, 2008; ISBN 978-3-540-32202-3.
147. Mondal, M.S.; Sharma, N.; Garg, P.K.; Kappas, M. Statistical Independence Test and Validation of CA Markov Land Use Land Cover (LULC) Prediction Results. *Egypt. J. Remote Sens. Space Sci.* **2016**, *19*, 259–272. [CrossRef]
148. Tran, D.X.; Pla, F.; Latorre-Carmona, P.; Myint, S.W.; Caetano, M.; Kieu, H.V. Characterizing the Relationship between Land Use Land Cover Change and Land Surface Temperature. *ISPRS J. Photogramm. Remote Sens.* **2017**, *124*, 119–132. [CrossRef]
149. Paul, S.; Ghosh, S.; Oglesby, R.; Pathak, A.; Chandrasekharan, A.; Ramsankaran, R. Weakening of Indian Summer Monsoon Rainfall Due to Changes in Land Use Land Cover. *Sci. Rep.* **2016**, *6*, 32177. [CrossRef]
150. Hovenga, P.A.; Wang, D.; Medeiros, S.C.; Hagen, S.C.; Alizad, K. The Response of Runoff and Sediment Loading in the Apalachicola River, Florida to Climate and Land Use Land Cover Change: Responses to climate and lulc change. *Earths Future* **2016**, *4*, 124–142. [CrossRef]
151. Alexandru, A.; Sushama, L. Impact of Land-Use and Land-Cover Changes on CRCM5 Climate Projections over North America for the Twenty-First Century. *Clim. Dyn.* **2016**, *47*, 1197–1209. [CrossRef]
152. Mora, B.; Tsendbazar, N.-E.; Herold, M. Towards Better Use of Global Land Cover Datasets and Improved Accuracy Assessment Practices. Available online: https://www.researchgate.net/profile/Brice-Mora/publication/312936617_Towards_Better_Use_of_Global_Land_Cover_Datasets_and_Improved_Accuracy_Assessment_Practices/links/588a22a092851c2779b2564f/Towards-Better-Use-of-Global-Land-Cover-Datasets-and-Improved-Accuracy-Assessment-Practices.pdf (accessed on 17 September 2021).
153. Jansen, L.J.M.; Badea, A.; Milenov, P.; Moise, C.; Vassilev, V.; Milenova, L.; Devos, W. The Use of the Land-Cover Classification System in Eastern European Countries: Experiences, Lessons Learnt and the Way Forward. In *Land Use and Land Cover Mapping in Europe*; Manakos, I., Braun, M., Eds.; Remote Sensing and Digital Image Processing; Springer: Dordrecht, The Netherlands, 2014; Volume 18, pp. 297–325. ISBN 978-94-007-7968-6.
154. Jansen, L.J.M. Harmonization of Land Use Class Sets to Facilitate Compatibility and Comparability of Data across Space and Time. *J. Land Use Sci.* **2006**, *1*, 127–156. [CrossRef]
155. Capucim, M.N.; Brand, V.S.; Machado, C.B.; Martins, L.D.; Allasia, D.G.; Homann, C.T.; Freitas, E.D.d.; Dias, M.A.F.D.S.; Andrade, M.F.; Martins, J.A. South America Land Use and Land Cover Assessment and Preliminary Analysis of Their Impacts on Regional Atmospheric Modeling Studies. *IEEE J. Sel. Top. Appl. Earth Obs. Remote Sens.* **2015**, *8*, 1185–1198. [CrossRef]
156. Rozenstein, O.; Karnieli, A. Comparison of Methods for Land-Use Classification Incorporating Remote Sensing and GIS Inputs. *Appl. Geogr.* **2011**, *31*, 533–544. [CrossRef]
157. Franklin, S.E.; Ahmed, O.S.; Wulder, M.A.; White, J.C.; Hermosilla, T.; Coops, N.C. Large Area Mapping of Annual Land Cover Dynamics Using Multitemporal Change Detection and Classification of Landsat Time Series Data. *Can. J. Remote Sens.* **2015**, *41*, 293–314. [CrossRef]
158. Yang, L.; Jin, S.; Danielson, P.; Homer, C.; Gass, L.; Bender, S.M.; Case, A.; Costello, C.; Dewitz, J.; Fry, J.; et al. A New Generation of the United States National Land Cover Database: Requirements, Research Priorities, Design, and Implementation Strategies. *ISPRS J. Photogramm. Remote Sens.* **2018**, *146*, 108–123. [CrossRef]
159. Mallupattu, P.K.; Reddy, J.R.S. Analysis of Land Use/Land Cover Changes Using Remote Sensing Data and GIS at an Urban Area, Tirupati, India. *Sci. World J.* **2013**, *2013*, 7. [CrossRef] [PubMed]
160. Gong, P.; Yu, L.; Li, C.; Wang, J.; Liang, L.; Li, X.; Ji, L.; Bai, Y.; Cheng, Y.; Zhu, Z. A New Research Paradigm for Global Land Cover Mapping. *Ann. GIS* **2016**, *22*, 87–102. [CrossRef]
161. Wentz, E.A.; Anderson, S.; Fragkias, M.; Netzband, M.; Mesev, V.; Myint, S.W.; Quattrochi, D.; Rahman, A.; Seto, K.C. Supporting Global Environmental Change Research: A Review of Trends and Knowledge Gaps in Urban Remote Sensing. *Remote Sens.* **2014**, *6*, 3879–3905. [CrossRef]
162. Jacobs, S.; Burkhard, B.; Van Daele, T.; Staes, J.; Schneiders, A. 'The Matrix Reloaded': A Review of Expert Knowledge Use for Mapping Ecosystem Services. *Ecol. Model.* **2015**, *295*, 21–30. [CrossRef]

land

Article

Urban Type Classification and Characteristic Analysis through Time-Series Environmental Changes for Land Use Management for 31 Satellite Cities around Seoul, South Korea

Jin-Hyo Kim [1], Oh-Sung Kwon [2,*] and Jung-Hwa Ra [3]

1 Division of Forestland Spatial Planning, Korea Forest Conservation Association, Daejeon 35262, Korea; gsjhk@kfca.re.kr
2 Department of Ecology and Environment, Gyeonggi Research Institute, Suwon 16207, Korea
3 Department of Landscape Architecture, Kyungpook National University, Daegu 41566, Korea; jhra@knu.ac.kr
* Correspondence: kos8282@gri.re.kr; Tel.: +82-31-250-3252

Abstract: The objective of the present study was to determine changes in land coverage for 31 satellite cities surrounding Seoul and changes in values of MSPA (Morphological Spatial Pattern Analysis) for a time period of about 30 years (from 1988 to 2018). Cities that showed similar environmental changes were grouped utilizing a hierarchical cluster analysis. The results of this study are summarized as follows: First, as a result of analyzing changes in land coverage, urbanized areas in all 31 cities greatly increased, whereas areas of forest, grassland, farmland, wetland, etc., greatly decreased. Second, as a result of carrying out MSPA for green areas in each city, the number of Cores, Islets as stepping-stone green areas, and Branches greatly decreased. As a result of analyzing factors in cluster analysis, 12 variables were classified into four groups. After performing a cluster analysis, the 31 cities were classified into six clusters. Cluster-6 showed the biggest decrease in wetland areas. These results could be used as basic data for establishing differentiated environmental policies for clusters of cities that show similar environmental changes, and for establishing policy priorities that break away from uniform environmental policies at the local level.

Keywords: land-cover change; MSPA; cluster analysis; land use management

Citation: Kim, J.-H.; Kwon, O.-S.; Ra, J.-H. Urban Type Classification and Characteristic Analysis through Time-Series Environmental Changes for Land Use Management for 31 Satellite Cities around Seoul, South Korea. *Land* **2021**, *10*, 799. https://doi.org/10.3390/land10080799

Academic Editor: Giuseppe Pulighe

Received: 29 June 2021
Accepted: 27 July 2021
Published: 29 July 2021

1. Introduction

Korea has achieved economic growth at a very high speed since the 1960s. Rapid urbanization and industrialization has particularly progressed around Seoul, the capital of Korea. As the traffic congestion problem in Seoul became severe in the late 1970s, a suburbanization phenomenon involving population movements into Gyeonggi-do, the outskirts of Seoul, quickly emerged to mitigate this problem [1–3]. As a result, the capital area including Seoul and Gyeonggi-do formed a typical metropolitan area. About 26 million people, close to half of the Korean population, live in this capital area.

Although this rapid growth of the city has been accompanied by changes in various aspects such as the natural environment and the human environment, it has had direct effects on the natural environment in particular. Environmental damage in the capital area is intensifying day by day. Such environment damage is occurring across Korea and all over the world. As numerous development plans are damaging the natural environment at an irrevocable level, efforts have been made to solve environmental problems in various aspects.

First, in relation to basic studies, European countries such as Germany, Italy, etc., where environmental damage is intensifying, have conducted studies based on landscape ecology, and many efforts have been made to create sustainable land environments by grafting diverse theories into actual land plans [4–7]. Similar studies are also actively conducted in North American countries such as the USA [8–11]. For example, Forman has arranged a theoretical basis for North American landscape ecological planning based on a landscape ecology [12].

Moreover, data about environmental changes (in land use and land coverage) can now be collected and analyzed within a short period of time thanks to the development of diverse technologies such as Geographic Information System (GIS) and Remote Sensing (RS) technology [13–18]. In particular, FRAGSTATS, a spatial pattern analysis program developed by McGarigal and Marks [19], is a core tool that can quantitatively determine structural patterns and change aspects of landscape elements. Utilizing FRAGSTATS, various studies have been actively conducted all over the world [20–23]. For example, Reddy et al. [24] investigated the effect of forest fragmentation in India by utilizing landscape indices such as Mean Patch Size (MPS) and Edge Density (ED).

In legal and institutional aspects, efforts have been steadily made to cope with environmental changes in accordance with "Sustainable Development", which has been widely used after the announcement of the Bryndtkand report of WCED in 1987 and the "Paris Climate Agreement" adopted in the UN Weather Change Conference in 2015. Korea is working on an interlocking between land plans and environmental plans based on Article 5 (Environment-friendly Land Management) of the Framework Act on the National Land, and the Framework Act on Environmental Policy (Responsibilities of the State and Local Governments). The objective of such an interlocking plan is to sustainably upbuild the land by minimizing the effect on the environment at a plan level in advance [25–27].

Considering that development plans are direct causes of environmental problems that take place mostly in cities, establishing environmental policies at a city level has a very important meaning. However, it may be more effective to solve housing problems, traffic problems, and, in particular, environmental problems caused by urban expansion by comprehensively bundling up several cities together rather than solving these problems within individual cities. Thus, it is necessary to create a new spatial unit that bundles up cities that show similar environmental changes that transcend the boundaries of administrative districts.

In relation to this, studies that group cities showing similar characteristics have been actively conducted through cluster analyses [28–33]. Targeting about 280 villages where mountain village development programs have been carried out, Ko et al. [34] categorized villages showing similar mountain village scenery by utilizing the altitude, forest ratio, farmland ratio, coniferous forest ratio, broadleaf forest ratio, and ecological and natural map rating ratios. Further, Kuo et al. [35] classified the impact of urban development on the natural environment into six clusters for Tainan, Taiwan. Based on this, an evaluation methodology was established to simulate and analyze the impact of urban growth.

Although diverse efforts have been made to solve environmental problems, as mentioned earlier, precedent studies have the following limitations: First, studies that utilized FRAGSTATS did not present analysis results as drawings. They had limitations in that it was difficult to correlate their results with drawings of development plans. Additionally, with respect to cluster analyses, cities were categorized by considering various aspects such as social and economic issues, transportation, land cover change, MSPA structure, water supply and demand, surface temperature, and surface runoff. Among the various factors for colonization, MSPA has been limitedly used mainly for studies related to eco-corridors such as wildlife passageways. In other words, there have been relatively few studies examining changes in MSPA values and diagnosing environmental problems based on the changes in green areas for each cluster due to urbanization.

Accordingly, in the present study, we categorized 31 cities in Gyeonggi-do adjacent to Seoul at a regional level based on results of time-sequential environmental changes to grasp characteristics of environmental changes by type. We believe that the results of the present study could be used as basic data for establishing differentiated environmental plans for cities with similarities or for establishing environmental policy priorities, breaking away from standardized policies related to the environment that appear at the regional level in particular.

2. Data and Methods

2.1. Research Area

The present study was conducted for 31 satellite cities in Gyeonggi-do adjacent to Seoul, the capital of Korea (Figure 1). The main status of each city is shown as follows. First, the city with the biggest area among these 31 cities was found to be Yangpyeong (877.8 km^2), followed by Gapyeong and Pocheon. We could see that Gwacheon, Gwangmyeong, and Gunpo were much smaller in size than the other cities. Looking at the number of populations by city, Suwon, Goyang, Yongin, Seongnam, and Hwaseong had the largest population in that order. On the other hand, Yeoncheon, Gwacheon, Gapyeong, and Dongducheon have low population numbers. In particular, Yeoncheon has about 44,000 people, showing the lowest population among all 31 cities. In addition, when we looked into the geographical characteristics of each city, Yeoncheon and Paju were playing an important role militarily as they are close to North Korea. Goyang, Seongnam, Bucheon, and Hanam were under high development pressure in the past as they were directly in contact with Seoul. Such development pressure still exists today. The area, population, and location characteristics of the 31 cities are presented in Appendix A.

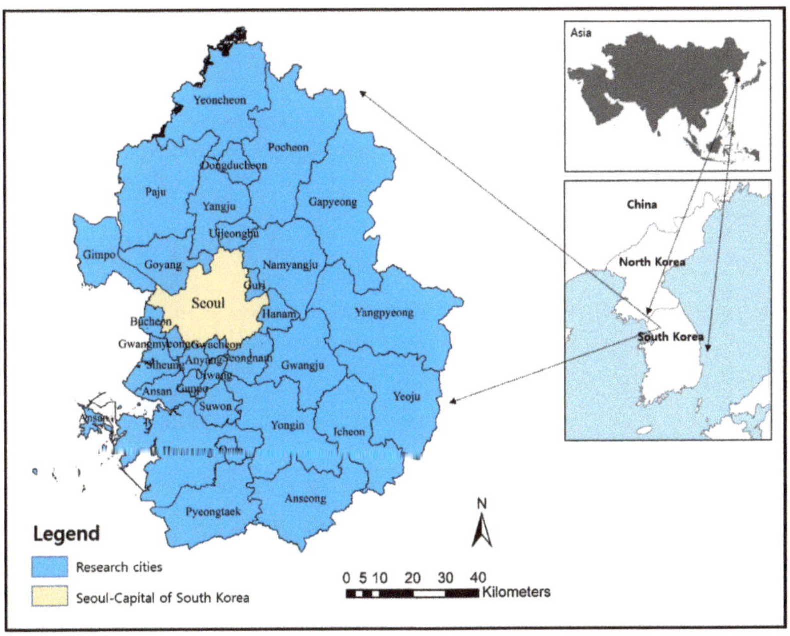

Figure 1. Location of research cities.

2.2. Procedure for Conductiong the Study

The procedure for conducting the present study is largely divided into three stages (Figure 2). First, an analysis of changes in land coverage for each of the 31 cities was performed based on land cover maps of 1988 and 2018. Second, an MSPA was performed for structural changes in each of the 31 cities. Third, a cluster analysis was performed to grasp the characteristics of each cluster. The detailed study method of each stage is shown below.

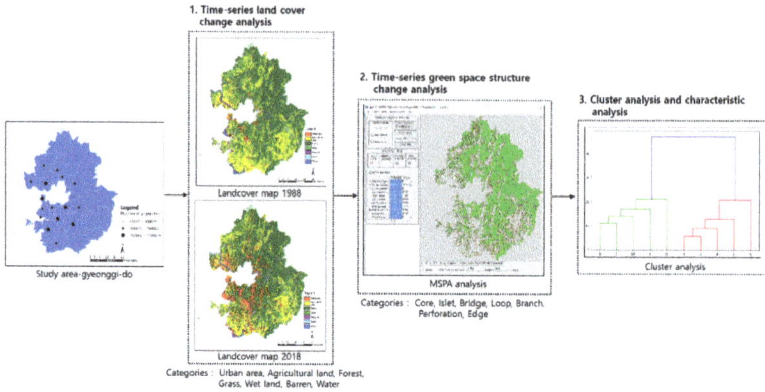

Figure 2. Research flow.

2.3. Analysis of Changes in Landcover Change in Each City

For the analysis of changes in the land coverage in each city, a Level I Land Cover Map provided by Environmental Geographic Information Service [36] was used. The major classification Level I Land Cover Map has been constructed every 10 years, starting in 1988. In this study, the most historical data, from 1988, and the most recent, from 2018, were used. The reason for this is that the first new town developments (Ilsan, Bundang, etc.) were carried out from 1989 to 1996, and the second new town developments (Seongnam, Hwaseong, etc.) were carried out from 2001 to 2017. In other words, out of the 31 cities, large-scale development plans were carried out for cities adjacent to Seoul until recently. Therefore, it was judged that the changes in land cover and the changes to MSPA structure according to the development plan for each city could be best understood if the data for 1988 and 2018 were used. A Level I Land Cover Map was divided into seven land cover types (urbanized areas, agricultural areas, forest areas, grassland, wetland, bare land, and waters). It was prepared based on a resolution of 30 m × 30 m. Changes in land coverage for each city were examined based on such land cover maps. Land cover maps utilized in the present study are as follows (Figure 3).

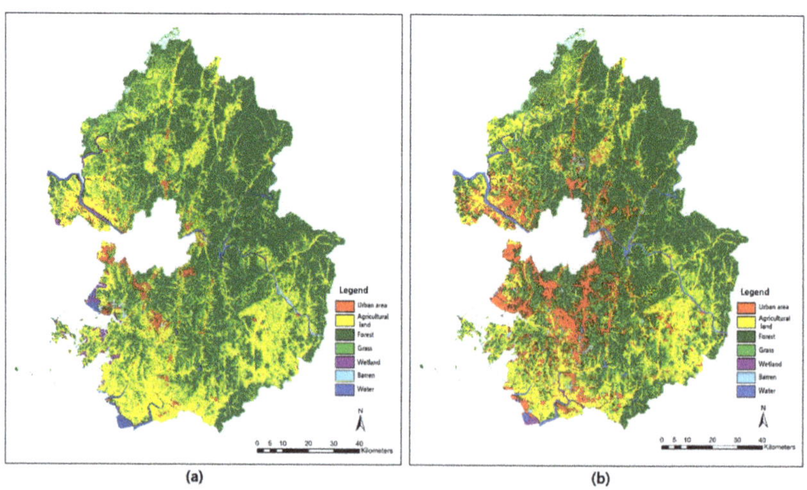

Figure 3. (**a**) Landcover map (1988) and (**b**) land cover map (2018).

2.4. MSPA Pattern Changes in Green Area of Each City

For MSPA of the structural changes in the green areas of each city, GUIDOS [37] was utilized for forest and grass areas among seven land cover types. The analysis was conducted after converting forest types extracted from land cover maps of 1988 and 2018 into Geo Tiff files and then applying a pixel size of 20 m [38].

GUIDOS (Graphic User Interface for the Description of Image Objects and their Shapes) is a program designed to overcome the limitations of numerical data presented by FRAGSTATS [39], and is an existing landscape pattern analysis program that can intuitively grasp changes in spatial forms. GUIDOS has been widely utilized in diverse fields recently [40–42].

In this study, MSPA (Morphological Spatial Pattern Analysis) analysis was used among various analysis methods of GUIDOS. MSPA has a high potential for being utilized during the establishment of various plans as it presents relations between diverse elements displayed on maps and their distribution patterns in the form of a diagram. In more detail, MSPA showed green areas on a drawing after dividing them into seven types (Core, Islet, Bridge, Loop, Branch, Perforation and Edge) depending on their forms. At the same time, the number of MSPA types of each city was calculated using the following formula (Figure 4). Accordingly, it is very useful for grasping changes in the form of green areas resulting from urbanization, as well as the extinction of dotted green spaces and strip green spaces on a small scale.

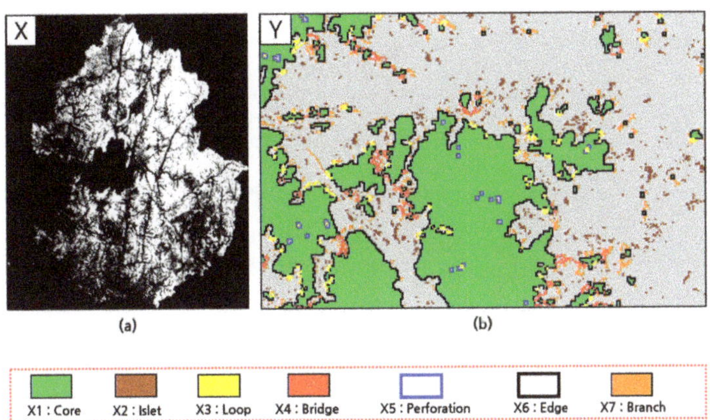

(c)

- X ---------- input binary image
- $X_1 = T_{d \geq s}[\mathrm{EDT}(X)]$ ---------- **core**
- $X_2 = X \setminus R^{\delta}_{X}(X_1)$ ---------- **islet**
- $A = \delta^{(s)}_{X \setminus X_1}[\mathrm{SKEL}_{X_1}(X)]$ ---------- connector
  - X_3 = pixels of A emanating from the same CC of X_1 ---------- **loop**
  - X_4 = pixels of A pixels emanating from different CCs of X_1 ---------- **bridge**
- $B = T_{0<d\leq s}[\mathrm{EDT}(X_1^c)] \cap [X \setminus (X_1 \cup X_2 \cup X_3 \cup X_4)]$ ---------- boundary
  - X_5 = pixels of B within a distance s to a hole of B ---------- **perforation**
  - $X_6 = B \setminus X_5$ ---------- **edge**
- $X_7 = X \setminus \{X_1 \cup X_2 \cup X_3 \cup X_4 \cup X_5 \cup X_6\}$ ---------- **branch**
- $Y = \{X_1, X_2, X_3, X_4, X_5, X_6, X_7\}$ ---------- segmented binary pattern

(d)

Figure 4. (**a**) Input binary image of research cities, (**b**) segmented binary pattern, (**c**) seven categories of MSPA and (**d**) equation for each step [40].

Characteristics of these seven MSPA types are shown as follows (Table 1). First, a Core is a park located in a large-scale forest or a city center. It is a type where the scale is big in comparison with other types. In other words, it is a type that functions as a key space for the inhabitation of biospecies. Next, a Bridge is a strip type that connects different Cores with each other and carries out a function similar to an eco-corridor. An Islet is a green space smaller than a Core. Although it plays an important role as a middle stopover during the movement of wild animals, it is also a type with a very high risk of extinction [43].

Table 1. Morphological spatial pattern analysis (MSPA) categories and ecological implication.

Categories	Description
Core	It can be used as the "source" of a variety of ecological processes, most of which are forest parks with large patch areas and large forest farms, etc., which are of great significance for species reproduction and biodiversity protection
Islet	Small patches, which are independent of each other and have low connectivity, are less likely to communicate with other patches in terms of materials and energy, and are mostly small green spaces in urban or rural areas. In addition, an islet is effective in enhancing connectivity by functioning as a stepping stone green area
Bridge	The narrow and long areas connecting the patches of different core areas; they have the characteristics of ecological corridors, which are mostly green belts, which are conducive to the migration of species and the connection of landscape within the territory
Loop	The internal channel of material and energy exchange in the same core area; they are shortcuts for material and energy exchange in the core area
Branch	Only one end is connected to the main patch; mainly an extension of the green space, which is the channel for species diffusion and energy exchange with the peripheral landscape
Perforation	As a transition region, the edge, etc., also exists between the core patch and its inner non-green space
Edge	The transition zone between the marginal zone of the core area and the peripheral non-green landscape area, which can reduce the impact brought by the external environment and human disturbance; usually the peripheral forest zone of forest parks and large forest farms

2.5. Cluster Analysis of 31 Cities

To effectively conduct categorizations by city through a cluster analysis, a factor analysis was conducted first. To enhance the accuracy of the analysis, 12 variables, excluding Grass and Waters, with low communality among a total of 14 selected variables (7 land cover types and 7 MSPA categories), were utilized for the analysis. Factors of these 12 selected variables were extracted through PCA (Principal Component Analysis) [44]. The Varimax Method of Orthogonal Rotation, widely utilized to simplify the characteristics of each factor, was utilized as a Factor Rotation Method. As the analysis was conducted without setting the number of clusters in advance during a cluster analysis based on factor scores, a hierarchical cluster analysis was performed. Hierarchical cluster analysis is a method where two objects close to each other start to form a cluster, and a dendrogram in the shape of a tree is formed through continuous clustering of the clusters adjacent to each other to determine the number of clusters. Ward's Method of hierarchical cluster analysis was used for grouping these 31 cities.

3. Results

3.1. Analysis of Landcover Changes by Each City

Changes in land coverage areas for each of the 31 cities are as follows (Table 2). First, in the case of Hwaseong, the increase in urbanized areas was found to be the biggest among the 31 cities. On the contrary, its sizes of agricultural areas, forest, wetland, and waters decreased. In particular, the sizes of its agricultural areas and wetland decreased by 7600 ha and 1705 ha, respectively. Decreases in the areas of these two land cover types were found to be the biggest among the 31 cities. Next, forest areas decreased by 2600 ha. The decrease in the area of natural space due to urbanization was noticeable in Hwaseong as a whole. In the case of Pyeongtaek, its urbanized areas increased by 6296 ha. It was found to be a city showing the second biggest increase in urbanized areas, following Hwaseong. Its farmland areas decreased by 7233 ha. It was found to be a city showing the second biggest decrease in farmland areas, following Hwaseong. Its forest areas decreased by 1325 ha. On the contrary, its grassland and wetland areas increased by 618 ha and 1737 ha, respectively.

In the case of Namyangju, its land cover types that showed increases in size were found to be urbanized areas, bare land, and waters (Figure 5). On the contrary, its land cover types that showed decreases in size were found to be agricultural areas, forest, grassland, and wetland. In particular, forest areas in Namyangju decreased by about 3900 ha due to urbanization. It was found to be a city showing the second biggest decrease in forest area, following Yongin. In the case of Siheung, its land cover types that showed increases in size were found to be urbanized areas and grassland. On the contrary, its land cover types that showed decreases in size were found to be agricultural areas, forest, wetland, and waters. In particular, its areas of wetland and waters decreased by 1286 ha and 1514 ha, respectively, due to large-scale reclamation projects that appeared in the areas close to the coast. Due to such reclamation projects, wetland and waters were converted to farmland and urbanized areas.

On the contrary, in the case of Gwacheon, Dongducheon, and Uiwang, urbanized areas increased by 193 ha, 503 ha and 601 ha, respectively. In particular, forest areas in Gwacheon and Dongducheon increased differently from other cities. This was because farmland and bare land located in the forests were converted into forests. When we put these results together, urbanized areas were found to have increased in all 31 cities. It was found that urbanization was concentrated in cities adjacent to Seoul, the capital. In addition, most of the forest, grassland, and wetland areas with high ecological values were found to show decreases in size due to urbanization.

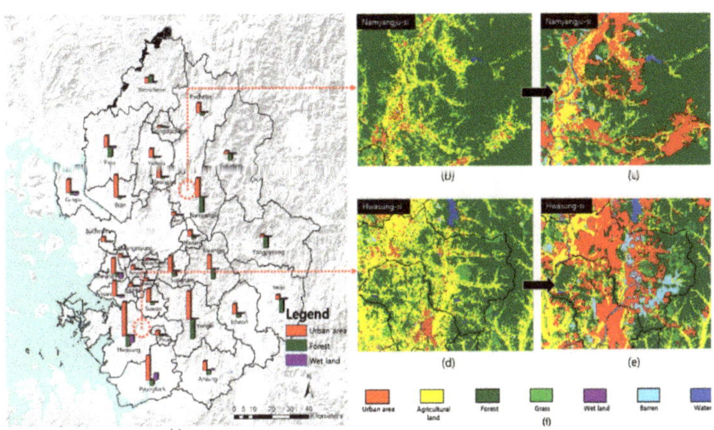

Figure 5. (**a**) Changes in major land cover by total city, (**b**) 1988 land cover map of Namyangju, (**c**) 2018 land cover map of Namyangju, (**d**) 1988 land cover map of Hwaseong, (**e**) 2018 land cover map of Hwaseong, and (**f**) legend of land cover.

Table 2. Changes in land cover type areas by city (2018–1988) Unit: ha.

Division	Urban Area	Agricultural Land	Forest	Grass	Wetland	Barren	Water
Gapyeong	695.24	−950.63	−1511.13	1044.13	−1.31	385.59	33799
Goyang	5688.11	−5789.74	−441.85	−185.05	−15.24	794.60	86.48
Gwacheon	193.32	−338.93	159.62	−115.88	2.01	100.42	−0.33
Gwangmyeong	668.63	−611.38	−209.67	111.89	24.41	17.31	−0.49
Gwangju	3565.62	−727.47	−2518.27	−522.29	−1.53	125.61	78.32
Guri	717.07	−736.47	−119.85	−191.99	−1.68	285.31	48.34
Gunpo	839.36	−417.37	−138.19	−317.95	4.88	19.85	9.45
Gimpo	3346.53	−4033.38	−477.35	660.85	661.73	638.35	525.64
Namyangju	4887.68	−1762.00	−3893.74	−603.03	−6.25	1275.71	101.63
Dongducheon	503.07	−534.42	165.59	−302.75	0.00	180.23	−11.73
Bucheon	1106.49	−1204.96	−97.41	191.61	−12.99	16.71	0.46
Seongnam	3737.96	−1967.57	−1390.05	−599.82	1.67	191.00	26.90
Suwon	3465.44	−2384.74	−590.55	−365.06	−23.54	−151.07	49.52
Siheng	3920.09	−1059.52	−920.43	182.25	−1286.49	677.93	−1514.39
Ansan	3868.27	−1045.76	−193.47	−292.40	−395.83	−1256.51	−684.71
Anseong	2328.93	−3439.41	−1072.45	1207.44	248.58	416.41	310.35
Anyang	1060.33	−767.78	71.54	−345.75	9.59	−26.93	−0.68
Yangju	2167.23	−2846.65	−192.94	−964.60	19.74	1785.28	31.97
Yangpyeong	451.60	1392.15	−2949.07	722.90	−3.79	339.36	47.25
Yeoju	1054.44	374.18	−3336.62	1129.63	128.86	400.46	249.24
Yeoncheon	1240.57	−1354.95	1755.13	−2062.68	76.26	128.85	221.89
Osan	1163.97	−944.33	−286.36	−135.66	8.49	183.58	10.31
Yongin	7130.08	−2978.75	−4976.87	−115.09	40.98	748.52	151.12
Uiwang	600.60	−406.09	−105.01	−252.72	0.33	145.49	17.39
Uijeongbu	1486.09	−1073.46	−292.48	−386.06	1.56	253.46	10.82
Icheon	2426.24	−1974.78	−976.91	−581.47	−0.96	1092.76	16.01
Paju	2990.49	1274.16	−2585.27	−3131.46	−276.22	1073.22	655.89
Pyeongtaek	6296.37	−7232.52	−1325.11	869.68	1737.38	1375.68	−1724.09
Pocheon	2416.37	−3083.93	−645.79	3.05	68.83	1088.78	153.31
Hanam	1402.77	−1506.13	−734.63	364.14	38.92	219.31	216.10
Hwaseong	8598.49	−7597.43	−2637.48	617.70	−1705.32	3221.19	−498.11

3.2. MSPA Pattern Changes in Green Area of Each City

Regarding MSPA changes for each of the cities, the results are as follows (Table 3). First, in the case of Yeoju, a city with the biggest number of Cores and Branches, it showed decreases compared to the data of the late 1988. In addition, the number of Bridges and Loops greatly decreased because most of its linear green areas with relatively small sizes ceased to exist due to the expansion of urbanized areas and increases in farmland areas. Moreover, Islets with small size green areas that could have played an important role as stepping-stone green areas greatly decreased in number.

In the case of Goyang, the number of Islets among the seven MSPA types decreased by 1322 when compared to that of 1988. Such a result was due to an increase in urbanized areas resulting from its geographical characteristic of being close to Seoul, as shown earlier in the land coverage changes. In addition, Cores known to play an important role as habitat spaces for biospecies also greatly decreased in number. In the case of Hwaseong (Figure 6), the number of Islets decreased by 3963, the biggest decrease among the 31 cities due to the urbanization of the last 30 years. Its Branches and Bridges also greatly decreased in number. Nevertheless, it is worth noting that its Cores increased in number because large scale green lands were fragmentized due to new constructions of roads and developments. Anseong was found to show changes in the spatial structure of green lands different from other cities. The result of MSPA showed that Cores, Loops, Bridges, and Branches increased in number in Anseong due to an increase in grassland areas among the land cover types. Next, in the case of Guri, Perforations belonging to the type arising from land use by humans in forests showed the biggest increase in number among the 31 cities. On the contrary, Perforations in Yangpyeong and Yongin decreased greatly in number due to the prohibition of farming activities, which took place in forests and forestation projects. MSPA drawings for 1988 and 2018 for all 31 cities are presented as Appendix A.

Table 3. Changes in MSPA categories by city (2018–1988).

Division	Core	Islet	Perforation	Edge	Loop	Bridge	Branch
Gapyeong	−183	105	−261	−342	109	−189	−944
Goyang	−127	−3122	−51	−536	−122	−167	−720
Gwacheon	3	−30	5	−20	5	−18	−34
Gwangmyeong	1	−258	−1	−49	−24	−7	−65
Gwangju	−138	128	−425	−878	−175	−235	−814
Guri	−42	−79	166	−109	−16	−25	−129
Gunpo	−6	−129	−12	−114	−17	−35	−83
Gimpo	−61	−1284	8	−155	−49	−30	−396
Namyangju	−284	793	−443	−1148	−123	−397	−900
Dongducheon	−70	−109	28	−172	−24	−72	−272
Bucheon	4	−263	2	28	−6	4	−30
Seongnam	−46	−151	−158	−432	−196	−57	−393
Suwon	−46	−979	−10	−223	−71	−62	−308
Siheng	−16	−336	−58	−332	−90	−79	−313
Ansan	0	−935	−15	−270	−90	−40	−356
Anseong	172	−990	−126	372	175	41	819
Anyang	−21	−216	−57	−120	−80	−31	−190
Yangju	−240	−547	−144	−1040	−107	−296	−979
Yangpyeong	−370	−236	−892	−1695	−113	−460	−1423
Yeoju	−514	−1575	−55	−2008	−350	−462	−2387
Yeoncheon	−313	−1004	38	−1093	2	−411	−1302
Osan	−63	−308	−6	−171	−38	−50	−348
Yongin	−43	−63	−475	−680	−231	−205	−539
Uiwang	−7	−51	−83	−136	−73	−24	−31
Uijeongbu	−11	−215	−30	−168	−64	−20	−144
Icheon	−45	−2143	−7	−74	38	−81	−375
Paju	−284	−1944	−337	−2040	−337	−484	−1712
Pyeongtaek	−124	−2190	−16	−275	7	−134	−493
Pocheon	−357	−2329	−55	−1302	−72	−301	−1705
Hanam	−18	−88	−41	−96	−26	−24	−188
Hwaseong	60	−3963	−39	−869	−419	−116	−824

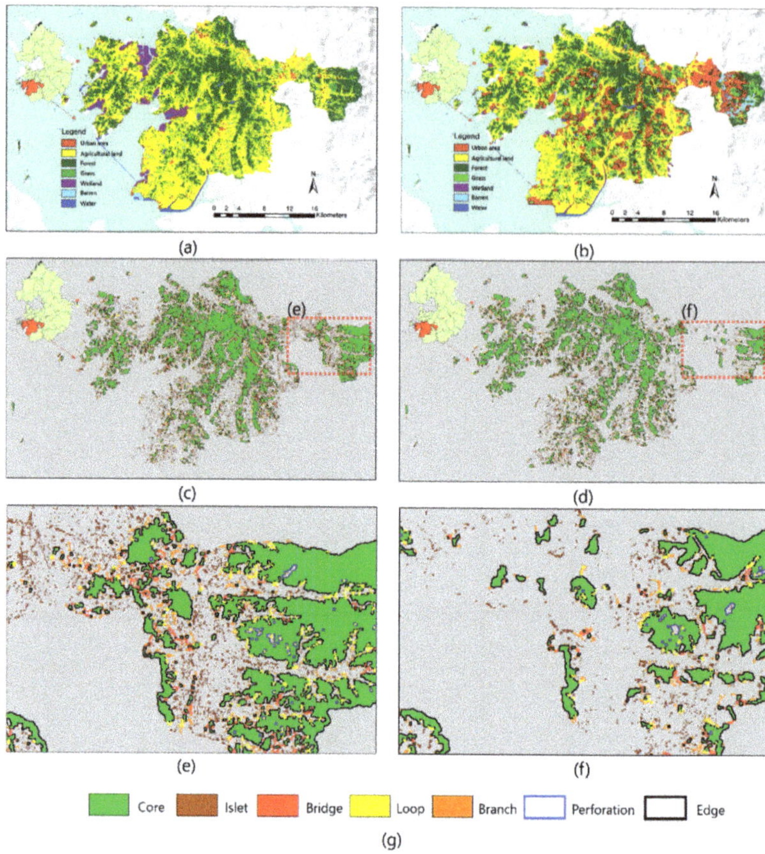

Figure 6. (**a**) 1988 land cover map of Hwaseong, (**b**) 2018 land cover map of Hwaseong, (**c**) 1988 MSPA map of Hwaseong (**d**) 2018 MSPA map of Hwaseong, (**e**) detailed view of map (**c**), (**f**) detailed view of map (**d**), and (**g**) legend of MSPA categories.

3.3. Cluster Analysis of 31 Cities

As a result of factor analysis, which utilized a total of 12 variables selected for categorization by city, factors were classified into a total of four (Table 4). These four factors converged a total of 25 times according to the Varimax Method. They explained about 88% of the total variance. The communality of the 12 variables was found to have high values (not smaller than 0.7). In addition, the KMO (Kaiser–Meyer–Olkin) value, which was the result of conducting a goodness of fit test of the measuring tool, was 0.575, meaning that there was no problem in selecting these variables. The goodness of fit of this model was found to be very high, because the probability value was found to be 0.00 in Bartlett's test of sphericity. Thus, there were significant differences, and commonality existed for the 12 selected variables.

The characteristics of each classified factor are shown as follows. First, the four variables belonging to Factor-1 were "Branch", "Core", "Bridge" and "Edge". They had an explanatory power of about 34% for the total variance. They were found to be variables with a high importance in the aspect of green networks. The reason these four variables are grouped un the same factor is because the annihilation of the Core directly affects the Branch, Bridge, and Edge. In fact, in the case of Yeoju, the number of Cores decreased the most, and at the same time, the number of Branches and Bridges existing in a linear form at the edge of the forest also decreased significantly compared to other cities.

Next, Factor-2 had an explanatory power of about 24%. It included "Agricultural Land", "Urban Area", "Barren", and "Islet". "Urban Area" and "Barren" showed negative values, differently from other variables. This was because "Urban Area" and "Barren" increased in size, contrary to the other 12 variables, for which the size or the number decreased when compared to those of 1988. Like this, the variables of Factor-2 were found to play an important role in clustering cities, with changes in the size of farmland due to the expansion of urbanized areas. Changes in Islets known to function as stepping-stone green areas were similar (i.e., highly urbanized cities). Factor-3 was made up of "Perforation" and "Forest". Its explanatory power was found to be about 17%. The reason these two variables are grouped as the same factor is because the disappearance of Perforation is directly related to the forest. As defined in Table 2, Perforation is a non-green space that appears in the forest, and is a variable that disappears when the forest is destroyed. In fact, looking at the cases of Yongin and Namyangju, it was found that the number of Perforations greatly decreased as the areas of forest decreased. Lastly, Factor-4 had an explanatory power of about 12%. This factor had a high effect on the grouping of cities in which time-sequential changes of "Wetland" and "Loop" were similar.

Table 4. Results of factor analysis.

Factor	Factor-1	Factor-2	Factor-3	Factor-4
Branch	0.965	0.112	0.100	0.113
Core	0.961	−0.080	0.101	−0.145
Edge	0.927	0.083	0.288	0.185
Bridge	0.920	0.036	0.316	−0.021
Agricultural land	−0.236	0.926	−0.061	−0.142
Urban area	0.083	−0.826	−0.400	−0.168
Barren	−0.234	−0.781	−0.135	−0.173
Islet	0.283	0.778	−0.318	0.195
Perforation	0.330	−0.146	0.856	−0.029
Forest	0.276	0.251	0.853	0.136
Wetland	−0.087	0.069	−0.018	0.942
Loop	0.471	0.317	0.323	0.609
Eigen value	4.173	2.971	2.050	1.463
Explained amount of total variance (%)	34.777	24.759	17.004	12.190
Cumulative explanation (%)	34.777	59.536	76.620	88.810

Based on these results of the factor analysis, 31 cities in Gyeonggi-do were classified into six clusters (Figure 7). Cluster-1 was made up of 15 cities, the biggest number of cities, which included Gwangmyeong, Gunpo and Anyang. Cluster-2 was a cluster made up of five cities, including Suwon, Gimpo and Goyang. Cluster-3 was made up of Gwangju, Namyangju, Yangpyeong and Yongin. Cluster-4 had five cities, Yeoju, Paju, Yangju, Pocheon and Yeoncheon, that showed similar environmental changes. Cluster-5 was an independent cluster formed by Pyeongtaek alone. Cluster-6, too, was an independent cluster formed by Hwaseong alone.

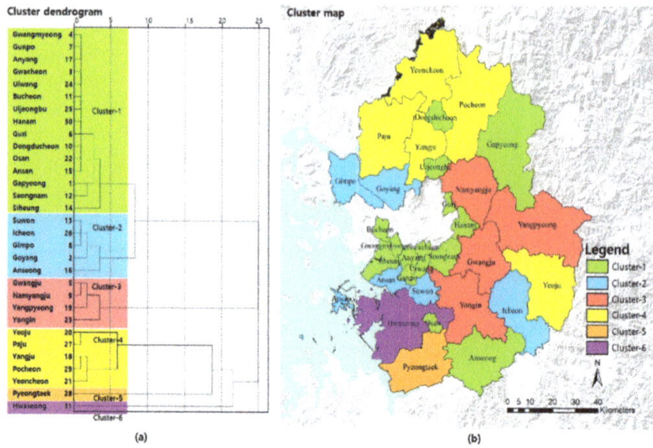

Figure 7. (**a**) Cluster dendrogram and (**b**) cluster map of research cities.

4. Discussion

4.1. Discussion and Limitation

In the present study, we determined changes in land coverage and changes in MSPA values in each of the 31 satellite cities surrounding Seoul for about 30 years, from 1988 to 2018. Cities that showed similar environmental changes were grouped. First, for changes in land coverage, urbanized areas in all 31 cities were found to have increased. In particular, urbanization was concentrated in cities adjacent to Seoul, the capital. For example, in the case of Goyang (Figure 8), most spaces that were farmland in the past were converted into urban areas due to such geographic characteristics. As a result, forest, grassland and wetland with high ecological values greatly decreased in size. Accordingly, for large cities such as Suwon, Icheon, Gimpo and Siheung, which showed environmental changes similar to those of Goyang, it is particularly important to preferentially search for a plan to effectively preserve natural spaces with high values in advance when establishing development plans.

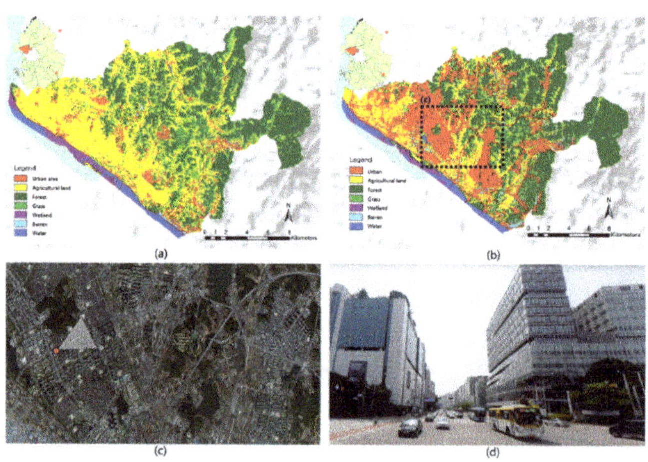

Figure 8. (**a**) 1988 land cover map of Goyang, (**b**) 2018 land cover map of Goyang, (**c**) satellite image of map (**b**,**d**) road view of map (**c**) Most of the space that used to be agricultural land has been converted into urbanization areas.

Next, when we look into the results after analyzing changes in MSPA category in each of the 31 cities, 103 Cores and 733 Islets were found to have decreased, on average, in these 31 cities. A total of 115 Perforations, 521 Edges, 83 Loops, 144 Bridges and 567 Branches decreased on average. As can be seen from such an analysis result, Cores that are key green lands and Islets that are stepping-stone green lands have ceased to exist in most cities due to the urbanization that has progressed for several decades, meaning that the green network function is greatly deteriorated at a whole city level. Accordingly, an environmental plan should be established with a focus on quantitatively/qualitatively securing green lands by creating additional green lands of diverse sizes and forms to recover the function of green networks.

To look into the characteristics of each of these six clusters in more detail, the average value of the factor of each cluster was calculated based on the characteristics examined (Table 5). In particular, as factors with a positive value and a negative value appeared in a mixture in Factor-2, among the four factors, the Agricultural Lands and Islets with decreased sizes and numbers were classified as Detailed Factor 2-1, and Urban Areas and Barrens with increased sizes were classified as Detailed Factor 2-2.

Table 5. Factor score average by cluster (the numbers in bold indicate the average value of the highest or lowest factor among the six clusters and can represent the characteristics of each cluster).

Factor	Cities	Factor-1	Factor-2		Factor-3	Factor-4
			2-1	2-2		
Cluster-1	Gwangmyeong, Gunpo, Anyang, Gwacheon, Uiwang, Bucheon, Uijeongbu, Hanam, Guri, Dongducheon, Osan, Ansan, Gapyeong, Seongnam, Siheung	**0.62999**	0.44254	−0.48388	0.39217	0.23549
Cluster-2	Suwon, Icheon, Gimpo, Goyang, Anseong	0.35109	−0.60356	0.08633	**0.42075**	−0.05106
Cluster-3	Gwangju, Namyangju, Yangpyeong, Yongin	−0.83973	0.63843	0.40742	**−1.96838**	−0.24216
Cluster-4	Yeoju, Paju, Yangju, Pocheon, Yeoncheon	**−1.60644**	−0.22083	0.11427	0.02580	−0.29516
Cluster-5	Pyeongtaek	0.11024	−2.03963	1.43641	0.13906	**2.08189**
Cluster-6	Hwaseong	0.07547	**−3.03028**	**3.18919**	−0.38086	**−2.91448**

First, Cluster-1 was found to be the cluster for which the score for Factor-1 was higher than that of any other cluster. That is, it is a cluster without a big environmental change. In reality, Cores, Branches, and Bridges were found to have decreased less in relevant cities than in cities of other clusters, although urbanization progressed. It was worth noting that the extinction of green lands was not high in Bucheon, Hanam, Guri, or Gwacheon, although these cities were very close to Seoul, the capital. Such a result is attributable to the restriction on development enforced by designating forests of relevant cities as Greenbelts to prevent the thoughtless expansion of Seoul. However, as Greenbelts are released gradually due to continuous development pressure, the emphasis should be put on maintaining the function of green networks by minimizing the extinction of small-scale green lands of diverse forms such as Branches, Bridges, Islets, etc., when development plans are established.

Cluster-2 was a cluster made up of five cities, including Suwon, Gimpo and Goyang, that were classified as relatively large cities. It showed the highest value for Factor-3 among the four Factors, meaning that its decrease in forest area was not high compared to the other clusters. However, it showed negative values for Factor-4 and Factor-2-1. Its spaces with high values such as wetland, stepping-stone green lands, etc., greatly decreased in number due to urbanization. In addition, as cities belonging to this cluster are still under high development pressure due to the continuous population increase and industrial development, emphasis should be put on the arrangement of a plan to preserve natural resources with high ecological values.

Cluster-3 was made up of Gwangju, Namyangju, Yangpyeong and Yongin. It was a cluster that showed a negative score for Factor-3. This cluster had a higher score for Factor -3 than the other clusters. Thus, it was a cluster with a large decrease in forest area. Further, as it had a big negative value for Factor-1, it was characterized by a big decrease in the number of linear connection green lands such as Branches, Bridges, etc. Accordingly, cities in the relevant cluster should put an emphasis on the minimization of additional damage to and the fragmentation of forests, while preserving small linear green lands at the same time.

Cluster-4 was a cluster made up of Yeoju, Paju, Yangju, Pocheon and Yeoncheon. It showed a higher negative value for Factor-1 than the other clusters. It was characterized by a big decrease in the actual number of large-scale green lands and linear green lands of diverse forms. As such a decrease in the number of green lands will lead to a decrease in the function of urban green networks, it is desirable to prioritize the creation of additional green lands to enhance green land connectivity when environmental plans are established.

Cluster-5 was an independent cluster formed by Pyeongtaek alone. It showed the highest positive value for Factor-4 among the four Factors. In fact, Pyeongtaek was the city with the largest increase in wetland areas. Accordingly, environmental plans should be established with an emphasis on the effective preservation of wetland and swamps that play an important role as habitats for biospecies.

Lastly, Cluster-6 was an independent cluster formed by Hwaseong alone, a city with the highest negative value for Factor-4. That is, this city had the biggest decrease in wetland areas. In addition, Cluster-6 showed larger changes for Factor 2-1 and 2-2 than the other clusters. As a whole, Cluster-6 showed a large environmental change due to urbanization. Accordingly, quantitative expansion of green spaces through the creation of additional green lands, and the establishment of a systematic preservation and management plan for wetland and swamps, should be preferentially considered for Hwaseong in order for it to function as an environmentally sustainable city.

From a small specific unit space to a large-scale city and even regional and territorial space, it is predicted that future land space will be more directly affected by numerous developments. Therefore, how will the various impacts caused by development change the future land space? By what means and methods will we be able to control the influencing factors and change processes in a natural environment and landscape-friendly way? In this respect, the results of this study are different from other related studies in that they involve clustered cities at the regional level and are based on the results of time-series environmental changes. The results thus present directions for establishing environmental policies for each cluster.

First, among studies examining environmental changes using landcover change [45–48], Kumar et al. [49] used land cover types in 1976, 1989, 2000 and 2014 for Usri watershed. They looked at the changes in the landscape matrix, and based on these analysis results, clustered land cover types showing similar values for each year. However, a limitations is that the scope of the space was limited to a specific watershed and that the analysis result could not be visually confirmed.

In addition, in the case of Korea, the consideration of the physical environment and the human psychological behavior aspect rather than the natural environment factor is considered more important in figuring out the value of the city. Lee's [50] research emphasizes the need for qualitative growth such as the quality of life rather than quantitative growth focusing on economic aspects. For this purpose, the 31 cities in Gyeonggi-do were grouped based on the Physical Environment Satisfaction and Neighborhood Satisfaction indices. Of the total of 14 indicators used for grouping, it was found that one of the indicators related to the natural environment was satisfaction with the park.

As evidenced by numerous studies, urban land use and environmental changes are inevitable phenomena in the process of urban development. Therefore, the Fragstats program, using the land cover map, has been used importantly in diagnosing environmental changes due to urbanization, but the fact that the analysis results cannot be looked at in drawings is a big limitation. In this aspect, the MSPA analysis used in this study is different in that the analysis results are presented in drawings. In addition, the results of this study are environmentally friendly in that they suggest a direction for establishing differentiated environmental policies for each city, such as the quantitative expansion of green areas, conservation of small-scale linear green areas, wetland conservation, and minimization of forest fragmentation in connection with land cover change. However, since a city is a space where humans and nature coexist, it is judged that it is desirable to comprehensively consider factors such as the physical environment and human psychological behavior, not just the natural environment aspect.

However, in order to increase the effectiveness of the study, it is necessary to additionally consider the following aspects in future studies. First, environmental changes in cities need to be determined using more detailed basic data. In Korea, a City Biotope Map was recently prepared with a high spatial resolution of a 5 m level on a national scale pursuant to Article 34-2 (Preparation/Utilization of City Biotope Map) of the Natural Environment Conservation Act. Accordingly, it is required to determine environmental changes by actively utilizing a more detailed basic map such as a City Biotope Map and to establish environmental plans based on it. Moreover, the present study had another limitation: the cities were clustered based on results of changes in land coverage and spatial structures of green lands. Accordingly, it is desirable to determine environmental changes in cities comprehensively by taking into account the values of diverse aspects such as soil, air, surface temperature and water quality and supply, as well as endangered species, etc., to find the appropriate direction to take when coping with such environmental changes based on the results of such determinations.

4.2. Policy Proposal

Although spatial plans for national land are established in Korea for each of the diverse spatial scales, such as Comprehensive National Territorial Plans, Metropolitan City Plans, Si/Gun Master Plans, etc., when we take into account the fact that concrete and detailed physical plans established under the premise of implementation are realized at a city level, we can see that establishing differentiated environmental plans for sustainable land use management based on time-sequential environmental changes at a city space level is very important.

However, most environmental policies established for many cities that belong to the same area (Gyeonggi-do, Gyeongsangbuk-do, etc.) put emphasis on the quantitative expansion of green lands. Accordingly, it is important to grasp time-sequential environmental changes and secure environmental drawings that can be connected to development plan drawings at the same time to establish differentiated environmental policies. In Korea, although a Biotope Map is built and renewed at five-year intervals based on Article 34-2 (Preparation/Utilization of Biotope Map) of the Natural Environment Conservation Act, there is no mandatory provision that forces its utilization when establishing development plans. Accordingly, the law/system is required to be reinforced to obtain basic data so that a Biotope Map or Ecosystem Service Map is mandatorily taken into account when establishing development plans.

Moreover, when we look into the budget of each ministry of the Korean government as of 2020, the budget of the Ministry of Environment was much smaller than that of the other ministries. It was about KRW 9 trillion, while that of the Ministry of Land, Infrastructure, and Transport was about KRW 50 trillion, that of the Ministry of National Defense was about KRW 50 trillion, and that of the Ministry of Employment and Labor was about KRW 30 trillion. Accordingly, to build diverse basic data and diagnose environmental changes in cities, the expansion of the budget is required. Each city is also required to use more of their budget for the preservation and management of the environment.

In addition, for sustainable urban development, the connection between disciplines that are being pursued in different perspectives, laws, and administrative organizations will be of utmost importance. For example, the natural environment conservation plan is mainly researched in the field of biology, the park and green space plan in the landscape field, and the artificial landscape plan in the field of architecture and civil engineering. However, as mentioned earlier, transdisciplinary cooperation to diagnose and solve environmental problems by integrating and synthesizing problems themselves for environmentally sustainable city development is more important than dealing with each environmental problem as a separate entity [51,52].

5. Conclusions

As pressure on the suburbs of large cities is increasing worldwide, this study examined land cover changes and MSPA changes over the past 30 years in 31 cities adjacent to the capital of Korea, Seoul. Cities with similar patterns were clustered based on the analysis results. Based on these results, this study is meaningful in that it suggests the characteristics of environmental change for each community and the direction of environmental planning based on these.

However, in the case of Korea, a number of development plans are scheduled for small and medium-sized cities that have not experienced significant environmental change compared to large cities. Therefore, prior to development plans, it is necessary to identify high-value environmental resources, key green areas for green network construction, and major habitats for species conservation, and to establish an urban development plan that prioritizes these spaces.

In addition, we believe that transdisciplinary cooperation is more important than ever. Therefore, experts in land use planning, landscape planning, forest planning, traffic planning, and climate change planning, which are all highly related to urban development planning, need to actively reflect their research results in urban development plans. It is judged that only if these efforts are supported, it will be possible to develop the national space more environmentally and sustainably.

Author Contributions: Conceptualization, J.-H.K. and O.-S.K.; methodology, J.-H.K.; validation, J.-H.K. and O.-S.K.; formal analysis, J.-H.R.; investigation, J.-H.K. and O.-S.K.; resources, J.-H.K.; data curation, O.-S.K. and J.-H.K.; writing, J.-H.K., J.-H.R.; supervision, J.-H.R. All authors have read and agreed to the published version of the manuscript.

Funding: This research received no external funding.

Institutional Review Board Statement: Not applicable.

Informed Consent Statement: Not applicable.

Data Availability Statement: Not applicable.

Conflicts of Interest: The authors declare no conflict of interest.

Appendix A

Table A1. Current status of 31 cities.

City Name	Area(km^2)	Population	Location and Characteristics
Gapyeong	834.4	62,605	Located in the northeast of Seoul
Goyang	267.4	1,076,406	Located in the north of Seoul and in direct contact with Seoul
Gwacheon	35.9	61,902	Located in the south of Seoul and in direct contact with Seoul
Gwangmyeong	38.5	308,678	Located in the southwest of Seoul and in direct contact with Seoul
Gwangju	431.8	379,480	Located in the southeast of Seoul
Guri	33.3	197,889	Located in the east of Seoul and in direct contact with Seoul
Gunpo	36.4	275,508	Located in the south of Seoul
Gimpo	276.6	458,505	Located in the west of Seoul
Namyangju	460.1	709,881	Located in the east of Seoul
Dongducheon	95.7	93,968	Located in the north of Seoul
Bucheon	53.5	824,865	Located in the east of Seoul and in direct contact with Seoul
Seongnam	141.8	940,966	Located in the south of Seoul and in direct contact with Seoul
Suwon	121.1	1,190,074	Located in the south of Seoul

Table A1. *Cont.*

City Name	Area(km^2)	Population	Location and Characteristics
Siheng	134.4	489,077	Located in the southwest of Seoul
Ansan	147.1	653,733	Located in the southwest of Seoul
Anseong	554.1	186,104	Located in the southeast of Seoul
Anyang	58.5	554,857	Located in the south of Seoul and in direct contact with Seoul
Yangju	310.2	229,052	Located in the north of Seoul
Yangpyeong	877.8	118,372	Located in the east of Seoul
Yeoju	607.9	111,438	Located in the southeast of Seoul
Yeoncheon	695.3	43,542	Located in the north of Seoul and on the border with North Korea
Osan	42.7	228,718	Located in the south of Seoul
Yongin	591.5	1,075,659	Located in the southeast of Seoul
Uiwang	54.0	162,751	Located in the southeast of Seoul
Uijeongbu	81.6	456,660	Located in the northwest of Seoul and bordering North Korea
Icheon	461.2	219,537	Located in the southeast of Seoul
Paju	672.6	459,158	Located in the northwest of Seoul and bordering North Korea
Pyeongtaek	452.1	527,166	Located in the south of Seoul
Pocheon	826.4	147,854	Located in the northeast of Seoul
Hanam	93.1	285,693	Located in the east of Seoul and in direct contact with Seoul
Hwaseong	688.1	842,864	Located in the south of Seoul

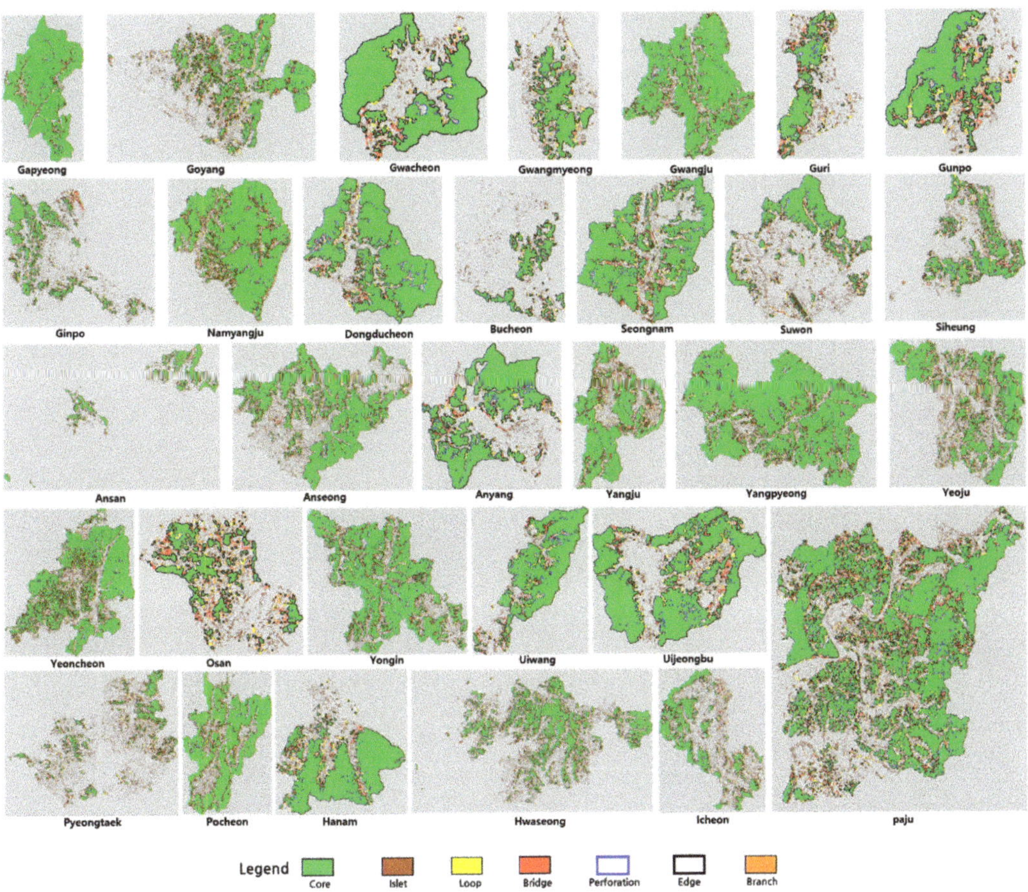

Figure A1. 1988 MSPA map of 31 cities.

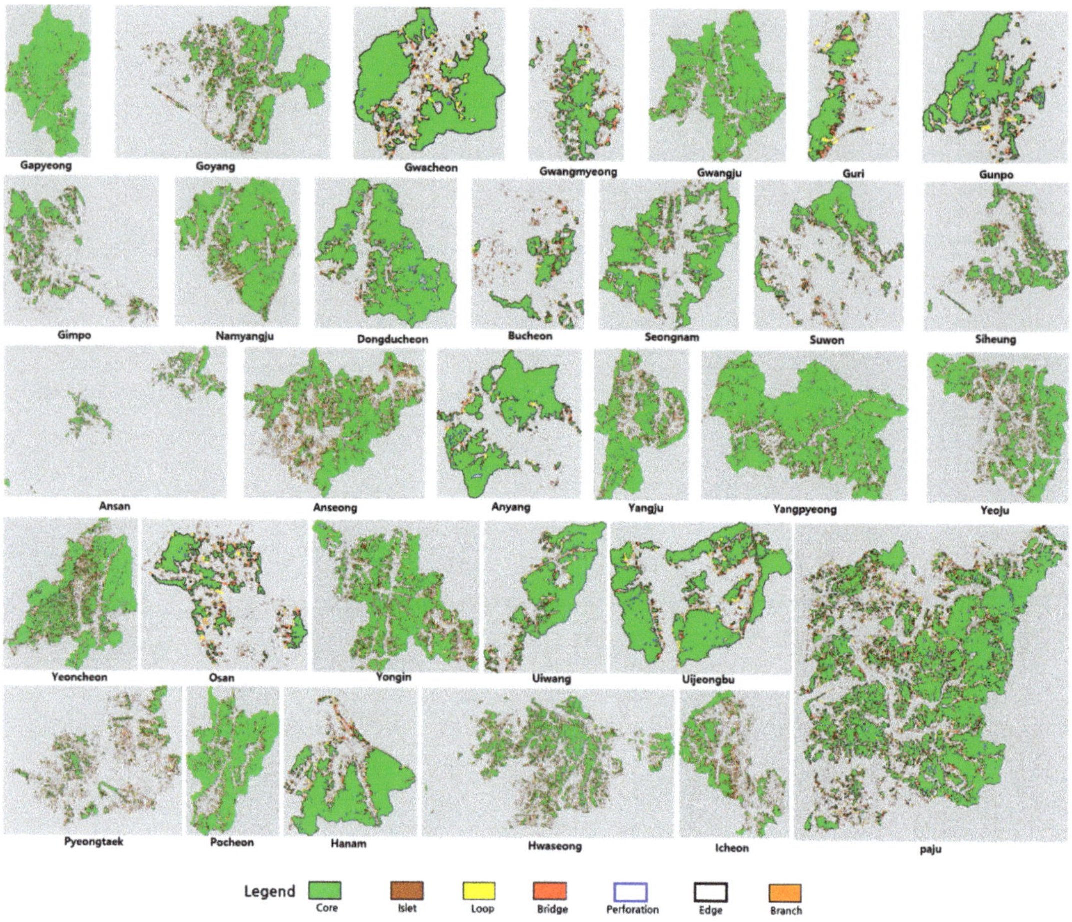

Figure A2. 2018 MSPA map of 31 cities.

References

1. Song, M.K.; Chang, H. Characterization of Cities in Seoul Metropolitan Area by Cluster Analysis. *J. Korean Soc. Geospat. Inform. Syst.* **2010**, *18*, 83–88.
2. Kim, B.S.; Yuh, H.K. A Study on Functional Features in Types of the Cities in Korea. *Geogr. J. Korea* **2010**, *44*, 537–552.
3. Kim, I.K. Socioeconomic concentration in the Seoul metropolitan area and its implications in the urbanization process of Korea. *Korean J. Soc.* **2010**, *44*, 111–128.
4. Zonneveld, I.S. The Land Unit-A Fundamental Concept in Landscape Ecology and Its Applications. *Landsc. Ecol.* **1989**, *3*, 67–86. [CrossRef]
5. Gobster, P.H. Landscape and urban planning at 100-Looking back moving forward. *Landsc. Urban Plan.* **2011**, *100*, 315–317. [CrossRef]
6. Bruns, D. Was kann Landschaftsplanung leisten? Alte und neue Funktionen der Landschaftsplanung. *Nat. Landsch.* **2003**, *35*, 114–118.
7. Wiens, J.A. Landscape ecology as a foundation for sustainable conservation. *Landsc. Ecol.* **2009**, *24*, 1053–1065. [CrossRef]
8. Thornton, D.H.; Branch, L.C.; Sunquist, M.E. The influence of landscape, patch, and within-patch factors on species presence and abundance—A review of focal patch studies. *Landsc. Ecol.* **2011**, *26*, 7–18. [CrossRef]
9. Lizee, M.H.; Manel, S.; Mauffrey, J.F.; Tatoni, T.; Cottin, M.D. Matrix configuration and patch isolation influences override the species-area relationship for urban butterfly communities. *Landsc. Ecol.* **2012**, *27*, 159–169. [CrossRef]
10. Downs, J.A.; Horner, M.W. Enhancing habitat connectivity in fragmented landscapes: Spatial modeling of wildlife crossing structures in transportation networks. *Ann. Am. Assoc. Geogr.* **2012**, *102*, 17–34. [CrossRef]

11. Forman, R.T.; Godron, M. Patches and structural components for a landscape ecology. *BioScience* **1981**, *31*, 733–740.
12. Forman, R.T.; Alexander, L.E. Roads and their major ecological effects. *Annu. Rev. Ecol. Systemat.* **1998**, *29*, 207–231. [CrossRef]
13. Parece, T.; McGee, J.A.; Campbell, J.B. *Remote Sensing with ArcGIS Pro. Virginia*; Geospatial Consortium: Wayland, MA, USA, 2019.
14. Gillanders, S.N.; Coops, N.C.; Wulder, M.A. Multitemporal remote sensing of landscape dynamics and pattern change: Describing natural and anthropogenic trends. *Prog. Phys. Geogr.* **2008**, *32*, 503–528. [CrossRef]
15. Chen, Y.H. Application of GIS and remote sensing technology in the production of image map. *Electron. Test.* **2016**, *11*, 107–108.
16. Yi, L.; Zhang, G. Object-oriented remote sensing imagery classification accuracy assessment based on confusion matrix. *Proc. Int. Conf. Geoinf.* **2012**, *10*, 1109.
17. Bruzzone, L. Detection of changes in remotely-sensed images by the selective use of multi –spectral information. *Int. J. Remote Sens.* **1997**, *18*, 3883–3888. [CrossRef]
18. Griffith, J.A. Geographic techniques and recent applications of remote sensing to landscape-water quality studies. *Water Air Soil Pollut.* **2002**, *138*, 181–197. [CrossRef]
19. McGarigal, K.; Marks, B.J. *FRAGSTATS: Spatial Pattern Analysis Program for Quantifying Landscape Structure*; Gen. Tech. Rep. PNW-GTR-351; Department of Agriculture, Forest Service, Pacific Northwest Research Station: Portland, OR, USA, 1995.
20. Kupfer, J.A. Landscape ecology and biogeography: Rethinking land-scape metrics in a post-FRAGSTATS landscape. *Prog. Phys. Geogr. Earth Environ.* **2012**, *36*, 400–420. [CrossRef]
21. McGarigal, K.; Cushman, S.A.; Ene, E. *Fragstats v4: Spatial Pattern Analysis Program for Categorical and Continuous Maps*; Computer Software Program; University of Massachusetts: Amherst, MA, USA, 2012.
22. McGarigal, K. 'Fragstats Help', University of Massachusetts, Amherst. 2015. Available online: https://www.umass.edu/landeco/research/fragstats/documents/fragstats_documents.html (accessed on 30 May 2021).
23. Mariana, B.; Barbara, R.; Luciana, G.; Rosa, L. The Extent Infrastructure Causing Fragmentation in the Hydrocarbon Basin in the Arid and Semi-Arid Zones of Patagonia (Argentina). *Sustainability* **2019**, *11*, 5956.
24. Reddy, C.S.; Sreelekshmi, S.; Jha, C.S.; Dadhwal, V.K. National Assessment of Forest Fragmentation in India: Landscape Indices as Measures of the Effects of Fragmentation and Forest Cover Change. *Ecol. Eng.* **2013**, *60*, 453–464. [CrossRef]
25. Korea Research Institute for Human Settlements. *A Study on the Urban Planning Measurement by Linking between National Land Plan and Environment Plan*; Korea Research Institute for Human Settlements: Anyang-si, Korea, 2015; pp. 61–85.
26. Heo, H.K.; Seong, H.C.; Lee, D.K.; Heo, M.J.; Park, J.H. A Study on Assessment Indicators for Integrated Management on Korea National Planning and Environmental Planning. *J. Korean Environ. Res. Technol.* **2018**, *21*, 27–45.
27. Heo, H.K.; Lee, D.K.; Seong, H.C.; Heo, M.J.; Park, J.H. Assessment of National and Regional Plans Using Integrated Management Index of Korea National Planning and Environmental Planning for Present Stats Evaluation. *J. Korean Environ. Res. Technol.* **2019**, *22*, 81–91.
28. de Oña, R.; López, G.; de los Rios, F.J.; de Oña, J. Cluster Analysis for Diminishing Heterogeneous Opinions of Service Quality Public Transport Passengers. *Procedia Soc. Behav. Sci.* **2014**, *162*, 459–466. [CrossRef]
29. Pirra, M.; Diana, M. Classification of Tours in the U.S. National Household Travel Survey through Clustering Techniques. *J. Transp. Eng.* **2016**, *142*, 04016021. [CrossRef]
30. Miriam, P.; Ruggero, G.P. Comparing Transport Quality Perception among Different Travelers in European Cities through Co Cluster Analysis. *Sustainability* **2019**, *11*, 7159.
31. Wang, C.; Wang, Z.; Li, Q. Emergence of urban clustering among U.S. cities under environmental stressors. *Sustain. Cities Soc.* **2020**, *63*, 102481. [CrossRef]
32. Noiva, K.; Fernandez, J.; Wescoat, L. Cluster analysis of urban water supply and demand: Toward large-scale comparative sustainability planning. *Sustain. Cities Soc.* **2016**, *27*, 484–496. [CrossRef]
33. Jia, Y.; Tang, L.; Xu, M.; Yang, X. Landscape pattern indices for evaluating urban spatial morphology—A case study of Chinese cities. *Ecol. Indicat.* **2019**, *99*, 27–37. [CrossRef]
34. Ko, A.R.; Lim, J.W.; Kim, S.H. Classification and Characteristic Analysis of Mountain Village Landscape Using Cluster Analysis. *J. Korean Soc. Rural Plan.* **2020**, *26*, 101–112. [CrossRef]
35. Kuo, H.; Tsou, K. Modeling and Simulation of the Future Impacts of Urban Land Use Change on the Natural Environment by SLEUTH and Cluster Anlysis. *Land* **2018**, *10*, 72. [CrossRef]
36. Environmental Geographic Information Service. Available online: https://egis.me.go.kr (accessed on 30 May 2021).
37. Vogiatzakis, I.N.; Mannion, A.M.; Griffiths, G.H. Mediterranean ecosystems: Problems and tools for conservation. *Prog. Phys. Geogr.* **2006**, *30*, 175–200. [CrossRef]
38. Murica, C. Edge Effects in Fragmented Forests: Implications for Conservation. *Trends. Ecol. Evol.* **1995**, *10*, 58–62.
39. McGarigal, K.; Cushman, S.A.; Neel, M.C.; Ene, E. FRAGSTATS: Spatial Pattern Analysis Program for Categorical Maps. Computer Software Program Produced by the Authors at the University of Massachusetts Amherst. 2002. Available online: https://www.umass.edu/landeco/research/fragstats/fragstats.html (accessed on 7 June 2021).
40. Vogt, P.; Riitters, K. Guidos Toolbox: Universal digital image object analysis. *Eur. J. Remote Sens.* **2017**, *50*, 352–361. [CrossRef]
41. Seo, H.J.; Choi, C.H.; Lee, K.J.; Woo, D.G. Landscape Characteristics Based on Effectiveness of Wildlife Crossing Structures in South Korea. *Sustainability* **2021**, *13*, 675. [CrossRef]
42. Soille, P.; Vogt, P. Morphological segmentation of binary patterns. *Pattern Recognit. Lett.* **2009**, *30*, 456–459. [CrossRef]

43. Ye, H.; Yang, Z.; Xu, X. Ecological Corridors Analysis Based on MSPA and MCR Model-A Case Study of the Tomur World Natural Heritage Region. *Sustainability* **2020**, *12*, 959. [CrossRef]
44. Carmen, C.V.; Romain, W.; Jeffrey, B. Clustering of European Smart Cities to Understand the Cities' Sustainability Strategies. *Sustainability* **2021**, *13*, 513.
45. Pauleit, S.; Duhme, F. Assessing the environmental performance of land cover types for urban planning. *Landsc. Urban Plan.* **2000**, *52*, 1–20. [CrossRef]
46. Lambin, E.F.; Meyfroidt, P. Land use transitions: Socio-ecological feedback versus socio-economic change. *Land Use Policy* **2010**, *27*, 108–118. [CrossRef]
47. Pauleit, S.; Ennos, R.; Golding, Y. Modeling the environmental impacts of urban land use and land cover change—A study in Merseyside, UK. *Landsc. Urban Plan.* **2005**, *71*, 295–310. [CrossRef]
48. Ding, H.Y.; Shi, W.Z. Land-use/land-cover change and its influence on surface temperature: A case study in Beijing City. *Int. J. Remote Sens.* **2013**, *34*, 5503–5517. [CrossRef]
49. Kumar, M.; Denis, D.M.; Singh, S.K.; Szabo, S.; Suryavanshi, S. Landscape Metrics for assessment of land cover change and fragmentation of a heterogeneous watershed. *Int. J. Remote Sens.* **2018**, *10*, 224–233. [CrossRef]
50. Lee, K.Y. Relationship between Physical Environment Satisfaction Neighborhood Satisfaction, and Quality of Life in Gyeonggi, Korea. *Land* **2021**, *10*, 663. [CrossRef]
51. Norris, P.E.; Rourke, M.O.; Mayer, A.S.; Halvorsen, K.E. Managing the wicked problem of transdisciplinary team formation in socio-ecological systems. *Landsc. Urban Plan.* **2016**, *154*, 115–122. [CrossRef]
52. Taylor, L.; Hochuli, D. Defining greenspace: Multiple uses across multiple disciplines. *Landsc. Urban Plan.* **2017**, *158*, 25–38. [CrossRef]

land

MDPI

Correction

Correction: Gu et al. Qualifying Land Use and Land Cover Dynamics and Their Impacts on Ecosystem Service in Central Himalaya Transboundary Landscape Based on Google Earth Engine. *Land* 2021, *10*, 173

Changjun Gu [1,2], Yili Zhang [1,2,3], Linshan Liu [1,*], Lanhui Li [4], Shicheng Li [5], Binghua Zhang [1,2], Bohao Cui [1,2] and Mohan Kumar Rai [1,2]

1 Key Laboratory of Land Surface Pattern and Simulation, Institute of Geographic Sciences and Natural Resources Research, CAS, Beijing 100101, China; gucj.18b@igsnrr.ac.cn (C.G.); zhangyl@igsnrr.ac.cn (Y.Z.); zhangbh.17b@igsnrr.ac.cn (B.Z.); cuibh.19b@igsnrr.ac.cn (B.C.); mkrai2019@igsnrr.ac.cn (M.K.R.)
2 College of Resources and Environment, University of Chinese Academy of Sciences, Beijing 100049, China
3 CAS Center for Excellence in Tibetan Plateau Earth Sciences, Beijing 100101, China
4 Fujian Key Laboratory of Pattern Recognition and Image Understanding, Xiamen University of Technology, Xiamen 361024, China; lilh.15b@igsnrr.ac.cn
5 Department of Land Resource Management, School of Public Administration, China University of Geosciences, Wuhan 430074, China; lisc@cug.edu.cn
* Correspondence: liuls@igsnrr.ac.cn

Citation: Gu, C.; Zhang, Y.; Liu, L.; Li, L.; Li, S.; Zhang, B.; Cui, B.; Rai, M.K. Correction: Gu et al. Qualifying Land Use and Land Cover Dynamics and Their Impacts on Ecosystem Service in Central Himalaya Transboundary Landscape Based on Google Earth Engine. *Land* 2021, *10*, 173. *Land* 2021, *10*, 506. https://doi.org/10.3390/land10050506

Received: 23 April 2021
Accepted: 30 April 2021
Published: 10 May 2021

Error in Figure

In the original article [1], there was a mistake in Figure 5. The distribution of ESVs in the KSL in 2000 and 2015. as published. Figure 5 and Figure 2 were same figure, and Figure 5 should be the presentation of the distributions of ESVs. The corrected Figure 5 appears below. The authors apologize for any inconvenience caused and state that the scientific conclusions are unaffected. The original article has been updated.

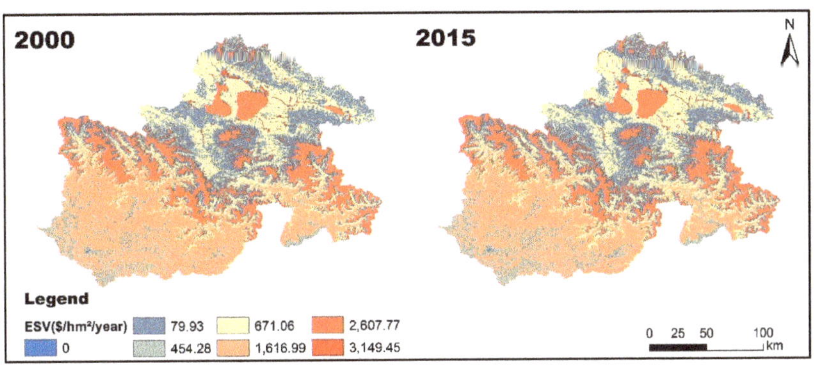

Reference

1. Gu, C.; Zhang, Y.; Liu, L.; Li, L.; Li, S.; Zhang, B.; Cui, B.; Rai, M.K. Qualifying Land Use and Land Cover Dynamics and Their Impacts on Ecosystem Service in Central Himalaya Transboundary Landscape Based on Google Earth Engine. *Land* **2021**, *10*, 173. [CrossRef]

land

MDPI

Article

Qualifying Land Use and Land Cover Dynamics and Their Impacts on Ecosystem Service in Central Himalaya Transboundary Landscape Based on Google Earth Engine

Changjun Gu [1,2], Yili Zhang [1,2,3], Linshan Liu [1,*], Lanhui Li [4], Shicheng Li [5], Binghua Zhang [1,2], Bohao Cui [1,2] and Mohan Kumar Rai [1,2]

1 Key Laboratory of Land Surface Pattern and Simulation, Institute of Geographic Sciences and Natural Resources Research, CAS, Beijing 100101, China; gucj.18b@igsnrr.ac.cn (C.G.); zhangyl@igsnrr.ac.cn (Y.Z.); zhangbh.17b@igsnrr.ac.cn (B.Z.); cuibh.19b@igsnrr.ac.cn (B.C.); mkrai2019@igsnrr.ac.cn (M.K.R.)
2 College of Resources and Environment, University of Chinese Academy of Sciences, Beijing 100049, China
3 CAS Center for Excellence in Tibetan Plateau Earth Sciences, Beijing 100101, China
4 Fujian Key Laboratory of Pattern Recognition and Image Understanding, Xiamen University of Technology, Xiamen 361024, China; lilh.15b@igsnrr.ac.cn
5 Department of Land Resource Management, School of Public Administration, China University of Geosciences, Wuhan 430074, China; lisc@cug.edu.cn
* Correspondence: liuls@igsnrr.ac.cn

Citation: Gu, C.; Zhang, Y.; Liu, L.; Li, L.; Li, S.; Zhang, B.; Cui, B.; Rai, M.K. Qualifying Land Use and Land Cover Dynamics and Their Impacts on Ecosystem Service in Central Himalaya Transboundary Landscape Based on Google Earth Engine. *Land* **2021**, *10*, 173. https://doi.org/10.3390/land10020173

Academic Editor: Claudia Capitani
Received: 28 December 2020
Accepted: 2 February 2021
Published: 8 February 2021

Abstract: Land use and land cover (LULC) changes are regarded as one of the key drivers of ecosystem services degradation, especially in mountain regions where they may provide various ecosystem services to local livelihoods and surrounding areas. Additionally, ecosystems and habitats extend across political boundaries, causing more difficulties for ecosystem conservation. LULC in the Kailash Sacred Landscape (KSL) has undergone obvious changes over the past four decades; however, the spatiotemporal changes of the LULC across the whole of the KSL are still unclear, as well as the effects of LULC changes on ecosystem service values (ESVs). Thus, in this study we analyzed LULC changes across the whole of the KSL between 2000 and 2015 using Google Earth Engine (GEE) and quantified their impacts on ESVs. The greatest loss in LULC was found in forest cover, which decreased from 5443.20 km^2 in 2000 to 5003.37 km^2 in 2015 and which mainly occurred in KSL-Nepal. Meanwhile, the largest growth was observed in grassland (increased by 548.46 km^2), followed by cropland (increased by 346.90 km^2), both of which mainly occurred in KSL-Nepal. Further analysis showed that the expansions of cropland were the major drivers of the forest cover change in the KSL. Furthermore, the conversion of cropland to shrub land indicated that farmland abandonment existed in the KSL during the study period. The observed forest degradation directly influenced the ESV changes in the KSL. The total ESVs in the KSL decreased from 36.53 $\times$ 10^8 USD y^{-1} in 2000 to 35.35 $\times$ 10^8 USD y^{-1} in 2015. Meanwhile, the ESVs of the forestry areas decreased by 1.34 $\times$ 10^8 USD y^{-1}. This shows that the decrease of ESVs in forestry was the primary cause to the loss of total ESVs and also of the high elasticity. Our findings show that even small changes to the LULC, especially in forestry areas, are noteworthy as they could induce a strong ESV response.

Keywords: land use and land cover; ecosystem service value; Google Earth Engine (GEE); forest fragmentation; transboundary landscape; Himalaya

1. Introduction

Ecosystem services can be defined as the benefits that humans gain from ecological processes that contribute to human well-being [1–4]. However, global ecosystem services have been altered by human activities over the past few centuries [5]. Anthropogenic activities can be found in almost every corner of the globe after the onset of the Anthropocene and have emerged as a global driver rapidly sculpturing the ecosystem [6–8].

According to Costanza et al. [9], 60% of worldwide ecosystem services have degraded over the past several decades. Land use and land cover (LULC) changes, mainly driven by human activities [10], are considered to be one of the greatest and most immediate threats affecting ecosystem services [11,12]. LULC changes have thus been considered an important research topic with regard to global environmental change and sustainable development [13–16]. Mountain ecosystems are rich sources of biodiversity [17] and host high plant endemism [18]. They also provide diverse ecosystem services [19]. On the other hand, mountain regions are fragile areas that are sensitive to external forces [20]. Human-driven LULC changes are considered to be among the greatest ecological pressures in mountain regions [21].

As a typical mountain system, the Hindu Kush Himalayan (HKH) region extends ca. four million square kilometers, encompassing eight countries: Afghanistan, Bangladesh, Bhutan (all), China, India, Myanmar, Nepal (all), and Pakistan [22,23]. It is the source of ten major river systems, which provide water, ecosystem services, and the basis for livelihoods to a population of around 210.53 million people in the region [24]. Harboring four of 36 global biodiversity hotspots [25], it provides habitats for numerous wild species but is deeply threatened. The region is extremely fragile in terms of land cover diversity and its association with variable terrain, climate, and sociodemographic interactions. The HKH region is significantly rich in terms of biodiversity but is also one of the least studied in the world [26,27]. The fourth and fifth reports of the Intergovernmental Panel on Climate Change (IPCC) explicitly pointed to the HKH as a data deficient area [28,29].

Even though 39% of the HKH region's land is divided into protected areas to support better conservation [30], the effectiveness of protected areas still faces challenges [31,32]. Almost one-third of the protected areas are transboundary and in these areas, as elsewhere in the HKH region, ecosystems and habitats extend across political boundaries [23]. When conservation policies meet with the administrative and political borders in the territory, the situation becomes more complex because of the nonconformity between natural ecological boundaries and administrative borders [33]. This means that landscape-level planning is necessary and management requires regional cooperation if the ecosystems or habitats are transboundary in nature [34]. For better conservation, seven transboundary landscapes have been identified across the HKH region—based on biodiversity significance, representation of ecoregions, cultural significance, and contiguity of ecosystems for conservation and sustainable development of the region [35]—and are being used to develop transboundary landscape-level planning and management approaches.

The Kailash Sacred Landscape (KSL) is one of the seven transboundary landscapes, named after the Mount Kailash, which is seen as the holiest shrine for several religions [36]. Three of Asia's great rivers have their sources in the landscape: the Indus, the Brahmaputra, and the Ganges River, which provide essential transboundary ecosystem goods and services, both locally and downstream [37]. However, increasingly frequent human activities, together with climate change, have caused rapid land use and land cover changes over the past decades. Uddin et al. [23] have shown the forest fragmentation in Nepal's Kailash Sacred Landscape from 1990 to 2009 and further predicted the future LULC in 2030. Duan et al. [38] assessed LULC changes in the Kailash Sacred Landscape of China from 1990–2008 and quantified driving forces. Singh et al. [39] studied the LULC changes in the Kailash Sacred Landscape of China from 1976–2011 and also pointed out forest fragmentation in the Indian part. All of these studies assessed the LULC changes in three countries using different data sources, study periods, classification systems, and methods. It is almost impossible to compare the differences in LULC changes among the three countries. In short, LULC data covering the entire area are still unclear.

A detailed and accurate knowledge of land cover is crucial for many scientific and operational applications and, as such, it has been identified as an Essential Climate Variable [40]. The development of remote sensing provided an important tool to explore historical and current land cover information at the local, national, regional, and global levels [41]. The complicated process of processing satellite imagery and the high cost of computing power has limited the relevant research. Google Earth Engine (GEE) provides a high-performance cloud-based platform and access for any researcher [42,43]. GEE houses a massive imagery data collection, including Landsat, MODIS (Moderate Resolution Imaging Spectroradiometer), and Sentinel that can be directly accessed using the JavaScript code within minutes, allowing users to interactively test and develop algorithms and preview results in real time without downloading any images [44]. Furthermore, GEE offers a packaged algorithm for image preprocessing and machine learning classifiers. The efficiency of GEE has been demonstrated by recent studies, including with regard to vegetation change detection [45,46], urban area mapping [47,48], agricultural land mapping [49], grassland monitoring [50,51], extraction of water bodies [52,53], and disaster monitoring [54].

Hence, we used satellite images and GEE to assess LULC changes and examine their impacts on ecosystem service values (ESVs) in the KSL between 2000 and 2015. Our main objectives were to explore: (1) the dynamics of LULC between 2000 and 2015; (2) the ESV changes based on LULC; and (3) their implications for landscape conservation and sustainable land use. This study is expected to provide insights into sacred landscape conservation for future land management.

2. Materials and Methods

2.1. Study Area

The Kailash Sacred Landscape is located between 79°40′ E–82°30′ E and 29°10′ N–31°20′ N (Figure 1). Mount Kailash, which is considered by multiple religions as the center of the universe, and Lake Manasarovar are the most prominent features in the KSL. There are two sacred lakes near Mount Kailash, Lake Manasarovar and Lake Rakshastal. The region covers an area of over 31,000 km^2, including parts of far-western Nepal, northern India, and Purang County, Tibet Autonomous Region of China [23,38,39]. The elevation drop from the highest mountain, Naimona'nyi, to the southern parts is over 7000 m. This loss in elevation causes abundant vegetation types, ranging from tropical broadleaved forest to alpine steppe. Diverse ecosystems provide habitats with rich biodiversity. The landscape is also home to 93 mammal species, 497 bird species, and 134 fish species, among other fauna, making it one of the ecologically richest areas in the western Himalayas [37].

Over a million people live within the landscape and most of this population resides in India and Nepal, with very few persons inhabiting the sparsely populated high-elevation areas on the Tibetan Plateau [37]. Local people rely heavily on the natural resources of this region. In KSL-China, grazing is the primary mode of utilization of grassland, often exerting pressures on fragile ecosystems. Agriculture accounts for a relatively small proportion of land use. In KSL-Nepal and KSL-India, forests cover large parts of these two regions and offer livelihoods to the local people while simultaneously supporting biodiversity conservation. Deforestation and fragmentation because of cropland expansion, infrastructure construction, and illegal timber harvesting have been reported in these regions. Forest cover loss and fragmentation are regarded as main causes of global ecosystem degradation [56]. Accordingly, human activities pose a serious threat to the fragile ecosystems in the KSL.

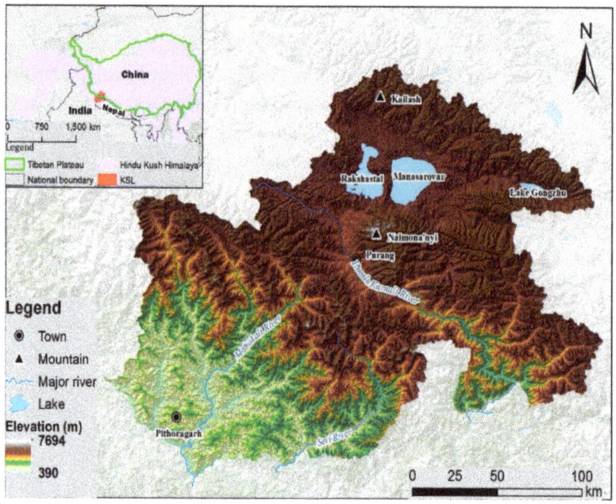

Figure 1. Map showing the location and topographic features of Kailash Sacred Landscape (KSL). The Hindu Kush Himalaya (HKH) boundary was obtained from https://rds.icimod.org/home/datadetail?metadataid=3924 (accessed on 8 February 2021) and the Tibetan Plateau boundary from Zhang et al. 2014 [55].

2.2. *Classification System and Training Data Collection*

Land cover classification systems have been defined separately in KSL-China, KSL-Nepal, and KSL-India. Duan et al. [38] classified land cover in KSL-China into ten types: barren land, cropland, desert, glacier, wetland, water bodies, built-up land, low coverage rangeland, medium coverage rangeland, and high coverage rangeland. Uddin et al. [23] divided the land cover in the KSL-Nepal into seven types: forest, shrub land, grassland, cropland, barren area, water bodies, and snow/glacier. Singh et al. [39] classified the land cover system for KSL-India into seven types: forest, settlement, water, agriculture, grassland, scrubland, and snow. The landscape in KSL-China differs from that in KSL-Nepal and KSL-India, thus resulting in different land cover systems. Even though there is a certain resemblance in landscape between KSL-Nepal and KSL-India, differences exist in the classification systems. Following previous frameworks [57–59], we defined our land cover classification system as shown in Table 1. Land cover classes were defined through visual interpretation of high-resolution imagery available from Google Earth, using Landsat images as a reference. Visual interpretation of reference imagery was based on elements that help identify land cover features such as location, size, shape, tone/color, shadow, texture, and pattern [60]. Furthermore, considering the time intervals defined in this study, the training points that were stable during the study period were selected. Finally, we obtained all the defined land cover types and training points shown in Table 1 and Figure S1.

Table 1. Classification system and list of training points.

Land Cover Code	Land Cover Type	Number of the Training Points
1	Water bodies	165
2	Snow/glacier	255
3	Forest	182
4	Built-up area	80
5	Shrub land	113
6	Cropland	194
7	Grassland	439
8	Barren land	285
9	Wetland	89

2.3. Preprocessing of the Landsat Images

The Landsat-5 Thematic Mapper (TM), Landsat-7 Enhanced Thematic Mapper Plus (ETM+), and Landsat-8 Operational Land Imager (OLI) top-of-atmosphere (TOA) reflectance products were used for land cover change analysis (available online: https://earthengine.google.com (accessed on 8 February 2021)). The Landsat datasets covering our study area were then imported as image collections into GEE, a cloud-based geospatial analysis platform, for subsequent preprocessing tasks. Preprocessing methods presented by Alban et al. (2018) [61] were modified and applied in this study. The main preprocessing functions, including cloud masking, shadow masking, adding spectral index, etc., were packaged together. Using pixel-based image compositing methods, the best available observations from multiple Landsat images were selected to generate high-quality Landsat image composites for 2000 and 2015 [62–65]. Users can define parameters according to their own requirements, including location of the study area, composite years, cloud cover threshold, etc. The detailed parameters used in this study can be found in the supplementary materials. The quality of the image always suffers from high cloud cover, resulting in empty pixels or scenes. To solve this problem, we combined two strategies. First, we set the combine year parameter to three years to obtain as many images as possible; then we applied the focal_mean function offered by GEE (available at: https://developers.google.com/earth-engine (accessed on 8 February 2021)), a morphological mean filter, to each band of an image using a custom kernel (Figure S2). The detailed parameters used in this study can be found here: https://code.earthengine.google.com/17f98d1e3fe5b7c0e6b432480a65dc9b (accessed on 8 February 2021).

2.4. Classification Features Input and Classifier

Multiple spectral indices have been developed to establish the relationship between the spectral and radiometric responses measured by remote sensors and the presence of various land covers, especially vegetation [66]. Huang et al. [67] used the B2–B7 and nominalized difference vegetation index (NDVI) bands as predicting bands for mapping land cover changes in Beijing; Teluguntla et al. [68] used the B2–B7 and NDVI bands as the classification features to map the 30-m cropland extent in Australia and China; and Xiong et al. [69] used B2–B4, B8, and NDVI bands as the predicting bands to acquire a 30-m resolution cropland extent map of continental Africa. Tsai et al. [44] mapped the LULC in Fanjingshan National Nature Reserve using the Landsat spectral band together with the NDVI, normalized difference blue and red (NDBR), normalized difference green and red (NDGR), normalized difference shortwave infrared and near-infrared (NDII), modified soil-adjusted vegetation index (MSAVI), and spectral variability vegetation index (SVVI).

To obtain the most suitable predicting bands, we added spectral bands as below: the B2–B7 and temp bands were selected as the main classification feature inputs, together with 15 spectral indices derived from the Landsat data, including the NDVI [70], the land surface water index (LSWI) [71], the nominalized difference snow index (NDSI) [72], the enhanced vegetation index (EVI) [73], the normalized difference tillage index (NDTI) [74], the normalized difference moisture index (NDMI) [75], the normalized burn ratio (NBR) [76], the vegetation index green (VIG) [77], tasseled cap transformation [78], and other spectral index (SI). These indices were defined as follows:

$$\mathrm{NDVI} = \frac{\rho NIR - \rho \mathrm{Red}}{\rho NIR + \rho \mathrm{Red}} \tag{1}$$

$$\mathrm{LSWI} = \frac{\rho \mathrm{NIR} - \rho \mathrm{SWIR1}}{\rho \mathrm{NIR} + \rho \mathrm{SWIR1}} \tag{2}$$

$$\mathrm{NDSI} = \frac{\rho \mathrm{Green} - \rho \mathrm{SWIR1}}{\rho \mathrm{Green} + \rho \mathrm{SWIR1}} \tag{3}$$

$$\text{EVI} = \frac{\rho NIR - \rho\text{Red}}{\rho NIR + 6 \times \rho\text{Red} - 7.5 \times \rho\text{Blue} + 1} \quad (4)$$

$$\text{NDTI} = \frac{\rho\text{SWIR1} - \rho\text{SWIR2}}{\rho\text{SWIR1} + \rho\text{SWIR2}} \quad (5)$$

$$\text{NDMI} = \frac{\rho NIR - \rho\text{SWIR1}}{\rho NIR + \rho\text{SWIR1}} \quad (6)$$

$$\text{NBR} = \frac{\rho NIR - \rho\text{SWIR2}}{\rho NIR + \rho\text{SWIR2}} \quad (7)$$

$$\text{VIG} = \frac{\rho Green - \rho\text{Red}}{\rho Green + \rho\text{Red}} \quad (8)$$

$$\text{SI} = \frac{\rho Red - \rho\text{Blue}}{\rho Red + \rho\text{Blue}} \quad (9)$$

where ρNIR, ρRed, ρGreen, ρSWIR1, ρSWIR2, and $\rho Blue$ represent the surface reflectance values of the near-infrared band (0.76–0.9 μm), the red band (0.63–0.69 μm), the green band (0.52–0.6 μm), the shortwave infrared band 1 (1.55–1.750 μm), the shortwave infrared band 2 (2.11–2.29 μm), and the blue band (0.45–0.52 μm) for a given pixel, respectively. Furthermore, we also took the topographical factors (slope, elevation, and aspects, available at: https://developers.google.com/earth-engine/datasets/catalog (available in EE as USGS/SRTMGL1_003) (accessed on 8 February 2021)) and nighttime data (available at: https://developers.google.com/earth-engine/datasets/catlog/NOAA_VIIRS_DNB_MONTHLY_V1_VCMSLCFG (accessed on 8 February 2021)) into consideration to better depict cropland and urban areas. We obtained a total of 24 features.

GEE provides 21 classifiers of which random forest (RF) is one of the most widely used as it yields higher classification accuracies, requires less model training time, and is less sensitive to training sample qualities compared to support vector machine (SVM) and artificial neural network (ANN) classifiers [79,80]. In this study, the RF classifier in GEE was trained using 70% of the training data randomly selected and extracted from the sets of image stacks, with the remaining 30% of the training data used for the model validation. A confusion matrix was implemented to assess the accuracy of the classified image with the independent set of ground truth points [81]. The overall accuracy was calculated in GEE, together with the producer's accuracy (PA) and user's accuracy (UA) of each land cover type. A previous study indicated that that the accuracy of a LULC map should higher than 85% for optimal interpretation and identification [82]. To deal with salt and pepper noise, classified images were postprocessed with a majority filter to smooth isolated pixels [83,84]. The overall levels of accuracy for 2000 and 2015 were 88.6% and 89.42%, respectively. The RF classifier produced overall acceptable levels of accuracy for the four classification points in time and the defined LULC types (Tables S1 and S2).

2.5. Detection of LULC Changes and Estimation of ESVs

The LULC changes can be calculated using Equation (10). To identify the main conversion directions and highlight the dominant dynamic events in land use/cover changes, we used ArcGIS (version 10.4) to generate the transfer matrix for each period and visualized the transfer process with a Sankey Diagram (available online at: https://sankey.csaladen.es (accessed on 8 February 2021)) [85,86]. The calculation is as follows:

$$\text{R} = \frac{L_t - L_{t-1}}{L_{t-1} * \Delta t} \times 100\% \quad (10)$$

where R represents the LULC change rate, L_t represents land cover type in year t, L_{t-1} represent land cover in the most recent time interval, and Δt denotes the time interval (15 in this study).

To better understand the consequences of the conversion from forest to other LULC types, we further assessed the forest fragmentation of the KSL in 2000 and 2015 following the method described by Vogt et al. [87]. The forest LULC type was divided into six classes, patch, edge, perforated, small core (SC) (<250 acres), medium core (MC) (250–500 acres), and large core (LC) (>500 acres), by computing the distance from forest pixels to non-forest pixels. We defined the edge width as 100 m in reference to previous studies [23].

Costanza et al. [1] presented a model to estimate global ecosystem service value. However, this estimation method is best suited for Western countries; Xie et al. [88] therefore grouped the ESVs into four types and nine subtypes specific to China on this basis and using data from [5]. Costanza et al. [9] further presented a new method for the estimation of global ESVs and found that the ESVs of certain land cover types increased (e.g., the ESVs of forest land cover increased by 2462 USD per hectare per year from 1997–2011) while the remaining land cover types remained stable. In this study, we adopted the same equivalent value as that used by Song et al. [89] (Table 2). The equations used to evaluate the KSL's ESVs and their changes are as follow:

$$\mathrm{ESV}_t = \sum_{i=1}^{n} Area_i \times ESV_i \tag{11}$$

$$C_{\Delta t} = \frac{E_{end} - E_{start}}{E_{Start}} \times 100\% \tag{12}$$

where ESV_t denotes the total ESV at time t (2000, 2005, 2010, 2015); $Area_i$ represents the area of land cover i, ESV_i represents the ESV of land cover I, and n denotes the total number of land cover types (seven types after reclassification in this study). C_t represents the changes in ESV within a time interval (e.g., 2000–2005) and E_{end} and E_{start} denote the ESVs at the end and start of the time interval, respectively.

Table 2. Ecosystem service values (ESVs) of land cover types defined in this study.

Land Cover Defined in This Study	Equivalent Biome (Song et al. 2017) [89]	ESVs Per Unit Area ($/hm^2/year)
Water bodies Snow/glacier	Water areas	2607.77
Forest Shrub land	Forestry areas	1616.99
Grassland	Grassland	671.06
Cropland	Cultivated land	454.28
Built-up area	Built-up areas	0
Barren land	Unused land	79.93
Wetland	Wetland	3149.45

2.6. *Elasticity of ESV Changes in Response to LULC Changes*

For the purpose of investigating the relation between LULC and ESVs, elasticity as defined by Song et al. [89] was applied in this study. The concept of elasticity is used to measure the sensitivity of a variable to change in another variable. Here, elasticity was used to measure the percentage change in ESV in relation to the percentage change in LULC, and thus can be described as follows:

$$\mathrm{EEI} = \left| \frac{(E_{\mathrm{end}} - E_{\mathrm{start}})/E_{\mathrm{start}} \times 100\%}{LCP} \right| \tag{13}$$

$$\mathrm{LCP} = \frac{\sum_{i=1}^{7} \Delta LUT_i}{\sum_{i=1}^{7} LUT_i} \tag{14}$$

where EEI represents the elasticity of ESV change in response to changes in LULC, E_{end} is the ESV at the end of the study period, E_{start} is the ESV at the beginning of the study period, LCP is the conversion percentage of land (which reveals both speed and degree of

LULC conversion), ΔLUT_i is the converted area of the i type of LULC, LUT_i is the area of the i type of LULC, and T is the time gap (in years) of the study period.

3. Results

3.1. The Spatial Distribution of LULC and Its Changes

As shown in Figure 2, there was a significant difference in land cover between the Himalayan northern slopes (China side) and southern slopes (Nepal and India side). Statistical results indicated that most land in the KSL was covered by grassland (23.98% in 2000, 25.74% in 2015) followed by barren land (21.34% in 2000, 21.98% in 2015), and forest (17.45% in 2000, 16.04% in 2015) (Table 3). Grassland was mainly distributed in KSL-China and widely distributed on the Tibetan Plateau (55.73% in 2000, 52.84% in 2015). Over 60% of barren land was distributed in KSL-China and forest land cover was mainly distributed in KSL-Nepal and KSL-India (60.69% and 39.31% in 2000, respectively). Snow/glacier accounted for more than 15.16% of the total area and over 53% of snow/glacier was distributed in KSL-Nepal (54.14% in 2000, 53.61% in 2015). Cropland and built-up areas were the main land cover types closely relevant to human activities and these were mainly distributed in KSL-Nepal and KSL-India.

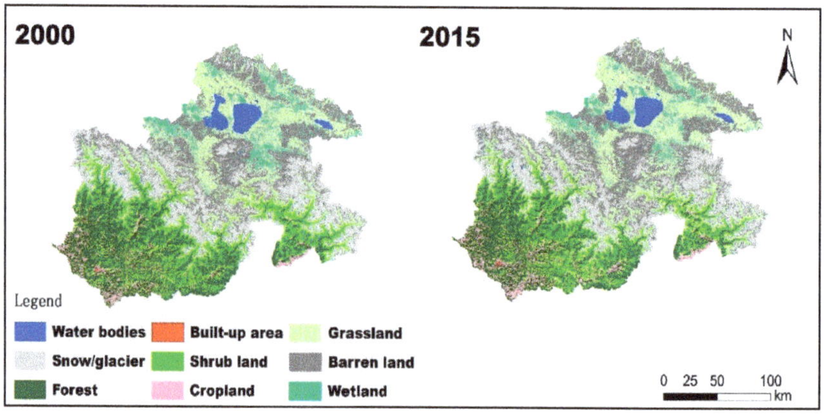

Figure 2. The distribution of different land cover types in KSL in 2000 and 2015.

Table 3. Land use and land cover (LULC) in the KSL during the period 2000–2015.

Land Cover	Area in 2000 (km^2)	%	Area in 2015 (km^2)	%	Changed Area (2000–2015)	Change Rate (2000–2015)
Water bodies	990.27	3.17	994.71	3.19	4.43	0.03
Snow/glacier	4728.51	15.16	4687.55	15.03	−40.96	−0.06
Forest	5443.20	17.45	5003.37	16.04	−439.82	−0.54
Built-up area	65.59	0.21	66.05	0.21	0.46	0.05
Shrub land	2917.78	9.35	2528.17	8.11	−389.61	−0.89
Cropland	1910.59	6.13	2257.50	7.24	346.90	1.21
Grassland	7479.89	23.98	8028.35	25.74	548.46	0.49
Barren land	6655.26	21.34	6854.46	21.98	199.20	0.20
Wetland	1000.04	3.21	770.98	2.47	−229.07	−1.53
Total	31,191.13	100	31,191.13	100		

Between 2000 and2015, four land cover types showed decreasing trends and the other five land cover types showed increasing trends (Table 3). The greatest loss was found for forest: a total of 439.82 km^2 forest cover loss was observed in the KSL. The decrease of forest cover in KSL-Nepal contributed 89.68% of the total forest loss during the research period. Shrub land also showed an obvious decreasing trend, with a total loss of 389.61 km^2 during the research period, decreasing at a rate of 0.89% per year. The decrease of shrub land in KSL-Nepal and KSL-India contributed 65.55% and 34.3%, respectively, to the total shrub land loss. During the research period, wetland and snow/glacier decreased by 229.07 km^2 and 40.96 km^2, respectively. Among the land cover types with increasing trends, the biggest gains were found in grassland: grassland increased by 548.46 km^2 during the research period. The increase in grassland in KSL-Nepal and KSL-India contributed 58.32% and 28.22% to the total gains in grassland in the KSL, respectively. From 2000–2015, cropland increased by 346.90 km^2 and at a rate of 1.21% per year. The biggest increase was observed in KSL-Nepal, where cropland increased by 247.94km^2 from 2000–2015. Barren land was found to increase from 6655.26 km^2 to 6854.46 km^2 between 2000 and 2015 and at a rate of 0.2% per year. Changes in water bodies and built-up areas were not obvious and only increased by 4.43 km^2 and 0.46km^2, respectively, during the research period. The results indicate that the area of land types with higher ecosystem service values (e.g., forest, shrub land, and wetland) decreased.

Forest, barren land and grassland were significantly converted to other land cover types in the period from 2000–2015 (Table 4 and Figure 3). A total of 857.81 km^2 of forest were converted to other land cover types, including 59.18% that were converted to shrub land and 34.19% that were converted to cropland. This indicates the forest fragmentation occurred between 2000 and 2015. About 1125.07 km^2 of barren land were converted into other land cover types during the research period, 39.96% of which were converted to snow/glacier. Snow/glacier mainly converted to barren land during the research period: a total of 526.78 km^2 of snow/glacier were converted to barren land. About 1150.91 km^2 of grassland were converted to other land cover types with 61.17% converted to barren land. During the study period, shrub land contributed most to the expansion of cropland: a total of 425.48 km^2 of shrub land was converted to cropland. Meanwhile, cropland mainly converted to shrub land and forest between 2000 and 2015: a total of 288.38 km^2 and 170.74 km^2 of cropland converted to shrub land and forest, respectively. The expansion of built-up areas was mainly at the cost of cropland. The results indicate that deforestation and cropland abandonment occurred in KSL-Nepal and KSL-India simultaneously.

Table 4. Transition matrix of different LULC types in the KSL during the period 2000–2015.

2015	2000								
	Water Bodies	**Snow/ Glacier**	**Forest**	**Built-up Area**	**Shrub Land**	**Cropland**	**Grassland**	**Barren Land**	**Wetland**
Water bodies	832.23	61.02	19.24	4.54	4.40	0.91	32.87	35.5	0.00
Snow/glacier	48.92	4068.60	2.66	2.43	3.55	0.27	76.05	526.78	0.70
Forest	7.87	0.09	4586.69	8.21	507.66	293.25	39.95	0.79	0.00
Built-up area	7.74	0.06	6.49	23.11	1.40	25.02	1.39	0.49	0.00
Shrub land	16.12	31.59	203.00	0.85	1641.12	425.48	588.14	11.69	0.00
Cropland	1.59	0.02	170.74	10.41	288.38	1420.75	17.31	1.62	0.24
Grassland	29.40	77.80	13.99	8.32	79.91	75.55	6329.91	704.02	161.93
Barren land	51.32	449.60	1.83	8.22	1.94	3.48	555.73	5531.32	52.95
Wetland	0.00	0.32	0.00	0.03	0.00	13.31	387.99	43.23	555.29

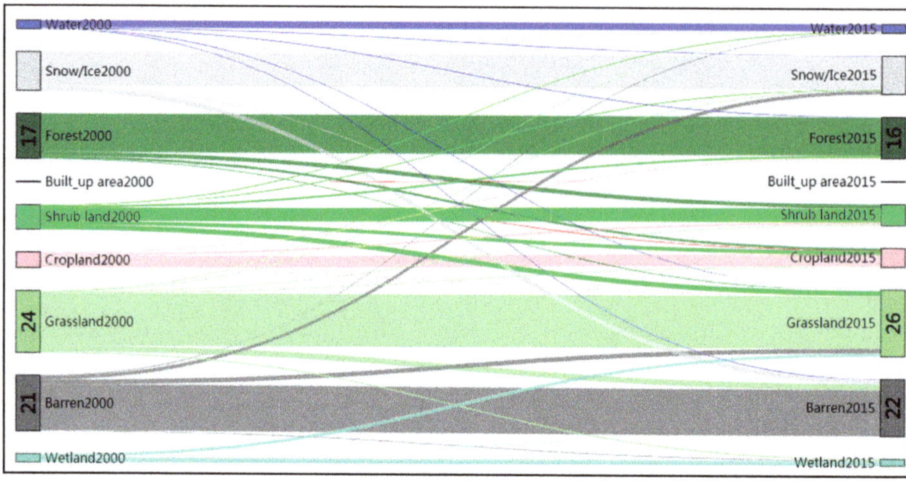

Figure 3. Sankey diagram of LULC transitions in the KSL during the period 2000–2015. The map was created using the tool at https://sankey.csaladen.es/#.

3.2. Forest Fragmentation in the KSL

A conversion of forest to other land cover types indicating forest fragmentation occurred in the KSL during the research period (Table 5). The distribution of and changes in forest fragmentation in 2000 and 2015 are depicted in Figure 4. During the study period, the forest fragmentation changed significantly. In 2000, core forest (>500 acres) covered 34.24% of the forest area, followed by edge forest covering 30.39%, perforated forest covering 18.44%, core forest (<250 acres) covering 7.72%, patch forest covering 5.95%, and core forest (250–500 acres) covering 2.89%. In 2015, edge forest covered 33.67% of the forest area, followed by core forest (>500 acres) covering 28.1%, perforated forest covering 17.22%, patch forest covering 8.69%, core forest (<250 acres) covering 8.53%, and core forest (250–500 acres) covering 3.79%. Core forest (>500 acres) decreased from 1883.90 km^2 to 1406.05 km^2, with a change rate of 25.36%. Meanwhile, patch forest increased from 323.81 km^2 to 434.83 km^2, with a change rate of 34.29%.

Table 5. Forest fragmentation and change in KSL between 2000 and 2015.

Type of Patches	2000 (km^2)	2015 (km^2)	2000–2015 (km^2)	Change Rate (%)
Patch	323.81	434.83	111.02	34.29
Edge	1654.24	1684.39	30.15	1.82
Perforated	1003.80	861.59	−142.21	−14.17
Core (<250 acres)	420.12	426.81	6.69	1.59
Core (250–500 acres)	157.33	189.71	32.37	20.58
Core (>500 acres)	1883.90	1406.05	−477.85	−25.36

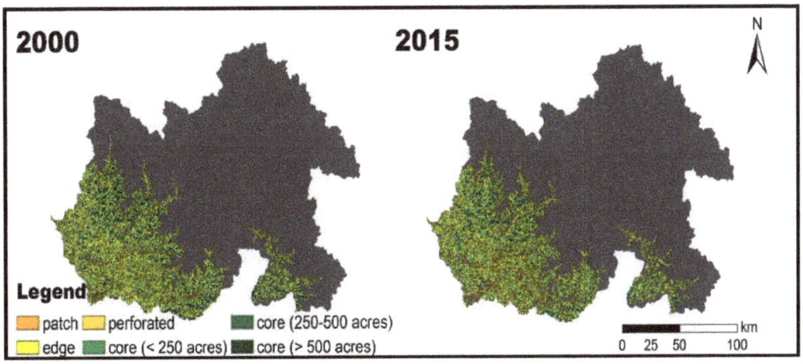

Figure 4. Forest fragmentation maps from 2000 and 2015. Core—relatively distant from the forest–non-forest boundary; patch—forests too small to be considered as core forest; perforated—boundaries between core forest and relatively small perforations; edge—boundaries of relatively large perforations and the exterior boundaries of core forest regions.

Forest land cover was mainly distributed in KSL-Nepal; this also true for the core forest (>500 acres), covering 57.68% of the total core forest (>500 acres) in KSL. In 2000, core forest (>500 acres) covered 32.90% of the forest area in KSL-Nepal, followed by edge forest covering 31.23%, perforated forest covering 18.28%, core forest (<250 acres) covering 8.24%, patch forest covering 6.12%, and core forest (250–500 acres) covering 3.23%. During the research period, core forest (>500 acres) decreased by 31.97% and patch forest increased by 38.62%. The rest of the forest land cover was distributed in KSL-India. In 2000, core forest (>500 acres) covered 37.25% of the forest area in KSL-India, followed by edge forest covering 29.09%, perforated forest covering 18.70%, core forest (<250 acres) covering 6.91%, patch forest covering 5.68%, and core forest (250–500 acres) covering 2.36%. From 2000–2015, core forest (>500 acres) decreased by 16.36% and core forest (250–500acrs) increased by 56.59%. The results suggest that deforestation and forest fragmentation occurred in KSL, especially in KSL-Nepal, during the research period.

3.3. The LULC Changes in KSL-China, KSL-Nepal, and KSL-India

In 2015, most land in the three countries was covered by different land cover types (Table 6). In KSL-China, barren land accounted for the largest proportion of land cover. During the research period, no evident changes were observed in barren land (increase of 2.43 km^2). As the second largest land cover type, grassland increased from 4168.41 km^2 to 4242.24 km^2 during the research period. Between 2000 and 2015, snow/glacier increased from 756.58 km^2 to 815.46 km^2, an increasing trend opposite to the broader picture for KSL snow/glacier. A great increase was observed in cropland, which increased by 73.07 km^2 between 2000 and 2015. Wetland, water bodies and shrub land showed decreasing trends during the research period, while the largest decrease was found in wetland (decreased by 203.29 km^2). In KSL-Nepal, grassland, cropland, and barren land contributed most to land cover increases. During the research period, the greatest gains were found in grassland, which increased by 319.87km^2, followed by cropland, which contributed the most to the cropland expansion in the KSL (increase of 247.94 km^2, accounting for over 70% of the total increase). Forest in KSL-Nepal decreased from 3303.37 km^2 to 2908.90 km^2 during the research period. Between 2000 and 2015, shrub land decreased by 255.43 km^2, second only to the loss of forest. Snow/glacier showed a decreasing trend and decreased by 47.31 km^2 during the research period. In KSL-India, the greatest gains were found for grassland, which increased by 154.76 km^2 between 2000 and 2015. Changes in cropland were not evident, with an increase from 796.86 km^2 to 822.83 km^2. Shrub land decreased from 905.79 km^2 to 772.13 km^2 during the research period. The most significant changes were observed for cropland expansion and forest loss, which were mainly distributed in KSL-Nepal.

Table 6. Dynamic changes in LULC types between 2000–2015.

Land Cover	KSL-China				KSL-Nepal				KSL-India			
	2000 (km^2)	2015 (km^2)	Change Area (km^2)	Change Rate (%)	2000 (km^2)	2015 (km^2)	Change Area (km^2)	Change Rate (%)	2000 (km^2)	2015 (km^2)	Change Area (km^2)	Change Rate (%)
Water bodies	754.55	748.52	−6.03	−0.05	136.44	150.07	13.64	0.67	99.45	96.30	−3.15	−0.21
Snow/glacier	756.58	815.46	58.89	0.52	2560.67	2513.36	−47.31	−0.12	1412.07	1359.45	−52.63	−0.25
Forest	0.00	0.00	0.00	0.00	3303.37	2908.90	−394.47	−0.80	2140.04	2094.67	−45.38	−0.14
Built-up area	0.15	1.83	1.69	75.37	32.92	32.26	−0.67	−0.14	32.60	32.04	−0.56	−0.11
Shrub land	0.63	0.05	−0.58	−6.17	2011.52	1756.09	−255.43	−0.85	905.79	772.13	−133.66	−0.98
Cropland	13.61	86.67	73.07	35.80	1100.41	1348.35	247.94	1.50	796.86	822.83	25.97	0.22
Grassland	4168.41	4242.24	73.83	0.12	2358.69	2678.55	319.87	0.90	952.65	1107.41	154.76	1.08
Barren land	4227.39	4229.81	2.43	0.00	1659.08	1801.75	142.67	0.57	767.51	821.74	54.23	0.47
Wetland	898.35	695.06	−203.29	−1.51	97.66	71.42	−26.24	−1.79	4.08	4.49	0.41	0.68

3.4. The Spatial Distribution of ESVs and Their Response to LULC Changes

The ESVs of the KSL in 2000 and 2015 were estimated (Figure 5 and Table 7). The total ESV of the KSL in 2000 was 36.53×10^8 USD y^{-1}. During the research period, the total ESV decreased by 1.17×10^8 USD y^{-1}. In general, water areas and forestry areas contributed most to the total ESV, accounting for about 77.83% in 2000 and 76.38% in 2015. In 2000, water areas contributed about 40.82% of the total ESV in the KSL and 41.91% in 2015. Forestry areas contributed the second most to the total ESV and decreased from 13.52×10^8 USD y^{-1} in 2000 to 12.18×10^8 USD y^{-1} in 2015. The ESVs of grassland, cultivated land, and unused land showed an increasing trend. The greatest gains were found in grassland: the ESV of grassland increased from 5.42×10^8 USD y^{-1} to 5.67×10^8 USD y^{-1} during the research period. With the expansion of cropland, the ESV of cropland increased by 0.16×10^8 USD y^{-1} from 2000–2015. Although wetland accounted for a small area in the KSL, the high ESV of wetland enlarged its influence on the total ESV. During the research period, the ESV of wetland decreased by 0.16×10^8 USD y$-^1$, which offset the increase of ESV of cropland. Since the ESV of built-up areas was zero, this kind of land cover contributed no ESV.

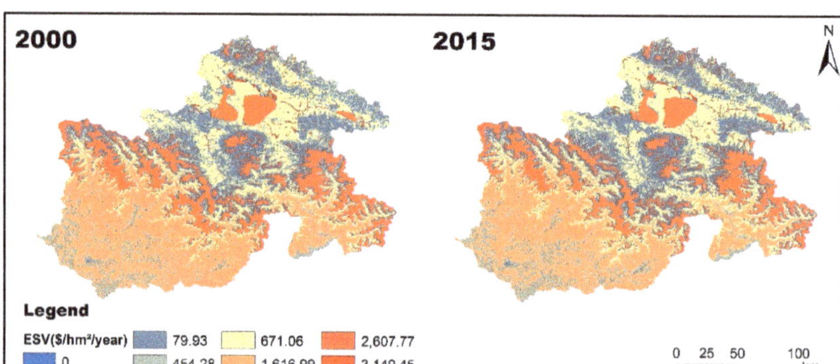

Figure 5. The distribution of ESVs in the KSL in 2000 and 2015.

Table 7. The ESVs of 11 LULC types in 2000 and 2015.

Land Cover	Value (10^8 USD y^{-1})		Change Value (10^8 USD y^{-1})	Change Rate (%)
	2000	2015	2000–2015	2000–2015
Water areas	14.91	14.82	−0.09	−0.6
Forestry area	13.52	12.18	−1.34	−9.91
Grassland	5.42	5.67	0.25	4.61
Cultivated land	0.87	1.03	0.16	18.39
Built-up areas	0	0	0	0
Unused land	0.53	0.55	0.02	3.77
Wetland	1.28	1.12	−0.16	−12.50
Total	36.53	35.35	−1.17	−3.20

In KSL-China, the total ESV was 8.44 × 10^8 USD y^{-1} in 2000, lower than the ESVs in KSL-Nepal (18.07× 10^8 USD y^{-1}) and KSL-India (10.03 × 10^8 USD y^{-1}). Water bodies contributed most to the total ESV in China, accounting for almost half of the total ESV in KSL-China, followed by Grassland (3.20 × 10^8 USD y^{-1}). From 2000–2015, the total ESV in KSL-China increased by 0.06 ×10^8 USD y^{-1}. The increase of water bodies by 0.13 × 10^8 USD y^{-1} contributed most to the increase of the total ESV in KSL-China. The decrease of the ESV of grassland was the main cause for the decrease of total ESV in the KSL. The ESV of wetland decreased from 0.95 × 10^8 USD y^{-1} in 2000 to 0.93 × 10^8 USD y^{-1} in 2015. In KSL-Nepal, the total ESV accounted for about 50% of the total ESV in the KSL. During the research period, ESV in KSL-Nepal decreased by 0.88 × 10^8 USD y^{-1}; the decrease of the ESVs of forestry areas was the main cause of the loss of total ESV in KSL-Nepal. In 2000, the ESVs of forestry areas accounted for 47.54% of the total ESV in KSL-Nepal. However, this number decreased to 43.87% in 2015. From 2000–2015, the ESV of forestry areas decreased by 1.05 × 10^8 USD y^{-1}. The ESVs of cultivated land and grassland increased by 0.11 × 10^8 USD y^{-1} and 0.21 × 10^8 USD y^{-1}, respectively, and offset a small part of the ESV loss. In KSL-India, the total ESV decreased from 10.03 × 10^8 USD y^{-1} in 2000 to 9.67 × 10^8 USD y^{-1} in 2015. The ESV of forestry areas contributed most to the total ESV in KSL-India, similar to KSL-Nepal, followed by water areas. The greatest loss was observed in forestry areas: the ESV of forestry areas decreased by 0.29 × 10^8 USD y^{-1}. From 2000 to 2015, the ESV of water areas decreased by 0.15 × 10^8 USD y^{-1}, second only to the loss in ESV of forestry areas. The greatest gains in ESV were found for grassland, which increased by 0.12 during the research period. The small changes of cropland in KSL-India made a relevant but small contribution to the changes in ESV in KSL-India.

The elasticity of ESV change with respect to LULC changes during the research period was 2.33, which indicates that a conversion of 1% of land area would result in an average change of 2.33% in the ESV. The elasticity of the ESVs in the three countries was further calculated. The results show that KSL-China had the highest elasticity at 5.27, indicating that a conversion of 1% of land area would result in an average change of 5.27% in the ESV. In KSL-Nepal, the elasticity was 4.34, higher than that of KSL-India (1.57).

4. Discussion

4.1. LULC Changes across the KSL

Detailed LULC research is of great significance for managing natural resource effectively [23]. In this study, we applied an RF algorithm to classify the LULC in the KSL in 2000 and 2015 using GEE. For a solution to the problem of the low-quality imagery caused by the high cloud cover in this region, we adopted a pixel-based image composite algorithm and filled the blank pixels using the focal_mean function. Furthermore, the spectral index, terrain factors, and nighttime light data were used to improve the accuracy of the classification. The entire process of LULC classification, except where otherwise noted, was accomplished in GEE. The overall accuracies of the LULC classification in 2000 and 2015 were 87.69% and 85.73%, respectively, indicating the good performance of our methods. Based on the LULC, we further estimated the ESVs of the KSL and qualified the responses of ESVs to LULC changes.

During the research period, the greatest land cover loss was found for forest cover, which decreased by 439.82 km^2. Forest area in Nepal and India decreased from 3303.37 km^2 to 2908.90 km^2 and from 2140.04 km^2 to 2094.67 km^2, respectively. The same phenomenon of forest cover decrease has also been found in other Himalayan regions [90,91]. The second greatest land cover loss was found for shrub land, which decreased by 389.61 km^2 between 2000 and 2015. These decreasing vegetated areas, especially the forest cover loss, may pose a threat to biodiversity conservation and livelihood [92,93]. Meanwhile, grassland and cropland areas significantly increased during the research period, which is consistent with the findings of Uddin et al.'s [23] and Singh et al. [39]. Cropland increased from 1910.59 km^2 to 2257.50 km^2, with a change rate of 1.21% per year. The three countries showed the same increasing trend. The largest growth was found in Nepal, where cropland increased by 247.94 km^2. Results from the transfer matrix show that expansions of cropland were mainly derived from forest and shrub land in KSL-India and KSL-Nepal. It has been previously shown that expansion of cropland is one of the major drivers of deforestation in the Himalayas [94]. Through statistical analysis of the LULC changes along the elevation, we further found that the increase of cropland was mainly distributed between 1000 and 2500 m in the KSL, accounting for 79.63% of the total increase (Figure S3). The largest growth was found at 1500–2000 m, accounting for 35.01%. The decrease of forest was mainly distributed between 1000 and 3500 m, accounting for 99.28% of the total loss. The most passive change of forest cover was between 1500 and 2000 m. An earlier study has shown that, in the final three decades of the 20th century, forest degradation mainly occurred in temperate oak forests at elevations of 1800–2800 m, with some forests also lost at lower elevations [95]. Lowland areas are considered more favorable for supporting human livelihood and thus result in more intense LULC changes [96].

The forest in the KSL is undergoing a process of fragmentation under the drivers of cropland expansion and illegal timber extraction [23,39]. As an important habitat for countless wild species, the decrease in forest cover along with forest fragmentation put wild life in danger. Sarker et al. (2018) assessed the habitat suitability and connectivity of the common leopard (Panthera pardus) in Kailash Sacred Landscape [97]. Their results show that the best forest connectivity for leopards lies between large forest patches situated at the middle elevational range of the landscape, associated with moderate to medium slopes and a high density of rivers and streams. The decrease in core forest cover may threaten the habitat of the common leopard. Increasing human activities (expansion of cropland and built-up areas) [98,99], together with climate change [100], have resulted in rapid changes in the Himalayan ecosystem [101]. Invasive species are another important issue to consider. Research has shown that species, including invasive species, tend to move to higher elevation regions in global warming contexts [102,103].

The conversion of cropland to forest and shrub land indicates that farmland abandonment occurred in the KSL. Between 2000 and 2015, 288.38 km^2 of cropland were converted to shrub land. A noticeable increasing trend in farmland abandonment has been reported all around the world, especially in mountain regions [104,105]. According to a previous study, the hill and mountain regions of the Nepal Himalayas are more prone to farmland abandonment because of labor migration [106–108]. Singh et al. (2015) also found that, in KSL-India, continuous migration forced the conversion of high-altitude agricultural lands into grasslands and scrublands [39]. From the perspective of ecosystems service, farmland abandonment itself has positive effects [109]; however, it also poses a threat to the food security of local livelihoods [107].

4.2. ESV Changes in Response to LULC Changes

LULC changes are generally accepted as one of the critical drivers of global change [110]. During the studied 15-year period, the total ESV of the KSL decreased from 36.53×10^8 USD y^{-1} to 35.35×10^8 USD y^{-1}, decreasing at a rate of 0.21%/year. The decrease of forestry areas was the primary cause for the loss of ESV. The largest ESVs were observed in KSL-Nepal, due to the large forestry areas, whereas KSL-China was responsible for the smallest proportion of the ESV. However, the ESVs in KSL-China showed an inverse trend compared to KSL-Nepal and KSL-India. Between 2000 and 2015, the total ESV of KSL-China increased by 0.06×10^8 USD y^{-1} thanks to the increase in water areas. On a global scale, the global terrestrial ESV decreased at a rate of 2.06%/year from 1997 to 2011 [9]. Hence, changes in ESVs in the KSL were more modest than those globally. On a national scale, the terrestrial ESVs in China decreased at a rate of 0.03% per year from 1988–2008. This indicates that the decreases in ESVs in KSL were much more rapid. However, there are large gaps in other Himalayan regions. Bhaskar et al. [111] assessed the ESVs in the Transboundary Karnali River Basin (KRB), Central Himalayas, and showed that they increased by 1.59×10^8 USD y^{-1} from 2000–2017. Increase of shrub/grassland contributed the most to the increase of ESVs in this region, followed by bare land. Raju et al. [112] estimated the ESVs in the Transboundary Gandaki River Basin (GRB), Central Himalayas, indicating that there was a 1.68×10^8 USD y^{-1} increase in ESVs from 1990–2015 due to the increase of cropland and forest cover. Zhao et al. [113] assessed the LULC changes and ESVs in the Koshi River Basin (KRB) and found that the latter decreased by 2.05×10^8 USD y^{-1} from 1990–2010 because of the decrease in forest and glacier cover. Even though large knowledge gaps are still present for different regions, the importance of forest land cover is obvious and changes to it directly affect regional ESVs.

With regard to the elasticity in the KSL, a result of 2.33 indicates that that the conversion of 1% of land area would result in average changes of 2.33% in ESVs. The region where changes in ESVs had the highest elasticity in relation to LULC changes was KSL-China, where the high elasticity of ESV change in relation to LULC changes was attributable to the concentrations of unused land, wetland, and water bodies, the LULC types with the highest and the lowest ESVs. In KSL-Nepal, deforestation was the main cause of the high elasticity. Forest cover in KSL-Nepal accounted for the largest proportion of this type of land cover in the KSL and decreased by 394.47 km^2 during the studied 15-year period. The elasticity in KSL-India was relatively small, mainly due to the small decrease of forest cover. High elasticity indicates that even small LULC changes would have serious effects on ESVs.

4.3. Uncertainty and Limitations of This Study

In this study, we failed to accurately extract the built-up areas in the KSL because of the limited resolution of Landsat images and relevant small buildings in the KSL mountain regions. To resolve this problem, we tried adding nighttime light data to improve accuracy for built-up areas. However, this approach only works in regions with night lights, such as Pithoragarh (Figure S4). Therefore, the changes to built-up areas in KSL-Nepal and KSL-India showed a slightly decreasing trend. Even so, LULC and ESV changes were not strongly affected due to the small proportion of built-up areas and their ESVs of zero. Long time-series LULC change monitoring can reveal more details behind these changes. Given the available images, we only studied the LULC changes from 2000–2015, and thus LULC change fluctuations may have been hidden. In this study, we adopted the benefit/value transfer method presented by Song et al. [89], though many critiques of the benefit/value transfer method remain unanswered. Biophysical models might be more helpful for analyzing complex ecological systems and their impacts.

5. Conclusions

In this study, we extended an LULC study to the entire KSL and further assessed the changes in ESVs between 2000 and 2015. During the study period, the KSL experienced significant LULC changes: forest and shrub land decreased by 439.82km^2 and 389.61km^2, respectively, whereas grassland and cropland increased by 548.46km^2 and 346.90km^2, respectively. The conversion of forestry areas to cropland was the main cause of cropland expansion. Meanwhile, the conversion of cropland to shrub land indicates that there was farmland abandonment in the KSL. The decrease of forestry areas may pose a threat to biodiversity and livelihoods there. During the studied 15-year period, the large core (>500 acre) forest type decreased by 25.36% and patch forest increased by 34.29%. Severe forest fragmentation was observed in the KSL, mainly distributed in KSL-Nepal, leading to a decrease in ESVs in the KSL. Between 2000 and 2015, the total ESV in the KSL decreased by 1.17×10^8 USD y^{-1} and the ESV of forestry areas decreased by 1.34×10^8 USD y^{-1}. The decrease of ESV in forestry areas contributed most to the loss of total ESV and also to the high elasticity. This study revealed that even small LULC changes can cause relevant high ESV changes in the KSL.

Supplementary Materials: The following are available online at https://www.mdpi.com/2073-445X/10/2/173/s1, Figure S1. Training points in Google Earth high-resolution image and Landsat5/7/8 false color composite image; Figure S2. Image with empty pixels (left) and processed using focal_mean function (right); Table S1. The confusion matrix in 2000; Table S2. The confusion matrix in 2015; Figure S3. The LULC changes along the elevation in KSL from 2000–2015; Figure S4. Built-up areas of Pithoragarh in 2000 and 2015.

Author Contributions: Y.Z. and L.L. (Linshan Liu) designed the study. C.G. and L.L. (Linshan Liu) conceived the research; C.G. studied and wrote the paper and mapped the land cover data and drew the figures; S.L., L.L. (Lanhui Li), B.Z. and B.C. revised the paper and polished the language. Y.Z. revised the paper and contributed to explanation of the results and the discussion. M.K.R. polished the manuscript. All authors have read and agreed to the published version of the manuscript.

Funding: This work was financially supported by the Second Tibetan Plateau Scientific Expedition and Research (Grant No. 2019QZKK0603), the Strategic Priority Research Program of the Chinese Academy of Sciences (Grant No. XDA20040201), and the National Natural Science Foundation of China (Grant No. 41761144081).

Institutional Review Board Statement: Not applicable.

Informed Consent Statement: Not applicable.

Data Availability Statement: Data used in this study will be available upon request from the first author.

Acknowledgments: We would like to express our thanks to anonymous reviewers for their helpful comments on our study. We would also like to express our special thanks to Wei Bo and Gong Dianqing in the Institute of Geographic Sciences and Natural Resources Research, CAS, for their timely help.

Conflicts of Interest: The authors declare no conflicts of interest.

References

1. Costanza, R.; DArge, R.; DeGroot, R.; Farber, S.; Grasso, M.; Hannon, B.; Limburg, K.; Naeem, S.; ONeill, R.V.; Paruelo, J.; et al. The value of the world's ecosystem services and natural capital. *Nature* **1997**, *387*, 253–260. [CrossRef]
2. Carpenter, S.R.; Mooney, H.A.; Agard, J.; Capistrano, D.; Defries, R.S.; Diaz, S.; Dietz, T.; Duraiappah, A.K.; Oteng-Yeboah, A.; Pereira, H.M.; et al. Science for managing ecosystem services: Beyond the Millennium Ecosystem Assessment. *Proc. Natl. Acad. Sci. USA* **2009**, *106*, 1305–1312. [CrossRef]
3. Bennett, E.M.; Peterson, G.D.; Gordon, L.J. Understanding relationships among multiple ecosystem services. *Ecol. Lett.* **2009**, *12*, 1394–1404. [CrossRef]
4. Li, S.; Zhang, H.; Zhou, X.; Yu, H.; Li, W. Enhancing protected areas for biodiversity and ecosystem services in the Qinghai–Tibet Plateau. *Ecosyst. Serv.* **2020**, *43*, 101090. [CrossRef]
5. Millennium Ecosystem Assessment. *Ecosystems and Human Well-Being*; Synthesis Island Press: Washington, DC, USA, 2005.
6. Randin, C.F.; Ashcroft, M.B.; Bolliger, J.; Cavender-Bares, J.; Coops, N.C.; Dullinger, S.; Dirnböck, T.; Eckert, S.; Ellis, E.; Fernández, N.; et al. Monitoring biodiversity in the Anthropocene using remote sensing in species distribution models. *Remote Sens. Environ.* **2020**, *239*, 111626. [CrossRef]
7. Ellis, E.C. Anthropogenic transformation of the terrestrial biosphere. *Philos. Trans. R. Soc. A Math. Phys. Eng. Sci.* **2011**, *369*, 1010–1035. [CrossRef]
8. Waters, C.N.; Zalasiewicz, J.; Summerhayes, C.; Barnosky, A.D.; Poirier, C.; Gałuszka, A.; Cearreta, A.; Edgeworth, M.; Ellis, E.C.; Ellis, M. The Anthropocene is functionally and stratigraphically distinct from the Holocene. *Science* **2016**, *351*, d2622. [CrossRef]
9. Costanza, R.; De Groot, R.; Sutton, P.; Van der Ploeg, S.; Anderson, S.J.; Kubiszewski, I.; Farber, S.; Turner, R.K. Changes in the global value of ecosystem services. *Glob. Environ. Chang.* **2014**, *26*, 152–158. [CrossRef]

10. Teixeira, Z.; Teixeira, H.; Marques, J.C. Systematic processes of land use/land cover change to identify relevant driving forces: Implications on water quality. *Sci. Total Environ.* **2014**, *470*, 1320–1335. [CrossRef] [PubMed]
11. Newbold, T.; Hudson, L.N.; Arnell, A.P.; Contu, S.; De Palma, A.; Ferrier, S.; Hill, S.L.; Hoskins, A.J.; Lysenko, I.; Phillips, H.R. Has land use pushed terrestrial biodiversity beyond the planetary boundary? A global assessment. *Science* **2016**, *353*, 288–291. [CrossRef]
12. Maxwell, S.L.; Fuller, R.A.; Brooks, T.M.; Watson, J.E. Biodiversity: The ravages of guns, nets and bulldozers. *Nat. News* **2016**, *536*, 143. [CrossRef] [PubMed]
13. Lambin, E.F.; Geist, H.J.; Lepers, E. Dynamics of land-use and land-cover change in tropical regions. *Annu. Rev. Environ. Resour.* **2003**, *28*, 205–241. [CrossRef]
14. Zomer, R.J.; Neufeldt, H.; Xu, J.; Ahrends, A.; Bossio, D.; Trabucco, A.; Van Noordwijk, M.; Wang, M. Global Tree Cover and Biomass Carbon on Agricultural Land: The contribution of agroforestry to global and national carbon budgets. *Sci. Rep.* **2016**, *6*, 1–12. [CrossRef] [PubMed]
15. Van Asselen, S.; Verburg, P.H. Land cover change or land-use intensification: Simulating land system change with a global-scale land change model. *Glob. Chang. Biol.* **2013**, *19*, 3648–3667. [CrossRef] [PubMed]
16. Liu, J.; Liu, M.; Tian, H.; Zhuang, D.; Zhang, Z.; Zhang, W.; Tang, X.; Deng, X. Spatial and temporal patterns of China's cropland during 1990–2000: An analysis based on Landsat TM data. *Remote Sens. Environ.* **2005**, *98*, 442–456. [CrossRef]
17. Brooks, T.M.; Mittermeier, R.A.; Da Fonseca, G.A.; Gerlach, J.; Hoffmann, M.; Lamoreux, J.F.; Mittermeier, C.G.; Pilgrim, J.D.; Rodrigues, A.S. Global biodiversity conservation priorities. *Science* **2006**, *313*, 58–61. [CrossRef]
18. Kruckeberg, A.R.; Rabinowitz, D. Biological aspects of endemism in higher plants. *Annu. Rev. Ecol. Syst.* **1985**, *16*, 447–479. [CrossRef]
19. Locatelli, B.; Lavorel, S.; Sloan, S.; Tappeiner, U.; Geneletti, D. Characteristic trajectories of ecosystem services in mountains. *Front. Ecol. Environ.* **2017**, *15*, 150–159. [CrossRef]
20. Wang, Y.; Dai, E.; Yin, L.; Ma, L. Land use/land cover change and the effects on ecosystem services in the Hengduan Mountain region, China. *Ecosyst. Serv.* **2018**, *34*, 55–67. [CrossRef]
21. Vigl, L.E.; Schirpke, U.; Tasser, E.; Tappeiner, U. Linking long-term landscape dynamics to the multiple interactions among ecosystem services in the European Alps. *Landsc. Ecol.* **2016**, *31*, 1903–1918. [CrossRef]
22. Gurung, J.; Chettri, N.; Sharma, E.; Ning, W.; Chaudhary, R.P.; Badola, H.K.; Wangchuk, S.; Uprety, Y.; Gaira, K.S.; Bidha, N.; et al. Evolution of a transboundary landscape approach in the Hindu Kush Himalaya: Key learnings from the Kangchenjunga Landscape. *Glob. Ecol. Conserv.* **2019**, *17*, e599. [CrossRef]
23. Uddin, K.; Chaudhary, S.; Chettri, N.; Kotru, R.; Murthy, M.; Chaudhary, R.P.; Ning, W.; Shrestha, S.M.; Gautam, S.K. The changing land cover and fragmenting forest on the Roof of the World: A case study in Nepal's Kailash Sacred Landscape. *Landsc. Urban Plan.* **2015**, *141*, 1–10. [CrossRef]
24. Molden, D.; Sharma, E.; Shrestha, A.B.; Chettri, N.; Pradhan, N.S.; Kotru, R. Advancing Regional and Transboundary Cooperation in the Conflict-Prone Hindu Kush–Himalaya. *Mt. Res. Dev.* **2017**, *37*, 502–508. [CrossRef]
25. Mittermeier, R.A.; Turner, W.R.; Larsen, F.W.; Brooks, T.M.; Gascon, C. *Global Biodiversity Conservation: The Critical Role of Hotspots*; Springer: Berlin, Germany, 2011; pp. 3–22.
26. Uddin, K.; Shrestha, H.L.; Murthy, M.S.R.; Bajracharya, B.; Shrestha, B.; Gilani, H.; Pradhan, S.; Dangol, B. Development of 2010 national land cover database for the Nepal. *J. Environ. Manag.* **2015**, *148*, 82–90. [CrossRef] [PubMed]
27. Sharma, E.; Chettri, N. ICIMOD's transboundary biodiversity management initiative in the Hindu Kush-Himalayas. *Mt. Res. Dev.* **2005**, 278–281. [CrossRef]
28. Mackay, A. Climate Change 2007: Impacts, Adaptation and Vulnerability. Contribution of Working Group II to the Fourth Assessment Report of the Intergovernmental Panel on Climate Change. *J. Environ. Qual.* **2008**, *37*, 2407. [CrossRef]
29. IPCC. *Climate Change 2014: Impacts, Adaptation, and Vulnerability. Part A: Global and Sectoral Aspects. Contribution of Working Group II to the Fifth Assessment Report of the Intergovernmental Panel on Climate Change*; Cambridge University Press: Cambridge, UK; New York, NY, USA, 2014; p. 1132.
30. Chettri, N.; Shakya, B.; Thapa, R.; Sharma, E. Status of a protected area system in the Hindu Kush-Himalayas: An analysis of PA coverage. *Int. J. Biodivers. Sci. Manag.* **2008**, *4*, 164–178. [CrossRef]
31. Oli, K.P.; Chaudhary, S.; Sharma, U.R. Are governance and management effective within protected areas of the Kanchenjunga landscape (Bhutan, India and Nepal). *Parks* **2013**, *19*, 25–36. [CrossRef]
32. Gu, C.; Zhao, P.; Chen, Q.; Li, S.; Li, L.; Liu, L.; Zhang, Y. Forest Cover Change and the Effectiveness of Protected Areas in the Himalaya since 1998. *Sustainability* **2020**, *12*, 6123. [CrossRef]
33. De Castro-Pardo, M.; Pérez-Rodríguez, F.; Martín-Martín, J.M.; Azevedo, J.C. Modelling stakeholders' preferences to pinpoint conflicts in the planning of transboundary protected areas. *Land Use Policy* **2019**, *89*, 104233. [CrossRef]
34. Sharma, E.; Chettri, N.; Gurung, J.; Shakya, B. *The Landscape Approach in Biodiversity Conservation*; ICIMOD: Kathmandu, Nepal, 2007.
35. Chettri, N.; Sharma, E.; Thapa, R. *Long Term Monitoring Using Transect and Landscape Approaches within Hindu Kush Himalayas*; ICIMOD: Kathmandu, Nepal, 2009; pp. 201–208.
36. Zomer, R.J.; Trabucco, A.; Metzger, M.; Oli, K.P. Environmental stratification of Kailash Sacred Landscape and projected climate change impacts on ecosystems and productivity. *ICIMOD* **2013**, *13*, 1–25.

37. Oli, K.P.; Zomer, R. *Kailash Sacred Landscape Conservation Initiative: Feasibility Assessment Report*; International Centre for Integrated Mountain Development (ICIMOD): Kathmandu, Nepal, 2011.
38. Duan, C.; Shi, P.; Song, M.; Zhang, X.; Zong, N.; Zhou, C. Land Use and Land Cover Change in the Kailash Sacred Landscape of China. *Sustainability* **2019**, *11*, 1788. [CrossRef]
39. Singh, G.; Sarkar, M.S.; Pandey, A.; Lingwal, S.; Rai, I.D.; Adhikari, B.S.; Rawat, G.S.; Rawal, R.S. Quantifying Four Decades of Changes in Land Use and Land Cover in India's Kailash Sacred Landscape: Suggested Option for Priority Based Patch Level Future Forest Conservation. *J. Indian Soc. Remote* **2018**, *46*, 1625–1635. [CrossRef]
40. Inglada, J.; Vincent, A.; Arias, M.; Tardy, B.; Morin, D.; Rodes, I. Operational high resolution land cover map production at the country scale using satellite image time series. *Remote Sens.* **2017**, *9*, 95. [CrossRef]
41. Foley, J.A.; DeFries, R.; Asner, G.P.; Barford, C.; Bonan, G.; Carpenter, S.R.; Chapin, F.S.; Coe, M.T.; Daily, G.C.; Gibbs, H.K. Global consequences of land use. *Science* **2005**, *309*, 570–574. [CrossRef]
42. Gorelick, N.; Hancher, M.; Dixon, M.; Ilyushchenko, S.; Thau, D.; Moore, R. Google Earth Engine: Planetary-scale geospatial analysis for everyone. *Remote Sens. Environ.* **2017**, *202*, 18–27. [CrossRef]
43. Kumar, L.; Mutanga, O. Google Earth Engine Applications Since Inception: Usage, Trends, and Potential. *Remote Sens.* **2018**, *10*, 1509. [CrossRef]
44. Tsai, Y.; Stow, D.; Chen, H.; Lewison, R.; An, L.; Shi, L. Mapping Vegetation and Land Use Types in Fanjingshan National Nature Reserve Using Google Earth Engine. *Remote Sens.* **2018**, *10*, 927. [CrossRef]
45. Zhou, B.; Okin, G.S.; Zhang, J. Leveraging Google Earth Engine (GEE) and machine learning algorithms to incorporate in situ measurement from different times for rangelands monitoring. *Remote Sens. Environ.* **2020**, *236*, 111521. [CrossRef]
46. Martín-Ortega, P.; García-Montero, L.G.; Sibelet, N. Temporal Patterns in Illumination Conditions and Its Effect on Vegetation Indices Using Landsat on Google Earth Engine. *Remote Sens.* **2020**, *12*, 211. [CrossRef]
47. Chakraborty, T.; Lee, X. A simplified urban-extent algorithm to characterize surface urban heat islands on a global scale and examine vegetation control on their spatiotemporal variability. *Int. J. Appl. Earth Obs.* **2019**, *74*, 269–280. [CrossRef]
48. Yu, Z.; Yao, Y.; Yang, G.; Wang, X.; Vejre, H. Strong contribution of rapid urbanization and urban agglomeration development to regional thermal environment dynamics and evolution. *Forest Ecol. Manag.* **2019**, *446*, 214–225. [CrossRef]
49. Duan, Y.; Li, X.; Zhang, L.; Liu, W.; Liu, S.; Chen, D.; Ji, H. Detecting spatiotemporal changes of large-scale aquaculture ponds regions over 1988–2018 in Jiangsu Province, China using Google Earth Engine. *Ocean Coast. Manag.* **2020**, *188*, 105144. [CrossRef]
50. Oliphant, A.J.; Thenkabail, P.S.; Teluguntla, P.; Xiong, J.; Gumma, M.K.; Congalton, R.G.; Yadav, K. Mapping cropland extent of Southeast and Northeast Asia using multi-year time-series Landsat 30-m data using a random forest classifier on the Google Earth Engine Cloud. *Int. J. Appl. Earth Obs.* **2019**, *81*, 110–124. [CrossRef]
51. Xie, Z.; Phinn, S.R.; Game, E.T.; Pannell, D.J.; Hobbs, R.J.; Briggs, P.R.; McDonald-Madden, E. Using Landsat observations (1988–2017) and Google Earth Engine to detect vegetation cover changes in rangelands-A first step towards identifying degraded lands for conservation. *Remote Sens. Environ.* **2019**, *232*, 111317. [CrossRef]
52. Yao, F.; Wang, J.; Wang, C.; Crétaux, J. Constructing long-term high-frequency time series of global lake and reservoir areas using Landsat imagery. *Remote Sens. Environ.* **2019**, *232*, 111210. [CrossRef]
53. Zhou, Y.; Dong, J.; Xiao, X.; Liu, R.; Zou, Z.; Zhao, G.; Ge, Q. Continuous monitoring of lake dynamics on the Mongolian Plateau using all available Landsat imagery and Google Earth Engine. *Sci. Total Environ.* **2019**, *689*, 366–380. [CrossRef]
54. Daldegan, G.A.; Roberts, D.A.; Ribeiro, F.D.F. Spectral mixture analysis in Google Earth Engine to model and delineate fire scars over a large extent and a long time-series in a rainforest-savanna transition zone. *Remote Sens. Environ.* **2019**, *232*, 111340. [CrossRef]
55. Zhang, Y.; Li, B.; Zheng, D. Datasets of the boundary and area of the Tibetan Plateau. *Acta Geogr. Sin.* **2014**, *69*, 164–168.
56. Newman, M.E.; McLaren, K.P.; Wilson, B.S. Assessing deforestation and fragmentation in a tropical moist forest over 68 years; the impact of roads and legal protection in the Cockpit Country, Jamaica. *Forest Ecol. Manag.* **2014**, *315*, 138–152. [CrossRef]
57. Xie, F.; Wu, X.; Liu, L.; Zhang, Y.; Paudel, B. Land use and land cover change within the Koshi River Basin of the central Himalayas since 1990. *J. Mt. Sci. Engl.* **2021**, *18*, 159–177. [CrossRef]
58. Wu, X.; Gao, J.; Zhang, Y.; Liu, L.; Zhao, Z.; Paudel, B. Land Cover Status in the Koshi River Basin, Central Himalayas. *J. Resour. Ecol.* **2017**, *8*, 10–19.
59. Zhang, Y.; Wu, X.; Zheng, D. Vertical differentiation of land cover in the central Himalayas. *J. Geogr. Sci.* **2020**, *30*, 969–987. [CrossRef]
60. Campbell, J.B.; Wynne, R.H. *Introduction to Remote Sensing*; Guilford Press: New York, NY, USA, 2011.
61. De Alban, J.; Connette, G.; Oswald, P.; Webb, E. Combined Landsat and L-Band SAR Data Improves Land Cover Classification and Change Detection in Dynamic Tropical Landscapes. *Remote Sens.* **2018**, *10*, 306. [CrossRef]
62. Griffiths, P.; van der Linden, S.; Kuemmerle, T.; Hostert, P. A Pixel-Based Landsat Compositing Algorithm for Large Area Land Cover Mapping. *IEEE J. Stars.* **2013**, *6*, 2088–2101. [CrossRef]
63. Hansen, M.C.; Loveland, T.R. A review of large area monitoring of land cover change using Landsat data. *Remote Sens. Environ.* **2012**, *122*, 66–74. [CrossRef]
64. Wulder, M.A.; Masek, J.G.; Cohen, W.B.; Loveland, T.R.; Woodcock, C.E. Opening the archive: How free data has enabled the science and monitoring promise of Landsat. *Remote Sens. Environ.* **2012**, *122*, 2–10. [CrossRef]

65. White, J.C.; Wulder, M.A.; Hobart, G.W.; Luther, J.E.; Hermosilla, T.; Griffiths, P.; Coops, N.C.; Hall, R.J.; Hostert, P.; Dyk, A.; et al. Pixel-Based Image Compositing for Large-Area Dense Time Series Applications and Science. *Can. J. Remote Sens.* **2014**, *40*, 192–212. [CrossRef]
66. Wilson, B.T.; Knight, J.F.; McRoberts, R.E. Harmonic regression of Landsat time series for modeling attributes from national forest inventory data. *ISPRS J. Photogramm.* **2018**, *137*, 29–46. [CrossRef]
67. Huang, H.; Chen, Y.; Clinton, N.; Wang, J.; Wang, X.; Liu, C.; Gong, P.; Yang, J.; Bai, Y.; Zheng, Y.; et al. Mapping major land cover dynamics in Beijing using all Landsat images in Google Earth Engine. *Remote Sens. Environ.* **2017**, *202*, 166–176. [CrossRef]
68. Teluguntla, P.; Thenkabail, P.S.; Oliphant, A.; Xiong, J.; Gumma, M.K.; Congalton, R.G.; Yadav, K.; Huete, A. A 30-m landsat-derived cropland extent product of Australia and China using random forest machine learning algorithm on Google Earth Engine cloud computing platform. *ISPRS J. Photogramm.* **2018**, *144*, 325–340. [CrossRef]
69. Xiong, J.; Thenkabail, P.; Tilton, J.; Gumma, M.; Teluguntla, P.; Oliphant, A.; Congalton, R.; Yadav, K.; Gorelick, N. Nominal 30-m Cropland Extent Map of Continental Africa by Integrating Pixel-Based and Object-Based Algorithms Using Sentinel-2 and Landsat-8 Data on Google Earth Engine. *Remote Sens.* **2017**, *9*, 1065. [CrossRef]
70. Tucker, C.J. Red and photographic infrared linear combinations for monitoring vegetation. *Remote Sens. Environ.* **1978**, *8*, 127–150. [CrossRef]
71. Fensholt, R.; Sandholt, I. Derivation of a shortwave infrared water stress index from MODIS near-and shortwave infrared data in a semiarid environment. *Remote Sens. Environ.* **2003**, *87*, 111–121. [CrossRef]
72. Riggs, G.A.; Hall, D.K.; Salomonson, V.V. *A Snow Index for the Landsat Thematic Mapper and Moderate Resolution Imaging Spectroradiometer;* IEEE: Piscataway, NJ, USA, 1994; Volume 4, pp. 1942–1944.
73. Xiao, X.; Hagen, S.; Zhang, Q.; Keller, M.; Moore, B., III. Detecting leaf phenology of seasonally moist tropical forests in South America with multi-temporal MODIS images. *Remote Sens. Environ.* **2006**, *103*, 465–473. [CrossRef]
74. Van Deventer, A.P.; Ward, A.D.; Gowda, P.H.; Lyon, J.G. Using thematic mapper data to identify contrasting soil plains and tillage practices. *Photogramm. Eng. Remote Sens.* **1997**, *63*, 87–93.
75. Goodwin, N.R.; Coops, N.C.; Wulder, M.A.; Gillanders, S.; Schroeder, T.A.; Nelson, T. Estimation of insect infestation dynamics using a temporal sequence of Landsat data. *Remote Sens. Environ.* **2008**, *112*, 3680–3689. [CrossRef]
76. Escuin, S.; Navarro, R.; Fernandez, P. Fire severity assessment by using NBR (Normalized Burn Ratio) and NDVI (Normalized Difference Vegetation Index) derived from LANDSAT TM/ETM images. *Int. J. Remote Sens.* **2008**, *29*, 1053–1073. [CrossRef]
77. Gitelson, A.A.; Kaufman, Y.J.; Stark, R.; Rundquist, D. Novel algorithms for remote estimation of vegetation fraction. *Remote Sens. Environ.* **2002**, *80*, 76–87. [CrossRef]
78. Crist, E.P.; Cicone, R.C. A Physically-Based Transformation of Thematic Mapper Data—The TM Tasseled Cap. *IEEE Trans. Geosci. Remote.* **1984**, *GE-22*, 256–263. [CrossRef]
79. Pal, M. Random forest classifier for remote sensing classification. *Int. J. Remote Sens.* **2005**, *26*, 217–222. [CrossRef]
80. Belgiu, M.; Drăguţ, L. Random forest in remote sensing: A review of applications and future directions. *ISPRS J. Photogramm.* **2016**, *114*, 24–31. [CrossRef]
81. Congalton, R.G. A review of assessing the accuracy of classifications of remotely sensed data. *Remote Sens. Environ.* **1991**, *37*, 35–46. [CrossRef]
82. Andersson, E.; McPhearson, T.; Kremer, P.; Gomez Baggethun, E.; Haase, D.; Tuvendal, M.; Wurster, D. Scale and context dependence of ecosystem service providing units. *Ecosyst. Serv.* **2015**, *12*, 157–164. [CrossRef]
83. Sun, Z.; Xu, R.; Du, W.; Wang, L.; Lu, D. High-resolution urban land mapping in China from sentinel 1A/2 imagery based on Google Earth Engine. *Remote Sens.* **2019**, *11*, 752. [CrossRef]
84. Kobayashi, T.; Umehara, T.; Satake, M.; Nadai, A.; Uratsuka, S.; Manabe, T.; Masuko, H.; Shimada, M.; Shinohara, H.; Tozuka, H. Airborne dual-frequency polarimetric and interferometric SAR. *IEICE Trans. Commun.* **2000**, *83*, 1945–1954.
85. Reid, R.S.; Kruska, R.L.; Muthui, N.; Taye, A.; Wotton, S.; Wilson, C.J.; Mulatu, W. Land-use and land-cover dynamics in response to changes in climatic, biological and socio-political forces: The case of southwestern Ethiopia. *Landsc. Ecol.* **2000**, *15*, 339–355. [CrossRef]
86. Cao, Y.; Zhou, W.; Wang, J.; Yuan, C. Spatial-temporal pattern and differences of land use changes in the Three Gorges Reservoir Area of China during 1975–2005. *J. Mt. Sci.-Engl.* **2011**, *8*, 551–563. [CrossRef]
87. Vogt, P.; Riitters, K.H.; Estreguil, C.; Kozak, J.; Wade, T.G.; Wickham, J.D. Mapping spatial patterns with morphological image processing. *Landsc. Ecol.* **2007**, *22*, 171–177. [CrossRef]
88. Xie, G.D.; Zhen, L.; Lu, C.X.; Xiao, Y. Expert knowledge based valuation method of ecosystem services in China. *J. Nat. Resour.* **2008**, *23*, 911–919.
89. Song, W.; Deng, X. Land-use/land-cover change and ecosystem service provision in China. *Sci. Total Environ.* **2017**, *576*, 705–719. [CrossRef]
90. Kanade, R.; John, R. Topographical influence on recent deforestation and degradation in the Sikkim Himalaya in India; Implications for conservation of East Himalayan broadleaf forest. *Appl. Geogr.* **2018**, *92*, 85–93. [CrossRef]
91. Munsi, M.; Malaviya, S.; Oinam, G.; Joshi, P.K. A landscape approach for quantifying land-use and land-cover change (1976–2006) in middle Himalaya. *Reg. Environ. Chang.* **2010**, *10*, 145–155. [CrossRef]
92. Hofer, T.; Messerli, B. *Floods in Bangladesh: History, Dynamics and Rethinking the Role of the Himalayas;* UNU Press: Tokyo, Japan; FAO: Roma, Italy, 2006.

93. Hansen, M.C.; Potapov, P.V.; Moore, R.; Hancher, M.; Turubanova, S.A.; Tyukavina, A.; Thau, D.; Stehman, S.V.; Goetz, S.J.; Loveland, T.R. High-resolution global maps of 21st-century forest cover change. *Science* **2013**, *342*, 850–853. [CrossRef]
94. Adnan, A.; QI-Jing, L.; M., N.S.; Abdul, M.; Sajjad, S. Carbon emission from deforestation, forest degradation and wood harvest in the temperate region of Hindukush Himalaya, Pakistan between 1994 and 2016. *Land Use Policy* **2018**, 781–790. [CrossRef]
95. Tambe, S.; Arrawatia, M.L.; Sharma, N. Assessing the priorities for sustainable forest management in the Sikkim Himalaya, India: A remote sensing based approach. *J. Indian Soc. Remote* **2011**, *39*, 555–564. [CrossRef]
96. Achard, F.; DeFries, R.; Eva, H.; Hansen, M.; Mayaux, P.; Stibig, H.J. Pan-tropical monitoring of deforestation. *Environ. Res. Lett.* **2007**, *2*, 45022. [CrossRef]
97. Sarkar, M.S.; Pandey, A.; Singh, G.; Lingwal, S.; John, R.; Hussain, A.; Rawat, G.S.; Rawal, R.S. Multiscale statistical approach to assess habitat suitability and connectivity of common leopard (Panthera pardus) in Kailash Sacred Landscape, India. *Spat. Stat.* **2018**, *28*, 304–318. [CrossRef]
98. Singh, J.S.; Pandey, U.; Tiwari, A.K. Man and forests: A Central Himalayan case study. *Ambio* **1984**, *13*, 80–87.
99. Smadja, J. Studies of climatic and human impacts and their relationship on a mountain slope above Salme in the Himalayan Middle Mountains, Nepal. *Mt. Res. Dev.* **1992**, *12*, 1–28. [CrossRef]
100. Xu, J.; Grumbine, R.E.; Shrestha, A.; Eriksson, M.; Yang, X.; Wang, Y.; Wilkes, A. The melting Himalayas: Cascading effects of climate change on water, biodiversity, and livelihoods. *Conserv. Biol.* **2009**, *23*, 520–530. [CrossRef] [PubMed]
101. Bawa, K.S.; Koh, L.P.; Lee, T.M.; Liu, J.; Ramakrishnan, P.S.; Douglas, W.Y.; Zhang, Y.; Raven, P.H. China, India, and the environment. *Science* **2010**, *327*, 1457–1459. [CrossRef]
102. Bertrand, R.; Lenoir, J.; Piedallu, C.; Riofrío-Dillon, G.; de Ruffray, P.; Vidal, C.; Pierrat, J.; Gégout, J. Changes in plant community composition lag behind climate warming in lowland forests. *Nature* **2011**, *479*, 517–520. [CrossRef]
103. Thapa, S.; Chitale, V.; Rijal, S.J.; Bisht, N.; Shrestha, B.B. Understanding the dynamics in distribution of invasive alien plant species under predicted climate change in Western Himalaya. *PLoS ONE.* **2018**, *13*, e0195752. [CrossRef]
104. Yan, J.; Yang, Z.; Li, Z.; Li, X.; Xin, L.; Sun, L. Drivers of cropland abandonment in mountainous areas: A household decision model on farming scale in Southwest China. *Land Use Policy* **2016**, *57*, 459–469. [CrossRef]
105. Paudel, B.; Wu, X.; Zhang, Y.; Rai, R.; Liu, L.; Zhang, B.; Khanal, N.R.; Koirala, H.L.; Nepal, P. Farmland abandonment and its determinants in the different ecological villages of the Koshi River Basin, Central Himalayas: Synergy of high-resolution remote sensing and social surveys. *Environ. Res.* **2020**, *188*, 109711. [CrossRef] [PubMed]
106. Jaquet, S.; Schwilch, G.; Hartung-Hofmann, F.; Adhikari, A.; Sudmeier-Rieux, K.; Shrestha, G.; Liniger, H.P.; Kohler, T. Does outmigration lead to land degradation? Labour shortage and land management in a western Nepal watershed. *Appl. Geogr.* **2015**, *62*, 157–170. [CrossRef]
107. Paudel, B.; Zhang, Y.; Yan, J.; Rai, R.; Li, L.; Wu, X.; Chapagain, P.S.; Khanal, N.R. Farmers' understanding of climate change in Nepal Himalayas: Important determinants and implications for developing adaptation strategies. *Clim. Chang.* **2020**, *158*, 485–502. [CrossRef]
108. Rai, R.; Zhang, Y.; Paudel, B.; Khanal, N. Status of Farmland Abandonment and Its Determinants in the Transboundary Gandaki River Basin. *Sustainability* **2019**, *11*, 5267. [CrossRef]
109. Bowen, M.E.; McAlpine, C.A.; House, A.P.; Smith, G.C. Regrowth forests on abandoned agricultural land: A review of their habitat values for recovering forest fauna. *Biol. Conserv.* **2007**, *140*, 273–296. [CrossRef]
110. Lambin, E.F.; Turner, B.L.; Geist, H.J.; Agbola, S.B.; Angelsen, A.; Bruce, J.W.; Coomes, O.T.; Dirzo, R.; Fischer, G.; Folke, C. The causes of land-use and land-cover change: Moving beyond the myths. *Glob. Environ. Chang.* **2001**, *11*, 261–269. [CrossRef]
111. Shrestha, B.; Ye, Q.; Khadka, N. Assessment of Ecosystem Services Value Based on Land Use and Land Cover Changes in the Transboundary Karnali River Basin, Central Himalayas. *Sustainability* **2019**, *11*, 3183. [CrossRef]
112. Rai, R.; Zhang, Y.; Paudel, B.; Acharya, B.K.; Basnet, L. Land use and land cover dynamics and assessing the ecosystem service values in the trans-boundary Gandaki River Basin, Central Himalayas. *Sustainability* **2018**, *10*, 3052. [CrossRef]
113. Zhao, Z.; Wu, X.; Zhang, Y.; Gao, J. Assessment of changes in the value of ecosystem services in the Koshi River Basin, Central High Himalayas based on land cover changes and the CA-Markov Model. *J. Resour. Ecol.* **2017**, *8*, 67–76.

land

MDPI

Article

Aboveground Biomass Distribution in a Multi-Use Savannah Landscape in Southeastern Kenya: Impact of Land Use and Fences

Edward Amara [1,2,3,*], Hari Adhikari [1,2], Janne Heiskanen [1,2], Mika Siljander [1,2], Martha Munyao [1,4], Patrick Omondi [4] and Petri Pellikka [1,2]

1 Department of Geosciences and Geography, University of Helsinki, P.O. Box 68, 00014 Helsinki , Finland; hari.adhikari@helsinki.fi (H.A.); janne.heiskanen@helsinki.fi (J.H.); mika.siljander@helsinki.fi (M.S.); martha.munyao@helsinki.fi (M.M.); petri.pellikka@helsinki.fi (P.P.)
2 Institute for Atmospheric and Earth System Research, Faculty of Science, University of Helsinki, 00014 Helsinki, Finland
3 Sierra Leone Agricultural Research Institute, Tower Hill, Freetown PMB1313, Sierra Leone
4 Biodiversity Research and Planning Directorate, Kenya Wildlife Service, P.O. Box 40241, Nairobi 00100, Kenya; pomondi@kws.go.ke
* Correspondence: edward.amara@helsinki.fi

Received: 18 August 2020; Accepted: 5 October 2020; Published: 9 October 2020

Abstract: Savannahs provide valuable ecosystem services and contribute to continental and global carbon budgets. In addition, savannahs exhibit multiple land uses, e.g., wildlife conservation, pastoralism, and crop farming. Despite their importance, the effect of land use on woody aboveground biomass (AGB) in savannahs is understudied. Furthermore, fences used to reduce human–wildlife conflicts may affect AGB patterns. We assessed AGB densities and patterns, and the effect of land use and fences on AGB in a multi-use savannah landscape in southeastern Kenya. AGB was assessed with field survey and airborne laser scanning (ALS) data, and a land cover map was developed using Sentinel-2 satellite images in Google Earth Engine. The highest woody AGB was found in riverine forest in a conservation area and in bushland outside the conservation area. The highest mean AGB density occurred in the non-conservation area with mixed bushland and cropland (8.9 Mg·ha^{-1}), while the lowest AGB density (2.6 Mg·ha^{-1}) occurred in overgrazed grassland in the conservation area. The largest differences in AGB distributions were observed in the fenced boundaries between the conservation and other land-use types. Our results provide evidence that conservation and fences can create sharp AGB transitions and lead to reduced AGB stocks, which is a vital role of savannahs as part of carbon sequestration.

Keywords: savannah; multifunctionality; protected areas; conservation; airborne laser scanning; aboveground woody biomass

1. Introduction

Savannahs are characterized by scattered tree cover and continuous coverage of grass-dominated herbaceous plants [1,2]. On the African continent, savannahs and woodlands play a particularly large role in the carbon cycle, and wildlife and biodiversity conservation, while providing livelihoods for a huge human population [3]. The area covered by savannahs is roughly three times larger than that of forests, corresponding to approximately 50% of the total area of the African continent. Savannahs therefore represent a major carbon stock in Africa despite having a lower carbon density compared to forests [4–6]. Another significant feature of the African carbon cycle is that emissions caused by land-use change are greater than fossil fuel emissions [7,8]. A large part of these emissions originates from land cover conversion of savannahs and woodlands to croplands while forests still remain an important sink [7]. Woody vegetation is mainly converted into agricultural land in response to rapid population growth [9]. In contrast to woody cover loss, widespread woody encroachment has also been observed in African savannahs [10–13]. Encroaching is particularly severe in the central interior of Africa in areas with moderate woody cover, e.g., Cameroon, the Central African Republic, South Sudan, and Uganda [12]. Species with the potential to fix nitrogen, such as *Vachelia tortillis* and *Senegalia mellifera* [11], are typical encroachers in African savannahs.

African savannahs often exhibit multi-use landscapes. They can be used for wildlife-based activities, pastoralism, subsistence agriculture, forestry, and fuelwood production, and provide other ecosystem services such as climate change regulation and water reservoirs [14]. Wildlife conservation in protected areas, such as national parks, national reserves, community conservancies, and wildlife sanctuaries, promote wildlife-based tourism [15,16], which is a significant source of income for many countries, e.g., Kenya. Through wildlife management, some savannahs have been transformed into game ranching areas with high economic growth, albeit at a significant cost to conservation [17]. On the other hand, in some cases these areas have provided funds for conservation efforts. Furthermore, savannah ecosystems are suitable for livestock grazing. Therefore, they support both wild and domestic herbivores and their potential predators [18], considering the nutritional suitability of the plants [19], and the structure, productivity, phenology, composition, and chemical attributes of the ecosystem. Uncontrolled domestic herbivore populations in protected areas threaten the conservation of wild herbivores [20]. In addition, communities in savannah areas and near conservation areas grow crops for their own use and as cash crops to support their livelihoods. Population growth and land-use policies support the expansion of agricultural activities [20] at the expense of biodiversity and wildlife conservation. Although the extraction of timber, fuelwood, and non-timber forest products contributes to the livelihood options of savannah landscape dwellers, these practices may also have a negative impact on woody vegetation structure and biodiversity.

Savannahs in Eastern Africa are extremely rich in biodiversity, with high numbers of threatened species that constitute part of the largest remaining populations of iconic wildlife left on the continent [21,22]. Many countries in this region have designated a significant portion of their terrestrial areas to biodiversity conservation, amongst them some of the world-famous national parks and reserves (e.g., Serengeti National Park in Tanzania, and Tsavo National Parks and Maasai Mara National Reserve in Kenya) [22]. Their management depends on the ownership and purpose of the conservation. A large portion of these sites are owned and managed by the government for tourism, biodiversity conservation, education, and research. Recently, private and community owned conservation areas, mainly for tourism, have increased [23]. The social and economic conditions that support their management are critical for the maintenance of wildlife within their boundaries [15]. This means that human-induced drivers have more influence on wildlife abundances than those affecting ecological processes such as changes in the size of a conservation area [15].

Megaherbivores (e.g., elephants) are often of disproportionate importance in motivating conservation actions [24]. These animals are sensitive to human impact and are most likely to survive in conservation areas. However, they impact ecosystem structure [25], shape ecosystem functions [26], and affect primary productivity and soil nutrient balance [27]. They impact habitats and the presence of other animals, even small ones such as termites [28,29]. Fences are used as conservation measures to reduce the impact of large herbivores on vegetation and human habitat [29–32]. Fencing can protect stands of dense vegetation [31,32] and mitigates human–wildlife conflicts [33]. Fences are also used to demarcate protected area boundaries. However, fencing can alter ecological processes, such as dispersal of wildlife and livestock and lead to differences in plant biomass densities in grazed and non-grazed areas [34]. The role of fencing in threatening biodiversity has been also stressed [33]. Cost associated with the construction and maintenance of fences and the conflicts occurring between protected area management and communities around fenced areas are further drawbacks [35]. Woody biomass in savannah landscapes is highly variable as a result of the various factors affecting vegetation structure. However, very little information currently exists on the biomass variations in African multi-use savannahs.

Remote sensing has a central role in understanding terrestrial carbon dynamics and in the implementation of national greenhouse gas (GHG) emission inventories and payments for ecosystem services schemes such as Reducing Emissions from Deforestation and Forest Destruction (REDD+) [36–39]. Remote sensing provides information on the extent and changes of the land-use and land cover (LULC) types, and on biomass and carbon densities. The former is typically based on LULC classification, and the latter is derived from aboveground biomass (AGB) maps. AGB maps also serve other purposes, such as natural resource management [40,41]. Optical satellite images are the most common data for LULC classification and are increasingly used in cloud computing platforms, particularly Google Earth Engine (GEE) [42]. On the other hand, airborne light detection and ranging (LiDAR, also known as airborne laser scanning, ALS) provides the most accurate remote sensing method for mapping the AGB of forests [43], but savannah, bushland, and cropland AGBs in Africa have remained less studied [44,45]. Therefore, more research on the feasibility of ALS data on AGB estimation outside forests in the African savannahs are needed.

In this study, our main objective was to assess the effect of land use and wildlife fences on woody AGB density and distribution patterns in a multi-use savannah landscape in southeastern Kenya. In this landscape, fences between conservation areas and other land-use regions are used to reduce human–wildlife conflict. More specifically, we (1) used ALS and other remote sensing data to map AGB distribution and land cover in the study area, (2) examined the effect of land use (wildlife conservation, livestock management, small-holder farming) and land cover types on AGB, and (3) studied the effect of wildlife fences on AGB patterns in the boundaries of land-use regions. We hypothesized that land use considerably affects the woody AGB distribution in the studied landscape because it drives the observed patterns of land cover, and each land cover type has a characteristic AGB density. Furthermore, fences affect the distributions and effects of wildlife and livestock, and hence, contribute to the observed woody AGB patterns.

2. Material and Methods

2.1. Study Area

The study area is located in the plains southwest of the Taita Hills (3°20′ S, 38°15′ E), in southeastern Kenya (Figure 1). The area belongs to Taita Taveta County. The county covers an area of 17,071 km^2 and has 340,670 inhabitants [46]. Typical lowland land uses include conservation in national parks, livestock management on ranches, mining, commercial sisal plantations, and dryland small-holder agriculture [6,46]. Lowland soil type is characterized by very deep, acidic, dark red, sandy clay soil (Ferralsols). Elevation ranges from 600–1000 meters above sea level (m a.s.l.) in the plains to the highest peak in the Taita Hills at 2208 m a.s.l. Average daily temperature ranges between 20 °C and 30 °C. Mean annual rainfall ranges from 500 mm to 1200 mm from the plains to the hills, and the rainfall pattern is bimodal with long rains in March–May and short rains in October–December [47,48]. Lowlands are much drier than highlands, e.g., the average yearly rainfall recorded at the Maktau weather station located within the study area was 483 mm in 2014–2016 [49].

Considerable variation in annual rainfall may also occur. A drought period occurred from 2007 to 2010 according to Voi meteorological station data at 580 m a.s.l., located 40 km east of the study area. The lowest annual rainfall (241 mm) was recorded in 2008 and the highest (553 mm) in 2009. The short rains in November–December 2008 resulted in only 35 mm of precipitation. The average annual precipitation was 563 mm from 2000 to 2018, while rainfall in 2006 and 2011 was 866 mm and 794 mm, respectively. As the Maktau weather station was established in October 2013 [50], we possess no rainfall data from the area of interest for the drought period, but the drought was evident. It caused a lack of water and forage for large mammals, such as elephants, which consequently caused a loss of woody vegetation, especially in conservation areas.

The Tsavo ecosystem, including Tsavo East and West National Parks (NP), cover ca. 62% of Taita Taveta County. In addition to Tsavo NPs, the Tsavo ecosystem consists of several other protected areas, namely Taita Hills Wildlife Sanctuary (THWS), Rukinga, and LUMO Community Wildlife Sanctuary, and gazetted forest patches in the Taita Hills and Kasigau Mountain. Wildlife populations (e.g., elephants, buffaloes, lions, antelopes, and giraffes) are large in the lowlands of the Tsavo ecosystem [51,52]. Cattle, elephants, and buffaloes constitute the most important herbivores and have increased from the late 1970s to date [53]. Wildlife densities may vary significantly during the wet and dry seasons. For example, 462 elephants were recorded in THWS in November 2013 during the dry season ground census, while only 17 were sighted during the wet season census in June 2013 [54]. Wildlife congregates in man-made waterholes, the Bura River, and riverine forests of THWS during the dry season, in search of water and fresh vegetation.

The study area (Figure 1) was defined by the extent of ALS data (see details in Section 2.3). The landscape includes typical lowland land-use and land cover types within THWS and a small part of Tsavo West National Park (TWNP) and LUMO Community Wildlife Sanctuary (LUMO). The three conservation areas are very different in their wildlife and livestock management. Tsavo West National Park is the largest of the three, covering ca. 9065 km^2, while LUMO and THWS are smaller. Although the conservation areas are managed exclusively for wildlife and wildlife-based tourism, large cattle herds may be found grazing seasonally within the boundaries. Within LUMO, the western part of Mramba (West Mramba) is preserved for livestock management, while the eastern part (East Mramba) is preserved for wildlife but is very often invaded by large cattle herds that may further invade the western plains of THWS. Cattle typically only graze in the eastern parts of THWS, while livestock occurs very seldom within TWNP. Mramba ranch holds 3500 heads of cattle and 2000 heads of goats. The entire Oza area has 3000 goats, 1500 cattle, and 130 camels, but numbers are smaller in our study site [55] and the number of livestock fluctuates between seasons and years.

Agriculture is practiced on single farms in West Mramba and in the eastern parts of THWS. Outside the conservation areas, the landscape consists of grazing land and dryland agriculture, for which the term 'matrix' is used here (Figure 1). The most common crops include cassava, maize, and legumes. Common woody species in the Acacia-Commiphora bushlands and thickets (Figure 2) include *Vachellia tortillis, Commiphora baluensis, Vachellia xanthophloea, Albizia antihelmintica, Commiphora schimperi, Maerua angolensis, Carres tomentosa, Commiphora trothe, Senegalia mellifera, Acacia brevispica, Acacia elata, Balanites aegyptica, Boscia coriacea, Newtonia hildebrantii, Delonix elata,* and *Grewia villosa*. The landscape is divided by the road from Voi to Taveta. A 33 km long electric wildlife fence constructed in 1999 separates the matrix and conservation areas (Figure 1).

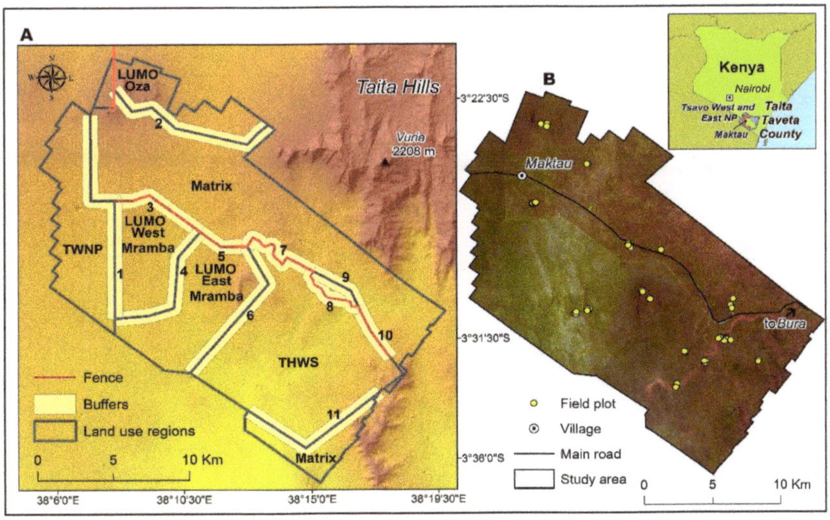

Figure 1. (**A**) Location and topography of the study area with land-use regions, fences, and buffers. Land-use regions: Taita Hills Wildlife Sanctuary (THWS), LUMO Community Wildlife Sanctuary (West Mramba and East Mramba), Tsavo West National Park (TWNP), and other land use (matrix). Numbers refer to buffers. (**B**) False color composite of Sentinel-2 satellite image showing positions of the field plots for woody aboveground biomass (AGB) estimation.

2.2. Field Data

The field data were collected between 15 and 22 August 2018 to estimate the AGB of woody plants (trees and shrubs). The sample plots were selected subjectively to cover variation in land-use and land cover type based on high resolution satellite imagery in Google Earth, and tree cover and tree height based on ALS data (Figure 1). In total, 49 sample plots were surveyed. The field plots were positioned using a Trimble GeoXH GNSS receiver with differential correction.

The sample plot design consisted of circular plots of different sizes. The main plot was 0.1 ha in size (radius 17.84 m) and was used for inventorying all the trees with a diameter at breast height (DBH, 1.3 m height from the ground) of more than 5 cm. Height (H) for the highest, median, and shortest tree were also measured at each plot using a hypsometer (Suunto). Tree species was identified for all of these trees. Furthermore, four "subplots" of 0.01 ha (radius 5.64 m) located within the main plot were used for inventorying shrubs with DBHs of 1–5 cm (see [56] for subplot locations), and four "micro plots" of 0.001 ha (radius 1.78 m) in the central points of the subplots for measuring shrubs with DBHs < 1 cm. Shrub measurements included count, DBH, basal diameter (BD), crown diameter (CD), and height for a median-sized shrub. The dominant woody species of each plot was also recorded.

Figure 2. Land-use and land cover types in the study area. (**A**) Riverine forest characterized by *Vachellia xanthophloea* trees along the Bura River in THWS (J. Heiskanen, 27.8.2018). (**B**) Partly grazed Acacia-Commiphora bushland characterized by *Vachellia tortillis* and *Commiphora baluensis* in the matrix in Maktau (P. Pellikka 26.2.2019). (**C**) Grassland in the THWS conservation area with a *Vachelia tortillis* tree (P. Pellikka, 29.9.2018). (**D**) A maize (*Zea mays*) field next to Maktau weather station with Taita Hills in the background (P. Pellikka, 5.1.2020). (**E**) Degraded grassland in the livestock management area of West Mramba in LUMO (P. Pellikka, 16.8.2018).

Aboveground biomass of trees with DBH > 5 cm (AGB_{trees}) was computed using pan-tropical biomass model [57] due to the absence of local, species-specific allometric equations. The model (Equation (1)) is based on DBH (cm), H (m), and wood-specific gravity (ρ, g/m^3). Wood densities were obtained from a species-specific list in the BIOMASS package [58] in the R software environment [59].

$$AGB_{trees} = 0.0673 \times \left(\rho DBH^2 H\right)^{0.976} \tag{1}$$

Aboveground shrub biomass (AGB_{shrubs}) was calculated using the equation in Conti et al. [60]. The model is based on BD (cm), CD (m), and H (cm) (Equation (2)). As BD, we used diameter at the 10 cm height (D10), which was calculated from the ground-level diameter using equation [61], as recommended in [60].

$$AGB_{shrubs} = \exp(-2.281 + 1.525\ln(BD) + 0.831\ln(CD) + 0.523\ln(H)) \quad (2)$$

Finally, we normalized AGB values per hectare and calculated the plot-level AGB as a sum of the tree and shrub AGB. Hereafter, by AGB, we refer to this aboveground biomass of woody plants unless specified otherwise.

2.3. Airborne Laser Scanning Data (ALS) and Biomass Mapping

Airborne laser scanning data were used to generate a reference canopy height model and to predict a high-resolution wall-to-wall AGB map for the study area. The scanning was conducted in late March 2014 and covered an area of 433 km^2. The sensor was a Leica ALS60 and a maximum of four returns per pulse were recorded. The pulse density was 1.04 pulses/m^2.

The data vendor (Ramani Geosystems, Kenya) pre-processed the ALS data, including filtering of the ground returns using Terrascan software (Terrasolid Oy, Finland). The data were delivered as georeferenced point clouds in the UTM/WGS84 coordinate system with ellipsoidal heights. The ground-classified returns were used for generating digital elevation models (DEM) at a 1-m cell size. The ALS point cloud elevations were normalized to height from the ground levels using DEM. Furthermore, buildings, power lines, and outliers (high points) were filtered using Terrascan, LAStools (Rapidlasso GmbH), and manual editing.

A 3.5-m height threshold provided the best model between ALS metrics and field biomass and was used to separate understory and ground returns from the canopy returns. Height metrics were calculated separately using first and last returns and canopy cover metrics using all returns (single, first, and last) (Table A1). The variables included all the variables available in the FUSION software [62] and ones used in our earlier study [63]. Square root transformation was applied to AGB, as it was found to improve the linear relationship between AGB and explanatory variables. The "regsubset" function in the "leaps" package [64] was used to fit multiple linear regression models between the ALS metrics calculated from the ALS point density clipped over the field plot and the AGB calculated from that field plot. The leave-one-out cross-validation root mean square error (RMSE) and the coefficient of determination (R^2) were used to select the best AGB model. The predictions were back-transformed (squared), and the square of the residual standard error was added to the predicted values to avoid back-transformation bias [45,65]. For AGB prediction at wall-to-wall level, spatial grids of ALS metrics were generated at a spatial resolution of 30 m × 30 m. Mean densities of AGB in each land-use and land cover class was calculated from the AGB map.

2.4. Satellite Imagery and Land Cover Mapping

We collected Sentinel-2 images (top of atmosphere reflectance) with cloud cover less than 20% in the images during the dry seasons [short dry season (January 1 to February 28) and long dry season (July 1 to September 30)] in 2017 and 2018, and pre-processed them in the GEE platform. In total, 103 Sentinel-2 images (bands with a resolution of 10 m and 20 m only) were used to calculate the median dry season image. Median dry season images were calculated for all bands in the blue to the shortwave infrared spectral range based on all available cloud-free pixels (Figure 1B). In addition, a normalized difference vegetation index (NDVI) [66], an enhanced vegetation index (EVI) [67], EVI2 [68], two variants of normalized difference infrared index (NDII-1 and NDII-2) [69], and an optimized soil-adjusted vegetation index (OSAVI) [70] were calculated from the median image.

Additionally, land cover classification was performed in the GEE platform. In addition to median dry season Sentinel-2 composite and vegetation indices, input data included an ALS-based canopy height model (CHM). The land cover in the landscape was classified into four land cover types (cropland, grassland, forest and bushland) according to the Land Degradation Surveillance Framework [71].

Cropland is cultivated land with annual or perennial crops, while grassland contains grasses and other herbs with less than 10% woody cover. Forest in our classification is made up of a continuous stand of trees with partly interlocking crowns, typically along the riverbeds. Bushland is made up of mixed trees and shrubs with a canopy cover of 40% or more, while thickets are closed stands of bushes and climbers usually between 2 m and 7 m tall and shrubland are open or closed stands up to 3 m tall. For this study, thickets and shrubland were incorporated into bushland because we had few field plots for those classes and the classes were similar in reflectance and vegetation characteristics.

In the first step, training data were collected through visual interpretation using ArcGIS 10.3 for the four land cover classes. The pixels for training the classifier were selected based on image interpretation and CHM. Classification and regression trees (CART) [72] were observed to obtain the highest overall accuracy among the classifiers in GEE and was thus selected for the classification. The reference data set for accuracy assessment included the 49 points surveyed in 2018 in the field, which were not used as training points in the classification. Finally, manual editing was performed in ArcGIS to address some of the apparent misclassification in the land cover map.

2.5. Wildlife and Livestock Data

Elephant, buffalo, and cattle data were taken from the Tsavo–Mkomazi large mammal census of 2014 to be comparable with the 2014 ALS data used. The wildlife census is conducted by the Kenya Wildlife Service (KWS) every three years to establish the status of key species in the Tsavo ecosystem. The census is carried out from fixed-wing aircrafts and the data collection procedure is described in detail in [73]. The animal spatial distribution and densities were further compared with AGB in the studied landscape (Figure 3).

2.6. Statistical Analyses of AGB Data

The plot-level AGB values were used for computing descriptive statistics (range, mean, median, and standard deviation) for the field data. The Kruskal–Wallis test was conducted to study whether differences in AGB were statistically significant between the land-use regions and land cover classes. Furthermore, median and mean values of the AGB per class were illustrated with a box plot for the different land-use regions and land cover classes. We also estimated the percentage area covered by each land cover in the respective land-use region. Finally, 500-m wide buffers were set in 11 segments of land-use region boundaries to assess local AGB differences (Figure 1A). The buffers were categorized into fenced and non-fenced segments to determine the effect of the fence on AGB. Pixel values were studied separately for two sides of the boundary by calculating the percentage of zero AGB pixels. Furthermore, medians of the non-zero AGB values were studied using the Wilcoxon test. All analyses were performed in the R statistical environment [74].

3. Results

3.1. Aboveground Biomass Estimates and Map

Woody AGB estimates based on the field plot measurements are summarized in Table 1. The maximum plot-level values are nearly 365 megagrams per hectare (Mg/ha) and were observed in the riverine forest. The plots with the lowest AGB had very little woody biomass and were located in the grassland areas.

Table 1. Summary of the aboveground biomass (AGB) values based on the field data according to the diameter at breast height (DBH) class (n = 49). AGB was estimated based on diameter at ground level for shrubs with a DBH < 1 cm. SD = standard deviation, IQR = interquartile range.

DBH Class	AGB (Mg/ha)					
	Mean	Min	Max	SD	IQR	Median
DBH > 5 cm	42.15	0.28	364.04	85.41	20.94	7.91
DBH 1–5 cm	3.69	0.00	19.46	4.98	4.03	2.07
DBH < 1 cm	0.52	0.00	2.56	0.60	0.49	0.35
Total	38.02	0.00	364.54	78.27	21.56	10.04

The final modeling results for mapping AGB using ALS data are shown in Figure 3. The model was based on two variables: CC.all (percentage of all returns above 3.5 m; $p < 0.001$) and Elev.min.fr (minimum elevation of the first returns above 3.5 m; $p < 0.001$). The model performed well in terms of model fit ($R^2 = 0.88$) although RMSE based on leave-one-out cross-validation was relatively large (26 Mg/ha, 75.6% of the mean AGB). However, the model did not show any signs of systematic over- or under-estimation (Figure 3).

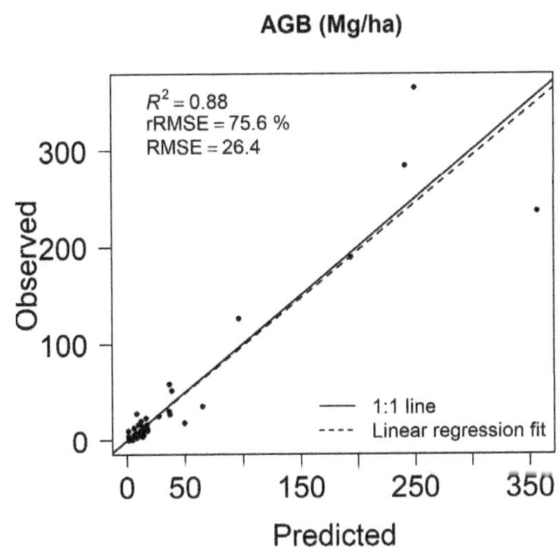

Figure 3. Airborne laser scanning- (ALS)-predicted vs. field-observed AGB based on leave-one-out cross-validation.

The AGB map shows predicted biomass density patterns at 30 m × 30 m resolution (Figure 4). The mean AGB in the study area was 5.9 Mg/ha. The Riverine forests in the southern and southeastern parts of the landscape within THWS had the largest AGB densities. We also observed relatively large AGB densities outside the protected areas towards the foothills of the Taita Hills, in the northeastern part of the landscape. Aboveground biomass spatial variations were also relatively large in the matrix and in LUMO Oza. On the other hand, the lowest AGB values were found in the nearly treeless grassland of THWS, LUMO East Mramba, LUMO West Mramba, and TWNP.

Wildlife (elephant and buffalos) and livestock (cattle) were highly evident in the conservation areas based on the 2014 KWS wildlife census. Elephants were present in LUMO Mramba East and THWS, and were absent in the matrix (Figure 4, Table 2). Cattle were found in all the land-use regions, except in the small portion of TWNP captured during the ALS campaign (Figure 4). Their density was highest in LUMO East Mramba (11.43 animals/km^2), a portion of the landscape secured for livestock grazing and was second highest in the matrix (4.40 animals/km^2), where agriculture is the most common land use. Buffalos were found in the conservation areas, showing the highest number per unit area in THWS (Table 2). We categorized the animals into three herd sizes, in which the number of animals per herd differed per animal species (Figure 4). We saw no elephants or buffaloes in the matrix during the 2014 wildlife census. Furthermore, no animals were visible in LUMO Oza.

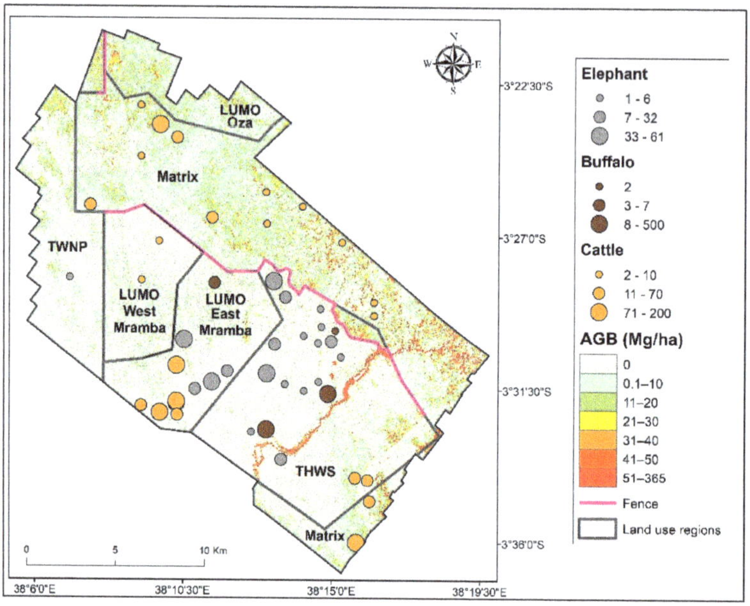

Figure 4. Biomass map showing the boundaries of the land-use regions: Taita Hills Wildlife Sanctuary (THWS), LUMO Community Wildlife Sanctuary (LUMO East Mramba, LUMO West Mramba, and LUMO Oza), Tsavo West National Park (TWNP), and other land use (matrix), and animal counts for elephants, buffalos, and cattle in 2014.

Table 2. Animal counts (animals) and densities (animals/km^2) per land-use region during the 2014 wildlife census by Kenya Wildlife Service. Land-use regions: Taita Hills Wildlife Sanctuary (THWS), LUMO Community Wildlife Sanctuary (LUMO East Mramba and LUMO West Mramba), Tsavo West National Park (TWNP), and other land use (matrix).

Animal	Land-Use Region (area)				
	TWNP (48.92 km^2)	LUMO East Mramba (47.24 km^2)	LUMO West Mramba (33.92 km^2)	THWS (101.50 km^2)	Matrix (141.63 km^2)
Elephant	5 (0.10)	137 (2.90)	0 (0)	237 (2.33)	0 (0)
Cattle	0 (0)	540 (11.43)	20 (0.59)	100 (0.99)	623 (4.40)
Buffalo	2 (0.04)	7 (0.15)	0 (0)	802 (7.90)	0 (0)

3.2. *Land Cover Classification*

The overall land cover classification accuracy was 88.78%. The producer's and user's accuracy are shown in Table A2. The land cover map shows the distribution of the land cover classes in the landscape (Figure 5). Bushland and cropland dominate the matrix in northern and northeastern parts of the landscape, while grassland that is representative of the savannah biome dominates the southern and southeastern parts (THWS, LUMO Mramba, TWNP). LUMO Oza is almost completely bushland as there is less agriculture and livestock management. Forest is the land cover type with the smallest area, located mostly in THWS along the Bura River. THWS also has relatively large patches of bushland in its eastern parts bordering the matrix. Cropland is also present in the eastern part of THWS, while it is not observed in the other conservation areas. The THWS wardens consider it a form of informal encroachment.

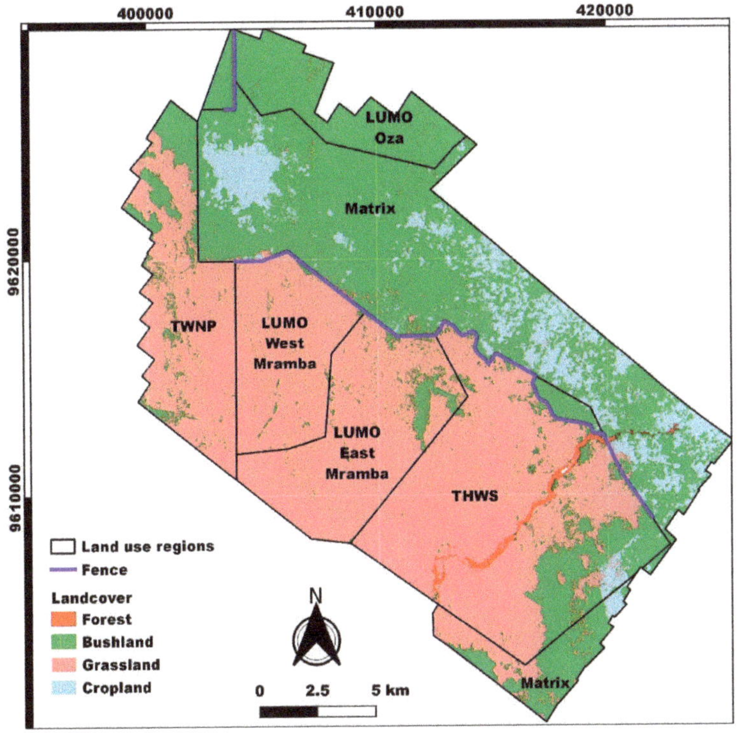

Figure 5. Land cover of the study area showing the boundaries of the land-use regions: Taita Hills Wildlife Sanctuary (THWS), LUMO Community Wildlife Sanctuary (LUMO East Mramba, LUMO West Mramba, and LUMO Oza), Tsavo West National Park (TWNP), and other land use (matrix).

3.3. *Effect of Land Cover and Land Use on Aboveground Biomass*

Aboveground biomass values for the land cover types are shown in Figure 6 and Table 3. The forest had the highest mean AGB (75.5 Mg/ha) followed by bushland (9.0 Mg/ha) and cropland (5.8 Mg/ha). Grassland had clearly the lowest mean AGB, as it is mostly treeless (mean 1.8 Mg/ha, median 0 Mg/ha). However, bushland, cropland, and grassland also had very high AGBs at certain locations (maximum values in Table 3). These areas correspond to "forest-like" bushland with trees and large shrubs. The highest values in the cropland were found in the fallowed fields and patches of bush and in the tree-covered areas next to the fields. In addition, certain farmers practice agroforestry, meaning that they grow trees for fruit and timber production and for providing shade for crops. Furthermore, the grasslands also have scattered large trees, e.g., in Figure 2C. We observed significant AGB differences among the land cover types ($p < 0.001$) according to the Kruskal–Wallis mean rank test. Furthermore, the Dunn test indicated a significant difference ($p < 0.05$) between all the land cover types (Figure 6).

When comparing the land-use regions, the mean AGB values in descending order were 8.9 Mg/ha in the matrix 8.8 Mg/ha in LUMO Oza, 4.8 Mg/ha in THWS, 3.8 Mg/ha in TWNP, 2.6 Mg/ha in LUMO West Mramba, and 2.4 Mg/ha in LUMO East Mramba (Table 3). According to the Kruskal–Wallis test, the AGB differences among land-use regions were significant ($p < 0.001$). These differences are mainly explained by dissimilarities in the land cover class distributions (Figure 7). The matrix has very little grassland with low AGB, but a large fraction of bushland with a relatively high AGB. The area also has some forest and cropland with high maximum values, which increase the mean AGB. LUMO Oza also mainly consists of higher AGB bushland, while lower AGB regions have larger fractions of grassland. This includes both the West Mramba grazing area and various protected areas. We conducted pairwise comparisons between the classes using the Dunn test, which indicates a significant difference ($p < 0.05$) between all the classes (Figure 7).

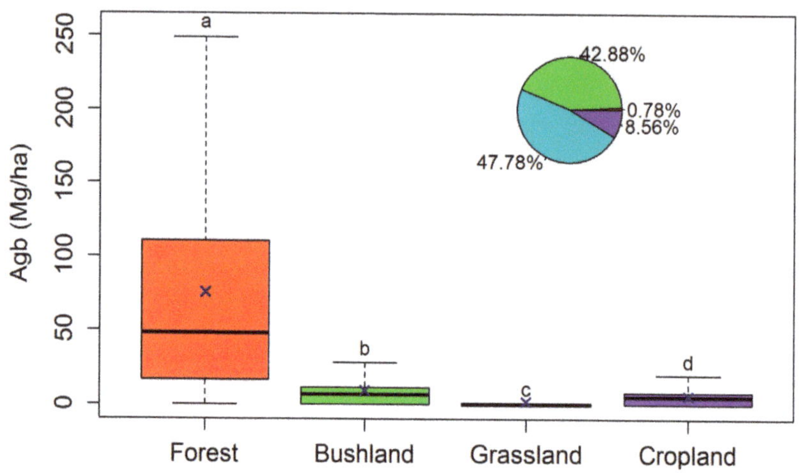

Figure 6. Aboveground biomass (AGB) distribution for the land cover types based on the AGB map. The pie chart shows the distributions of the types in the study area. Letters indicate significant differences ($p < 0.05$) according to the Dunn test. The outliers in each box plot are not shown. The "x" on each box plot represents the means and the whiskers represent confidence intervals.

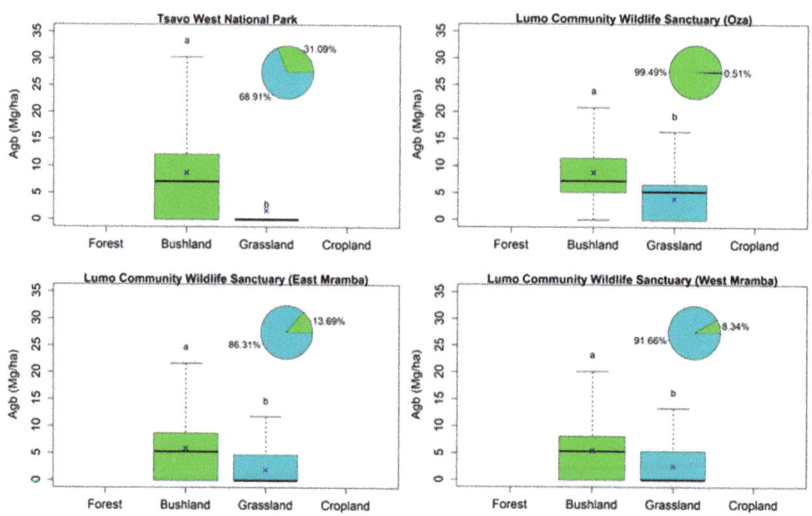

Figure 7. *Cont.*

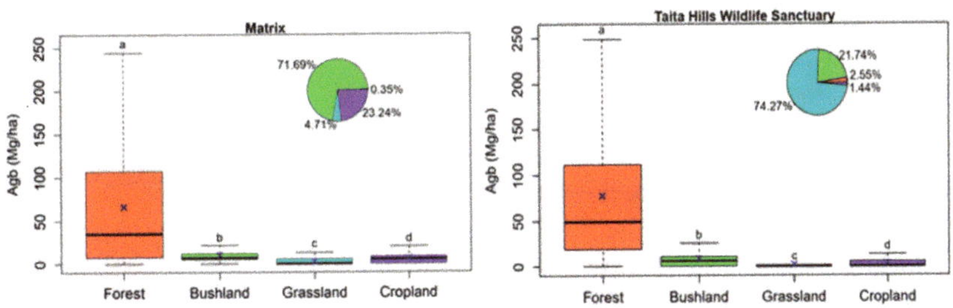

Figure 7. Land cover class-wise distribution of aboveground biomass (AGB) for each land-use region. Letters indicate significant differences ($p < 0.05$) according to the Dunn test. The outliers in each box plot are not shown. The pie chart shows the distribution of the types in the study area. The "x" on each box plot represents the means and the whiskers represent confidence intervals.

Table 3. Aboveground biomass (AGB) statistics for land-use regions and land cover types based on AGB and land cover maps. IQR = interquartile range. Land-use regions: Taita Hills Wildlife Sanctuary (THWS), LUMO Community Wildlife Sanctuary (LUMO East Mramba, LUMO West Mramba, and LUMO Oza), Tsavo West National Park (TWNP), and other land use (matrix).

Land-Use Region	Land Cover	AGB (Mg/ha)					
		Mean	Min	Max	SD	IQR	Median
LUMO Oza	Bushland	8.8	0.0	82.3	7.9	6.3	7.3
	Grassland	3.9	0.0	20.1	4.4	6.6	5.2
	All	8.8	0.0	82.3	7.9	6.2	7.2
LUMO East Mramba	Bushland	5.9	0.0	106.2	7.6	8.7	5.3
	Grassland	2.0	0.0	50.9	3.8	4.7	0.0
	All	2.4	0.0	106.2	4.6	5.1	0.0
LUMO West Mramba	Bushland	5.4	0.0	104.4	6.3	8.1	5.3
	Grassland	2.6	0.0	55.6	4.2	5.3	0.0
	All	2.6	0.0	104.4	4.4	5.4	0.0
THWS	Forest	77.4	0.0	353.0	79.0	92.5	49.0
	Bushland	8.1	0	159.3	11.9	10.1	5.5
	Grassland	1.4	0	237.3	4.2	0	0
	Cropland	3.0	0.0	100.3	7.0	5.1	0.0
	All	4.8	0.0	353.0	18.6	5.1	0.0
TWNP	Bushland	8.8	0	71.8	8.3	12.2	7.1
	Grassland	1.7	0	100.9	3.4	0	0
	All	3.8	0.0	100.9	6.3	6.2	0.0
Matrix	Forest	66.0	0.0	346.7	73.8	99.3	35.0
	Bushland	9.9	0.0	353	13.3	6.6	6.9
	Grassland	2.2	0.0	54.5	4.0	5.1	0.0
	Cropland	6.0	0.0	241.3	9.4	8.1	5.2
	All	8.9	0.0	353	13.5	10.5	6.3
All Land cover	Forest	75.5	0.0	353.0	78.3	94.2	47.8
	Bushland	9.2	0	353	11.9	11.2	6.7
	Grassland	1.8	0	237.3	4	0	0
	Cropland	5.8	0.0	241.3	9.3	7.9	5.1
	All	5.9	0.0	353.0	13.1	7.6	0.0

3.4. *Effect of Fences on Aboveground Biomass*

Lastly, we compared AGB values in the fenced and non-fenced boundaries of the land-use regions (see Figure 1A for buffer numbers). Table 4 reports the fraction of zero AGB pixels for two sides of the buffer and the Wilcoxon test results for the non-zero AGB values.

The largest differences in the percentage of zero AGB occurred in the fenced boundaries (buffers 3, 5, 7 and 8). Most of the non-fenced boundaries (buffers 1, 2, 4, 6 and 11) showed only small differences.

However, a greater difference was observed in the non-fenced buffer 9, which corresponds to the boundary between bushland part of THWS and the cropland-dominated matrix. Buffer 10 showed relatively small difference in the presence of zero AGB although there is a fence. This boundary is between THWS and the matrix in the eastern part of the study area.

The medians of the non-zero AGB values differed most substantially in the fenced buffers 7, 8 and 10 (all differences highly significant according to the Wilcoxon test) (Table 4). Although percentage zero AGB was substantially higher in LUMO West and East Mramba than in the matrix in the fenced buffers 3 and 5, median AGB did not differ significantly ($P > 0.05$). However, smaller but highly significant differences were also observed in the non-fenced buffers 1, 6 and 9. Buffer 1 is located in the non-fenced boundary between TWNP and LUMO West Mramba, where bushland in the northern part of TWNP has a relatively high AGB compared to grassland-dominated West Mramba. Buffer 6 corresponds to the boundary between two conservation areas, LUMO East Mramba and THWS.

Table 4. Percentage of zero woody aboveground biomass (AGB) and median AGB for non-zero AGB pixels. P value refers to the Wilcoxon test results made for the non-zero AGB values. Numbers in the end of land-use region names refer to the numbers of buffers in Figure 1A.

Side 1	Side 2	Fence	Percentage Zero AGB		Median for Non-Zero AGB		
			Side 1	Side 2	Side 1	Side 2	*P* Value
TWNP_1	LUMO West Mramba_1	No	57.6	52.9	7.1	6.6	<0.001
LUMO Oza_2	Matrix_2	No	20.7	17.0	8.2	8.2	>0.05
LUMO West Mramba_3	Matrix_3	Yes	84.3	44.1	6.5	6.4	>0.05
LUMO East Mramba_4	LUMO West Mramba_4	No	63.2	58.2	6.7	6.8	>0.05
Matrix_5	LUMO East Mramba_5	Yes	32.6	62.6	7.2	7.2	>0.05
LUMO East Mramba_6	THWS_6	No	85.9	84.9	6.5	6.9	<0.001
THWS_7	Matrix_7	Yes	75.2	22.0	6.9	8.8	<0.001
THWS1_8	THWS2_8	Yes	5.3	72.0	12.6	8.0	<0.001
Matrix_9	THWS_9	No	31.0	10.6	9.1	10.0	<0.001
Matrix_10	THWS_10	Yes	56.2	61.9	6.7	9.7	<0.001
THWS_11	Matrix_11	No	69.9	56.4	6.6	6.8	>0.05

4. Discussion

4.1. Remote Sensing—Based Biomass and Land Cover Maps

We used field data and ALS metrics to create a wall-to-wall high-resolution AGB map. The model fit and accuracy were similar [75,76] or compared favorably with previous studies in temperate and tropical forests [65,77–79]. Our model was based on two predictors: minimum elevation of the first returns above 3.5 m and percentage of all returns above 3.5 m. These variables characterize canopy height and cover, both of which are related to AGB. Similar combinations of height and cover variables have also been used in previous studies in sub-Saharan Africa [78,80,81]. Field-measured AGB included both shrubs and trees. According to the field data, shrubs (DBH 1–5 cm) can make an important contribution to woody AGB. However, as a height threshold of 3.5 m was used to separate canopy and ground returns, woody vegetation less than 3.5 m in height does not affect the ALS variables. Therefore, areas where shrubs are less than 3.5 m in height appear as zero AGB in the map. We selected the height threshold from the tested values, as it provided the most accurate predictions. Further research should be conducted to map AGB variations in the smallest shrubs and grasses.

We used Sentinel-2 satellite images and the CART algorithm in the GEE platform for creating the LULC map. We achieved a good overall accuracy of 88.78% when using dry season composites. Previous studies have shown that the dry season is best suited for separating variations in woody AGB [82,83]. One topic for further research would include classifying various grassland types within the study area.

Spatially explicit AGB and LULC maps offer additional knowledge of AGB variations across the savannah landscapes compared to spatially limited field inventories. In this study, maps demonstrated the link between LULC and AGB, and sharp AGB gradients in certain boundaries of the land-use regions. Furthermore, maps enable geospatial analyses of the AGB patterns, e.g., together with wildlife and livestock inventories, and can inform land management interventions [45]. In our study, maps showed that grassland concentrated in the wildlife conservation areas, where AGB was reduced due to the browsing effect on trees [83]. As there are fewer large mammals outside the conservation areas, their negative impact on woody vegetation is less in these areas. Therefore, wildlife and livestock frequency in the multi-use landscape contributes to the low biomass densities in the region.

4.2. Effect of Land Cover and Land Use on Aboveground Biomass

Our results reveal a significant difference ($P < 0.05$) in woody AGB among the land cover and land-use classes in the studied landscape. In general, AGB is concentrated in areas with larger tree densities. According to the field data, shrubs and smaller trees can also have a considerable effect on woody AGB density. We observed the highest AGB densities in the forest along the Bura River Valley and towards the foothills of the Taita Hills, while grassland had the lowest AGB densities. As the forest class only occupies a small area, other land cover classes contributed more to the total AGB stock at the landscape level. This emphasizes that a greater amount of AGB is stored in open savannah and bushland than in the forest. Bushland occupies more than half of the total area, and therefore contributes the most to the total AGB stock. The contribution of cropland to the total AGB in the landscape is due to agroforestry (i.e., trees growing on cropland). The mean AGB densities in the landscape were low compared to montane forest, exotic plantation, and woodland in the higher altitudes of the Taita Hills [65,84]. Furthermore, biomass in the bushland was comparable to the leaf biomass of sisal (*Agave sisalana*) in a commercially owned plantation established in the savannah landscape in Taita Taveta [85]. Low precipitation [39,86], small-scale farming by resource-poor farmers [87], low CO_2 concentrations in arid and semiarid regions [88], and disturbance from fire and herbivores [89,90] are among factors responsible for the generally low AGB in the savannah landscape.

We categorized the multi-use savannah landscape into conservation (TWNP, THWS, and LUMO) and non-conservation areas (matrix) based on land use. Furthermore, the conservation types were categorized based on ownership and management. The TWNP, LUMO, and THWS are government, community, and privately owned and managed, respectively. The AGB differences between land-use regions are driven by the land cover differences. We observed the highest woody AGB densities in non-conservation areas (matrix), which are mainly bushland and cropland, while LUMO West Mramba and LUMO East Mramba, community owned and managed wildlife sanctuaries that are mainly grassland, showed the lowest mean AGB densities. THWS and TWNP had similar mean AGB densities, while LUMO Oza had a much higher AGB compared to grassland-dominated regions because of its larger fraction of bushland.

Our results support the hypothesis that there is a link between the land use (conservation and non-conservation) and dominant land cover type, which affect the observed AGB patterns. Presence of wildlife is important for grassland to remain sparsely wooded, and hence, wildlife conservation contributes to open grassland with relatively low woody AGB. Furthermore, ranches for livestock contribute to the low AGB. THWS and LUMO West Mramba serve as a migratory corridor for elephants moving between Tsavo West and Tsavo East NPs in search of food and water [51]. Contrary to the February 2011 elephant census [73], the elephant density in the region increased from less than 0.5 elephants/km^2 in 2011 to > 2 elephants/km^2 in February 2014. This considerable increase in elephant population contributes to low AGB densities in the region. Habitat improvements through water supplementation in the protected areas also attract wildlife and further create pressure on the vegetation. Waterholes attract large congregations of herbivores particularly during the dry season [73]. Williams et al. [91] have also suggested the presence of surface water acts as a determinant of the distribution of water-dependent wildlife species. The wildlife and livestock census data also showed that private (THWS) and government (TWNP) owned conservation areas had more wildlife (elephants and buffalos) while the community owned conservation areas attract more livestock. This could be associated with the management strategies employed by the respective agencies. Therefore, policies and management strategies geared towards woody vegetation protection should be introduced into wildlife conservation management plan in order to reduce AGB decline in conservation areas.

Recent studies in the same region show that conversion of bushland to treeless cropland [92] increases land surface temperatures and decreases evapotranspiration, and low tree canopy cover areas cause higher land surface temperatures and higher temperatures in general [93]. Together with increasing proportion of agricultural areas, conservation areas have a negative contribution to the local climate, and furthermore, to the regional climate. Furthermore, bushland protection is vital for the conservation of flora and fauna, and for habitat conservation [91,94]. Furthermore, high AGB bushland supports, for example, the mitigation of wildfire, poor water quality, soil erosion, soil PH, air temperature and other ecosystem services of importance to the ecology, climate and wildlife [91,92,95]. Restoration of degraded areas by fencing, enrichment planting of woody plants and translocation of wildlife (browsers) to high biomass areas, agroforestry, and sustainable environmental regulation are some ways to mitigate these effects. Therefore, the trade-offs between the wildlife conservation and benefits of woody vegetation should be considered carefully in the conservation area management and land-use planning.

Although not addressed in this study, in addition to land use, natural factors, such as soil type, ground water table level, and rainfall, may contribute to land cover and AGB patterns. The soil type is typically red laterite, but parts of the landscape are characterized by sedimentary carbonites, which are drier and less fertile soils, thus introducing sparser woody vegetation. The water table level is high, especially along the Bura River Valley, enabling better tree growth. Furthermore, rainfall and mist emergence in topographically higher areas, such as Maktau Hill in LUMO Oza, may increase tree cover and height. Further studies should aim to clarify the roles of land use and natural factors on land cover and AGB in the study area.

4.3. *Effect of Wildlife Fences on Biomass Distribution and Density*

Fencing conservation areas is primarily done to prevent wildlife from intruding into surrounding communities and farmlands, in other words, to reduce human–wildlife conflicts [96–99]. Fences additionally help minimize wildlife poaching and the illegal extraction of other vital resources from protected areas [33] and hinder the transmission of vector-borne diseases between livestock and wildlife, as production animals and wildlife are kept separate. In Kenya, 60% of all protected areas are fully or partially fenced [35].

The ecosystem in the Taita Hills lowlands faces challenges, including livestock incursion, poaching, drought, land-use change, human–wildlife conflict, unprescribed fires, invasive species, and vegetation damage by elephants [100]. Electric and non-electric fences have therefore been constructed on the borders of the protected areas to minimize some of these challenges. The fence from Maktau to Bura Village was built in 1999 [33]. It restricts the movement of wildlife from conservation areas and hinders unauthorized access to the areas [99]. Fences also protect degraded habitats and support forest regeneration trials. Furthermore, fences around farms restrict wildlife and livestock from entering the farms.

According to our analysis of the AGB variation in the boundaries of the land-use regions (buffers), the largest differences in the percentage of zero AGB and median AGB occurred in the fenced boundaries. In the buffers 3, 5 and 7, which correspond to the boundary between the conservation areas (LUMO West and East Mramba, THWS) and the matrix, the percentage of zero AGB was considerably higher in the conservation area sides of the fence. The zero AGB pixels refer to pixels without any woody AGB, which may indicate large pressure from herbivores on woody vegetation. This is supported by the high density of wildlife close to the fence in LUMO East Mramba and THWS (Figure 4). In the buffer 7, matrix side also had higher median AGB but buffers 3 and 5 did not show significant difference in median AGB. The latter suggest that although woody vegetation is considerably less in LUMO West and East Mramba sides of the buffers, woody vegetation in both sides has similar character and median AGB. Buffer 8 matches the fenced boundary in the northern part of THWS and it is associated with a sharp transition from grassland to relatively dense bushland. This explains both the larger fraction of zero AGB pixels and the lower median AGB in the grassland side. Furthermore, buffer 10 is located in the fenced boundary between THWS and the matrix. THWS side of this boundary had slightly higher percentage of zero AGB than matrix side, similar to other conservation areas but the difference was smaller. However, THWS side of the buffer had significantly higher median AGB. This can be explained by the presence of riverine forest in that side of the boundary with greater AGB. Furthermore, the matrix in this area lies on a flood plain dominated by cropland interspersed with bushland in contrast to grassland in the THWS side, which may explain this difference.

Among the non-fenced boundaries, buffer 9 had the most apparent difference between the two sides of the boundary. This buffer corresponds to the northern boundary of THWS with rapid change from bushland to cropland-dominated area within the matrix. Matrix-side had clearly more zero AGB pixels corresponding to cropland and lower median AGB. Although not fenced, this boundary follows a road, which makes it clearly visible. Furthermore, the fence south of the area protects it from herbivores in THWS. In addition, statistically significant differences in median AGB were observed in the non-fenced buffers 1 and 6. In the boundary between TWNP and LUMO West Mramba (buffer 1), bushland in the northern part of TWNP has a relatively high AGB compared with grassland-dominated West Mramba, which explains higher median AGB in the TWNP side. Buffer 6 corresponds to the boundary between two LUMO East Mramba and THWS. Slightly higher AGB in the THWS side could relate to higher grazing pressure in LUMO East Mramba. However, differences in these two unfenced boundaries are very small in comparison to the fenced boundaries with obvious differences.

Our results support our hypothesis that fences play a role in the distribution of wildlife and livestock, and woody AGB patterns in the landscape. This creates sharp land cover transitions to the fenced boundaries of the land-use regions. The conservation and grassland sides of the buffers 5, 7 and 8 experience high pressure from wildlife and cattle while pressure is particularly low in the matrix sides of buffers 7 and 8 with fewer cattle (Figure 4). In buffers 3 and 10, the difference in herbivore density between the conservation areas and the matrix were not as evident at the time of counting. However, free ranging wildlife are constantly moving based on resource conditions.

In general, fencing can increase the wildlife population in the conservation areas and enhance biodiversity conservation [101–103]. However, an increased abundance of (mega)herbivores [104] reduce biomass densities due to tree mortality caused by browsing. The browsers suppress woody plant recruitment in the grassland and have a long-term impact on their growth and mortality rates [105]. This is particularly true for non-selective feeders, such as elephants, who debark trees and thus suppress recruitment and vegetation generation. According to Ogutu et al. [52], the landscape experienced a moderate growth in elephant density between 1977 and 2016. Similar pressure on woody plants was observed during 1970–1973, when the elephant population was large [55]. The problem is further aggravated by the fence, which restricts wildlife dispersal, and hence, reduces the ecosystem's resilience [98]. Thus, fencing combined with heavy browsing may reduce the biomass in conservation areas.

5. Conclusions

Taita Hills lowland savannah landscape, similar to other typical African savannah biomes, exhibits multi-use functionality, which results in heterogeneous land cover. We combined field data with ALS metrics to predict a woody AGB map in the study area and created a land cover map using Google Earth Engine. AGB densities in the region were comparatively low and influenced by wildlife conservation. The highest AGB densities were observed in the forest class (riverine forest) in THWS in the conservation area. Greater AGB densities were also found in the bushland in the matrix, LUMO Oza, and southern parts of THWS. The western parts of the landscape dominated by grassland and influenced by wildlife conservation and livestock grazing had a lower woody AGB density. Wildlife and livestock densities in the conservation area are high compared to the matrix. Bushland and cropland dominate the matrix, which support the livelihood of community members through farming and other livelihood options (fuelwood, etc.). The electric fence restricts the movement of wildlife, creating grassland within protected areas and contributing to the low densities of woody AGB. In addition to human–wildlife conflict mitigation, fencing also influences the spatial distribution and density of woody AGB in a multi-use savannah landscape. Further investigating the effect of wildlife and livestock fencing on land cover and biomass (including grass biomass) in multi-use savannah landscapes at various spatial and temporal scales is important. Furthermore, our results need to be scaled up and contributions of livestock management and conservation areas to climate change require investigation. The impact of wildlife conservation on land cover change, and plant species diversity and composition also deserve further investigation.

Author Contributions: E.A., H.A., and J.H. planned the study, analyzed the data, and wrote the manuscript. J.H. and H.A. collected and processed the data. P.P. supervised and commented on the manuscript. M.M. and P.O. commented on the manuscript. M.S. contributed by improving maps and commented on the manuscript. All authors have read and agreed to the published version of the manuscript.

Funding: This research was funded by the Building Bio-Carbon and Rural Development in West Africa (Biodev) program funded by the Ministry for Foreign Affairs of Finland, the SMARTLAND project (Environmental sensing of ecosystem services for developing climate smart landscape framework to improve food security in East Africa) funded by the Academy of Finland (Decision number 318645), the TAITAGIS project funded by the Ministry for Foreign Affairs of Finland and coordinated by the Finnish National Agency for Education HEI ICI Programme, and the Earth Observation and Environmental Sensing for Climate-Smart Sustainable Agropastoral Ecosystem Transformation in East Africa (ESSA project) funded by the Development of Smart Innovation through Research in Agriculture program (DeSIRA) of European Commission, DG Development.

Acknowledgments: The Faculty of Science at the University of Helsinki is acknowledged for supporting ALS data acquisition for the study area. We also acknowledge the Taita Research Station of the University of Helsinki in Kenya and its great staff for logistical support during the field inventory campaign. A special thanks to Darius Kimuzi, Hanna Hirvonen, Pinja Tolvanen, Mwadime Mjomba, and Zhipeng Tang for assistance in the field inventory campaign. Open access funding provided by University of Helsinki.

Conflicts of Interest: The authors declare no conflict of interest.

Appendix A

Table A1. Summary of airborne laser scanning metrics computed using Fusion [62,63].

Predictor	Description
H.p01, H.p05, H.p10, H.p20, H.p25, H.p30, H.p40, H.p50, H.p60, H.p70, H.p75, H.p80, H.p90, H.p95, H.p99	1st, 5th, 10th … and 99th percentile of return height > 3.5 m
H.L1, H.L2, H.L3, H.L4	L-moments 1–4 of return height > 3.5 m
H.L.cv	L-moments coefficient of variation of return height > 3.5 m
H.L.skewness	L-moments skewness of return height > 3.5 m
H.L.kurtosis	L-moments kurtosis of return height > 3.5 m
H.max	Maximum of return height > 3.5 m
H.mean	Mean of return height > 3.5 m
H.min	Minimum of return height > 3.5 m
H.mode	Mode of return height > 3.5 m
H.cv	Coefficient of variation of return height > 3.5 m
H.v	Variance of return height > 3.5 m
H.stdev	Standard deviation of return height > 3,5 m
H.skewness	Skewness of return height > 3.5 m
H.kurtosis	Kurtosis of return height > 3.5 m
H.IQ	75th percentile minus 25th percentile for cell

Table A1. *Cont.*

Predictor	Description
CC.first	First returns > 3.5 m/Total first returns * 100
CC.all	All returns > 3.5 m/Total all returns * 100
CC.all.first	All returns > 3.5 m/Total first returns * 100
CC.first.mean	First returns above mean/Total first returns * 100
CC.all.mean	All returns above mean/Total all returns * 100
CC.all.mean.first	All returns above mean/Total first returns * 100
CC.first.mode	First returns above mode/Total first returns * 100
CC.all.mode	All returns above mode/Total all returns * 100
CC.all.mode.first	All returns above mode/Total first returns * 100

All height variables (beginning with 'H') were calculated separately using first and last pulse returns, which are indicated by the prefix 'FR_' or 'LR_', respectively. All canopy variables (beginning with "CC") were calculated using all returns only.

Table A2. Errors of omission and commission per class in the land cover classification.

	Forest	Bushland	Grassland	Cropland	Row Total	Producer's Accuracy
Forest	47	5	0	0	52	95.91
Bushland	1	59	4	2	66	76.62
Grassland	0	6	100	3	109	92.59
Cropland	1	7	4	55	67	91.66
Column total	49	77	108	60	294	
User's accuracy	90.38	89.39	91.74	82.08		

References

1. Lehmann, C.E.R.; Archibald, S.A.; Hoffmann, W.A.; Bond, W.J. Deciphering the distribution of the savanna biome. *New Phytol.* **2011**, *191*, 197–209. [CrossRef] [PubMed]
2. Kaya, F.; Bibi, F.; Žliobaite, I.; Eronen, J.T.; Hui, T.; Fortelius, M. The rise and fall of the Old World savannah fauna and the origins of the African savannah biome. *Nat. Ecol. Evol.* **2018**, *2*, 241–246. [CrossRef] [PubMed]
3. Cawthra, H. Emergence of the African savannah. *Nat. Geosci.* **2019**, *12*, 588–589. [CrossRef]
4. Barlow, J.; França, F.; Gardner, T.A.; Hicks, C.C.; Lennox, G.D.; Berenguer, E.; Castello, L.; Economo, E.P.; Ferreira, J.; Guénard, B.; et al. The future of hyperdiverse tropical ecosystems. *Nature* **2018**, *559*, 517–526. [CrossRef]
5. Gaston, G.; Brown, S.; Lorenzini, M.; Singh, K.D. State and change in carbon pools in the forests of tropical Africa. *Glob. Chang. Biol.* **1998**, *4*, 97–114. [CrossRef]
6. Glenday, J. Carbon storage and emissions offset potential in an African dry forest, the Arabuko-Sokoke Forest, Kenya. *Environ. Monit. Assess.* **2008**, *142*, 85–95. [CrossRef]
7. Valentini, R.; Arneth, A.; Bombelli, A.; Castaldi, S.; Cazzolla Gatti, R.; Chevallier, F.; Ciais, P.; Grieco, E.; Hartmann, J.; Henry, M.; et al. A full greenhouse gases budget of Africa: Synthesis, uncertainties, and vulnerabilities. *Biogeosciences* **2014**, *11*, 381–407. [CrossRef]
8. Pellikka, P.; Hakala, E. Climate change. In *Megatreds in Africa*; Vastapuu, L., Mattlin, M., Hakala, E., Pellikka, P., Eds.; Ministry for Foreign Affairs of Finland: Helsinki, Finland, 2019; pp. 7–13.
9. Brink, A.B.; Eva, H.D. Monitoring 25 years of land cover change dynamics in Africa: A sample based remote sensing approach. *Appl. Geogr.* **2009**, *29*, 501–512. [CrossRef]
10. Devine, A.P.; McDonald, R.A.; Quaife, T.; Maclean, I.M.D. Determinants of woody encroachment and cover in African savannas. *Oecologia* **2017**, *183*, 939–951. [CrossRef]
11. Stevens, N.; Lehmann, C.E.R.; Murphy, B.P.; Durigan, G. Savanna woody encroachment is widespread across three continents. *Glob. Chang. Biol.* **2017**, *23*, 235–244. [CrossRef]
12. Venter, Z.S.; Cramer, M.D.; Hawkins, H.J. Drivers of woody plant encroachment over Africa. *Nat. Commun.* **2018**, *9*, 1–7. [CrossRef] [PubMed]
13. Mitchard, E.T.A.; Flintrop, C.M. Woody encroachment and forest degradation in sub-Saharan Africa's woodlands and savannas 1982–2006. *Philos. Trans. R. Soc. B Biol. Sci.* **2013**, *368*, 1–7. [CrossRef] [PubMed]
14. DeVries, J. *Shrublands in California: Literature Review and Research Needed for Management*; California Water Resources Center, University of California: Davis, CA, USA, 1984.

15. Barnes, M.D.; Craigie, I.D.; Harrison, L.B.; Geldmann, J.; Collen, B.; Whitmee, S.; Balmford, A.; Burgess, N.D.; Brooks, T.; Hockings, M.; et al. Wildlife population trends in protected areas predicted by national socio-economic metrics and body size. *Nat. Commun.* **2016**, *7*, 1–9. [CrossRef] [PubMed]
16. Myers, N. National parks in Savannah Africa. *Science* **1972**, *178*, 1255–1263. [CrossRef]
17. Pitman, R.T.; Fattebert, J.; Williams, S.T.; Williams, K.S.; Hill, R.A.; Hunter, L.T.B.; Slotow, R.; Balme, G.A. The Conservation Costs of Game Ranching. *Conserv. Lett.* **2017**, *10*, 402–412. [CrossRef]
18. Craigie, I.D.; Baillie, J.E.M.; Balmford, A.; Carbone, C.; Collen, B.; Green, R.E.; Hutton, J.M. Large mammal population declines in Africa's protected areas. *Biol. Conserv.* **2010**, *143*, 2221–2228. [CrossRef]
19. Wilmshurst, J.F.; Fryxell, J.M.; Colucci, P.E. What constrains daily intake in Thomson's gazelles? *Ecology* **1999**, *80*, 2338–2347. [CrossRef]
20. Fynn, R.W.S.; Augustine, D.J.; Peel, M.J.S.; de Garine-Wichatitsky, M. Strategic management of livestock to improve biodiversity conservation in African savannahs: A conceptual basis for wildlife-livestock coexistence. *J. Appl. Ecol.* **2016**, *53*, 388–397. [CrossRef]
21. Peters, M.K.; Hemp, A.; Appelhans, T.; Becker, J.N.; Behler, C.; Classen, A.; Detsch, F.; Ensslin, A.; Ferger, S.W.; Frederiksen, S.B.; et al. Climate–land-use interactions shape tropical mountain biodiversity and ecosystem functions. *Nature* **2019**, *568*, 88–92. [CrossRef]
22. Niskanen, L. *Situation Analysis—Conservation Areas & Species Diversity*; Conservation Areas and Species Diversity Programme, IUCN Eastern and Southern Africa Programme Office: Gland, Switzerland, 2011.
23. Vento, E. Divergent Objectives of Protected Area Management: Impact on Territorial Tourism Development in Taita Taveta County, Kenya. Master's Thesis, University of Helsinki, Helsinki, Finland, October 2017.
24. Bond, W.J.; Loffell, D. Introduction of giraffe changes acacia distribution in a South African savanna. *Afr. J. Ecol.* **2001**, *39*, 286–294. [CrossRef]
25. Bergstorm, R. Browse characteristics and impact of browsing on trees and shrubs in African savannas. *J. Veg. Sci.* **1992**, *3*, 315–324. [CrossRef]
26. Schmitz, O.J. Herbivory from Individuals Herbivory from Individuals to Ecosystems. *Annu. Rev. Ecol. Evol. Syst.* **2016**, *39*, 133–152. [CrossRef]
27. Augustine, D.J.; McNaughton, S.J. Interactive effects of ungulate herbivores, soil fertility, and variable rainfall on ecosystem processes in a semi-arid savanna. *Ecosystems* **2006**, *9*, 1242–1256. [CrossRef]
28. Vesala, R.; Harjuntausta, A.; Hakkarainen, A.; Rönnholm, P.; Pellikka, P.; Rikkinen, J. Termite mound architecture regulates nest temperature and correlates with species identities of symbiotic fungi. *PeerJ* **2019**, *2019*, 1–20. [CrossRef]
29. Vesala, R.; Niskanen, T.; Liimatainen, K.; Boga, H.; Pellikka, P.; Rikkinen, J. Diversity of fungus-growing termites (Macrotermes) and their fungal symbionts (Termitomyces) in the semiarid Tsavo Ecosystem, Kenya. *Biotropica* **2017**, *49*, 402–412. [CrossRef]
30. Massey, A.L.; King, A.A.; Foufopoulos, J. Fencing protected areas: A long-term assessment of the effects of reserve establishment and fencing on African mammalian diversity. *Biol. Conserv.* **2014**, *176*, 162–171. [CrossRef]
31. Xu, L.; Nie, Y.; Chen, B.; Xin, X.; Yang, G.; Xu, D.; Ye, L. Effects of fence enclosure on vegetation community characteristics and productivity of a degraded temperate meadow steppe in Northern China. *Appl. Sci.* **2020**, *10*, 2952. [CrossRef]
32. Nagaike, T.; Ohkubo, E.; Hirose, K. Vegetation Recovery in Response to the Exclusion of Grazing by Sika Deer (Cervus nippon) in Seminatural Grassland on Mt. Kushigata, Japan. *ISRN Biodivers.* **2014**, *2014*, 1–6. [CrossRef]
33. Packer, C.; Loveridge, A.; Canney, S.; Caro, T.; Garnett, S.T.; Pfeifer, M.; Zander, K.K.; Swanson, A.; MacNulty, D.; Balme, G.; et al. Conserving large carnivores: Dollars and fence. *Ecol. Lett.* **2013**, *16*, 635–641. [CrossRef]
34. Li, Q.; Zhou, D.W.; Jin, Y.H.; Wang, M.L.; Song, Y.T.; Li, G. Di Effects of fencing on vegetation and soil restoration in a degraded alkaline grassland in northeast China. *J. Arid Land* **2014**, *6*, 478–487. [CrossRef]
35. Pekor, A.; Miller, J.R.B.; Flyman, M.V.; Kasiki, S.; Kesch, M.K.; Miller, S.M.; Uiseb, K.; van der Merve, V.; Lindsey, P.A. Fencing Africa's protected areas: Costs, benefits, and management issues. *Biol. Conserv.* **2019**, *229*, 67–75. [CrossRef]

36. Erb, K.H.; Kastner, T.; Plutzar, C.; Bais, A.L.S.; Carvalhais, N.; Fetzel, T.; Gingrich, S.; Haberl, H.; Lauk, C.; Niedertscheider, M.; et al. Unexpectedly large impact of forest management and grazing on global vegetation biomass. *Nature* **2018**, *553*, 73–76. [CrossRef] [PubMed]
37. Lynch, J.; Maslin, M.; Balzter, H.; Sweeting, M. Choose satellites to monitor deforestation. *Nature* **2013**, *496*, 293–294. [CrossRef] [PubMed]
38. McNicol, I.M.; Ryan, C.M.; Mitchard, E.T.A. Carbon losses from deforestation and widespread degradation offset by extensive growth in African woodlands. *Nat. Commun.* **2018**, *9*. [CrossRef] [PubMed]
39. Xu, L.; Saatchi, S.S.; Shapiro, A.; Meyer, V.; Ferraz, A.; Yang, Y.; Bastin, J.F.; Banks, N.; Boeckx, P.; Verbeeck, H.; et al. Spatial Distribution of Carbon Stored in Forests of the Democratic Republic of Congo. *Sci. Rep.* **2017**, *7*, 1–12. [CrossRef]
40. Kumar, L.; Mutanga, O. Remote sensing of above-ground biomass. *Remote Sens.* **2017**, *9*, 935. [CrossRef]
41. Karlson, M.; Ostwald, M.; Reese, H.; Sanou, J.; Tankoano, B.; Mattsson, E. Mapping tree canopy cover and aboveground biomass in Sudano-Sahelian woodlands using Landsat 8 and random forest. *Remote Sens.* **2015**, *7*, 10017–10041. [CrossRef]
42. Gorelick, N.; Hancher, M.; Dixon, M.; Ilyushchenko, S.; Thau, D.; Moore, R. Google Earth Engine: Planetary-scale geospatial analysis for everyone. *Remote Sens. Environ.* **2017**, *202*, 18–27. [CrossRef]
43. Zolkos, S.G.; Goetz, S.J.; Dubayah, R. A meta-analysis of terrestrial aboveground biomass estimation using lidar remote sensing. *Remote Sens. Environ.* **2013**, *128*, 289–298. [CrossRef]
44. Egberth, M.; Nyberg, G.; Næsset, E.; Gobakken, T.; Mauya, E.; Malimbwi, R.; Katani, J.; Chamuya, N.; Bulenga, G.; Olsson, H. Combining airborne laser scanning and Landsat data for statistical modeling of soil carbon and tree biomass in Tanzanian Miombo woodlands. *Carbon Balance Manag.* **2017**, *12*. [CrossRef]
45. Heiskanen, J.; Adhikari, H.; Piiroinen, R.; Packalen, P.; Pellikka, P.K.E. Do airborne laser scanning biomass prediction models benefit from Landsat time series, hyperspectral data or forest classification in tropical mosaic landscapes? *Int. J. Appl. Earth Obs. Geoinf.* **2019**, *81*, 176–185. [CrossRef]
46. Kenya National Bureau of Statistic (KNBS). *Kenya Population and Housing Census*; KNBS: Nairobi, Kenya, 2019; Volume I, ISBN 9789966102096.
47. *County Integrated Development Plan 2018–2022*; County Government of Taita Taveta: Taita Taveta County, Kenya, 2018.
48. Pellikka, P.K.E.; Clark, B.J.F.; Gosa, A.G.; Himberg, N.; Hurskainen, P.; Maeda, E.; Mwang'ombe, J.; Omoro, L.M.A.; Siljander, M. Chapter 13—Agricultural expansion and its consequences in the Taita Hills, Kenya. In *Developments in Earth Surface Processes*; Paolo Paron, D.O.O., Christian Thine, O., Eds.; Elsevier: Amsterdam, The Netherlands, 2013.
49. Tuure, J.; Korpela, A.; Hautala, M.; Rautkoski, H.; Hakojärvi, M.; Mikkola, H.; Duplissy, J.; Pellikka, P.; Petäjä, T.; Kulmala, M.; et al. Comparing plastic foils for dew collection: Preparatory laboratory-scale method and field experiment in Kenya. *Biosyst. Eng.* **2020**, *196*, 145–158. [CrossRef]
50. Räsänen, M.; Aurela, M.; Vakkari, V.; Beukes, J.; Tuovinen, J.-P.; Josipovic, M.; Siebert, S.; Laurila, T.; Kulmala, M.; Laakso, L.; et al. The effect of rainfall amount and timing on annual transpiration in grazed savanna grassland. Hydrol. *Earth Syst. Sci. Discuss.* **2020**, 1–31. [CrossRef]
51. Smith, R.J.; Kasiki, S.M. *A Spatial Analysis of Human-Elephant Conflict in the Tsavo Ecosystem. Kenya*; African Elephant Specialist Group, Human-Elephant Conflict Task Force of IUCN: Gland, Switzerland, 2000.
52. Ogutu, J.O.; Piepho, H.P.; Said, M.Y.; Ojwang, G.O.; Njino, L.W.; Kifugo, S.C.; Wargute, P.W. Extreme wildlife declines and concurrent increase in livestock numbers in Kenya: What are the causes? *PLoS ONE* **2016**, *11*, 1–46. [CrossRef] [PubMed]
53. Ogutu, J.; Piepho, H.; Kuloba, B.; Kanga, E. Wildlife population dynamics in human-dominated landscapes under community-based conservation: The example of Nakuru Wildlife Conservancy, Kenya. *PLoS ONE* **2017**, *12*. [CrossRef]
54. Muteti, D.; Maloba, M. Taita Hills Wildlife Sanctuary: Dry season total ground wildlife census. 2013; unpublished.
55. Ngene, S.; Lala, F.; Nzisa, M.; Kimitei, K.; Mukeka, J.; Kiambi, J.; Davidson, Z.; Bakari, S.; Lyimo, E.; Khayale, C.; et al. *Aerial Total Count of Elephants, Buffalo and Giraffe in the Tsavo-Mkomazi Ecosystem*; Kenya Wildlife Service, Kenya and Tanzania Wildlife Research Institute: Arusha, Tanzania, 2017.
56. Vågen, T.-G.; Winowiecki, L.; Walsh, M.G.; Tamene Desta, L.; Tondoh, J.E. *The Land Degradation Surveillance Framework (LDSF): Field Guide*; World Agroforestry Centre (ICRAF): Nairobi, Kenya, 2010.

57. Chave, J.; Réjou-Méchain, M.; Búrquez, A.; Chidumayo, E.; Colgan, M.S.; Delitti, W.B.C.; Duque, A.; Eid, T.; Fearnside, P.M.; Goodman, R.C.; et al. Improved allometric models to estimate the aboveground biomass of tropical trees. *Glob. Chang. Biol.* **2014**, *20*. [CrossRef]
58. Réjou-Méchain, M.; Tanguy, A.; Piponiot, C.; Chave, J.; Hérault, B. Biomass: An R Package for Estimating Above-Ground Biomass and Its Uncertainty in Tropical Forests. *Methods Ecol. Evol.* **2017**, *8*, 1163–1167. [CrossRef]
59. Core Team. *R: A Language and Environment for Statistical Computing*; R Foundation for Statistical Computing: Vienna, Austria; Available online: https://www.R-project.org/ (accessed on 11 August 2018).
60. Conti, G.; Gorné, L.D.; Zeballos, S.R.; Lipoma, M.L.; Gatica, G.; Kowaljow, E.; Whitworth-Hulse, J.I.; Cuchietti, A.; Poca, M.; Pestoni, S.; et al. Developing allometric models to predict the individual aboveground biomass of shrubs worldwide. *Glob. Ecol. Biogeogr.* **2019**, *28*, 961–975. [CrossRef]
61. Paul, K.I.; Roxburgh, S.H.; Chave, J.; England, J.R.; Zerihun, A.; Specht, A.; Lewis, T.; Bennett, L.T.; Baker, T.G.; Adams, M.A.; et al. Testing the generality of above-ground biomass allometry across plant functional types at the continent scale. *Glob. Chang. Biol.* **2016**, *22*, 2106–2124. [CrossRef]
62. McGaughey, R. *FUSION/LDV: Software for LIDAR Data Analysis and Visualization*; US Department of Agriculture, Forest Service, Pacific Northwest Research Station: Seattle, WA, USA, 2016.
63. Adhikari, H.; Valbuena, R.; Pellikka, P.K.E.; Heiskanen, J. Mapping forest structural heterogeneity of tropical montane forest remnants from airborne laser scanning and Landsat time series. *Ecol. Indic.* **2020**, *108*, 105739. [CrossRef]
64. Package 'Leaps'. Available online: https://cran.r-project.org/web/packages/leaps/leaps.pdf (accessed on 16 January 2020).
65. Adhikari, H.; Heiskanen, J.; Siljander, M.; Maeda, E.; Heikinheimo, V.; Pellikka, P.K.E. Determinants of aboveground biomass across an afromontane landscape Mosaic in Kenya. *Remote Sens.* **2017**, *9*, 827. [CrossRef]
66. Rouse, J.W.; Hass, R.H.; Schell, J.A.; Deering, D.W. Monitoring vegetation systems in the Great Plains with ERTS. *Third Earth Resour. Technol. Satell. Symp.* **1973**, *1*, 309–317.
67. Huete, A.R.; Liu, H.Q.; Batchily, K.; Van Leeuwen, W. A comparison of vegetation indices over a global set of TM images for EOS-MODIS. *Remote Sens. Environ.* **1997**, *59*, 440–451. [CrossRef]
68. Jiang, Z.; Huete, A.R.; Didan, K.; Miura, T. Development of a two-band enhanced vegetation index without a blue band. *Remote Sens. Environ.* **2008**, *112*, 3833–3845. [CrossRef]
69. Serrano, L.; Ustin, S.L.; Roberts, D.A.; Gamon, J.A.; Peñuelas, J. Deriving water content of chaparral vegetation from AVIRIS data. *Remote Sens. Environ.* **2000**, *74*, 570–581. [CrossRef]
70. Clevers, J.G.P.W.; Gitelson, A.A. Remote estimation of crop and grass chlorophyll and nitrogen content using red-edge bands on sentinel-2 and-3. *Int. J. Appl. Earth Obs. Geoinf.* **2013**, *23*, 344–351. [CrossRef]
71. Vågen, T.-G.; Winowiecki, L.A.; Tamene Desta, L.; Tondoh, J.E. *Land Degradation Surveillance Framework (LDSF)—Field Guide*; World Agroforestry Center (ICRAF): Nairobi, Kenya, 2013.
72. Breiman, L.; Friedman, J.H.; Stone, C.J.S.; Olshen, R.A. *Classification and Regression Trees*, 1st ed.; Wadsworth & Brooks/Cole Advanced Books & Software: Monterey, CA, USA, 1984; 358p.
73. Ngene, S.; Njumbi, S.; Nzisa, M.; Kimitei, K.; Mukeka, J.; Muya, S.; Ihwagi, F.; Omondi, P. Status and trends of the elephant population in the Tsavo–Mkomazi ecosystem. *Pachyderm* **2013**, *53*, 38–50.
74. Nyangala, E.B. *Livestock Officer, Mwatate Sub County*, Taita Taveta County, Kenya, 2019; unpublished.
75. Asner, G.P.; Mascaro, J.; Muller-Landau, H.C.; Vieilledent, G.; Vaudry, R.; Rasamoelina, M.; Hall, J.S.; van Breugel, M. A universal airborne LiDAR approach for tropical forest carbon mapping. *Oecologia* **2012**, *168*, 1147–1160. [CrossRef]
76. Mascaro, J.; Detto, M.; Asner, G.P.; Muller-Landau, H.C. Evaluating uncertainty in mapping forest carbon with airborne LiDAR. *Remote Sens. Environ.* **2011**, *115*, 3770–3774. [CrossRef]
77. Asner, G.P.; Flint Hughes, R.; Varga, T.A.; Knapp, D.E.; Kennedy-Bowdoin, T. Environmental and biotic controls over aboveground biomass throughout a tropical rain forest. *Ecosystems* **2009**, *12*, 261–278. [CrossRef]
78. Chen, Q.; Vaglio Laurin, G.; Valentini, R. Uncertainty of remotely sensed aboveground biomass over an African tropical forest: Propagating errors from trees to plots to pixels. *Remote Sens. Environ.* **2015**, *160*, 134–143. [CrossRef]

79. Gonzalez, P.; Asner, G.P.; Battles, J.J.; Lefsky, M.A.; Waring, K.M.; Palace, M. Forest carbon densities and uncertainties from Lidar, QuickBird, and field measurements in California. *Remote Sens. Environ.* **2010**, *114*, 1561–1575. [CrossRef]
80. Asner, G.P.; Clark, J.K.; Mascaro, J.; Vaudry, R.; Chadwick, K.D.; Vieilledent, G.; Rasamoelina, M.; Balaji, A.; Kennedy-Bowdoin, T.; Maatoug, L.; et al. Human and environmental controls over aboveground carbon storage in Madagascar. *Carbon Balance Manag.* **2012**, *7*, 1–13. [CrossRef]
81. Mauya, E.W.; Ene, L.T.; Bollandsås, O.M.; Gobakken, T.; Næsset, E.; Malimbwi, R.E.; Zahabu, E. Modelling aboveground forest biomass using airborne laser scanner data in the miombo woodlands of Tanzania. *Carbon Balance Manag.* **2015**, *10*, 28. [CrossRef] [PubMed]
82. Heiskanen, J.; Liu, J.; Valbuena, R.; Aynekulu, E.; Packalen, P.; Pellikka, P. Remote sensing approach for spatial planning of land management interventions in West African savannas. *J. Arid Environ.* **2017**, *140*, 29–41. [CrossRef]
83. Forkuor, G.; Benewinde Zoungrana, J.-B.; Dimobe, K.; Ouattara, B.; Vadrevu, K.P.; Tondoh, J.E. Above-ground biomass mapping in West African dryland forest using Sentinel-1 and 2 datasets—A case study. *Remote Sens. Environ.* **2020**, *236*, 111496. [CrossRef]
84. Venter, F.J.; Naiman, R.J.; Biggs, H.C.; Pienaar, D.J. The evolution of conservation management philosophy: Science, environmental change and social adjustments in Kruger National Park. *Ecosystems* **2008**, *11*, 173–192. [CrossRef]
85. Pellikka, P.K.E.; Heikinheimo, V.; Hietanen, J.; Schäfer, E.; Siljander, M.; Heiskanen, J. Impact of land cover change on aboveground carbon stocks in Afromontane landscape in Kenya. *Appl. Geogr.* **2018**, *94*, 178–189. [CrossRef]
86. Vuorinne, I. Geoinformatics Assessing Agave Sisalana Biomass from Leaf to Plantation Using Field Measurements and Multispectral Satellite Imagery. Master's Thesis, University of Helsinki, Helsinki, Finland, 2020.
87. Davidson, E.A.; De Araüjo, A.C.; Artaxo, P.; Balch, J.K.; Brown, I.F.; Mercedes, M.M.; Coe, M.T.; Defries, R.S.; Keller, M.; Longo, M.; et al. The Amazon basin in transition. *Nature* **2012**, *481*, 321–328. [CrossRef]
88. Paruelo, J.M.; Burke, I.C.; Lauenroth, W.K. Land-use impact on ecosystem functioning in eastern Colorado, USA. *Glob. Chang. Biol.* **2001**, *7*, 631–639. [CrossRef]
89. Morgan, J.A.; Lecain, D.R.; Pendall, E.; Blumenthal, D.M.; Kimball, B.A.; Carrillo, Y.; Williams, D.G.; Heisler-White, J.; Dijkstra, F.A.; West, M. C4 grasses prosper as carbon dioxide eliminates desiccation in warmed semi-arid grassland. *Nature* **2011**, *476*, 202–205. [CrossRef] [PubMed]
90. Scheiter, S.; Higgins, S.I.; Beringer, J.; Hutley, L.B. Climate change and long-term fire management impacts on Australian savannas. *New Phytol.* **2015**, *205*, 1211–1226. [CrossRef] [PubMed]
91. Williams, H.F.; Bartholomew, D.C.; Amakobe, B.; Githiru, M. Environmental factors affecting the distribution of African elephants in the Kasigau wildlife corridor, SE Kenya. *Afr. J. Ecol.* **2018**, *56*, 244–253. [CrossRef]
92. Abera, T.A.; Heiskanen, J.; Pellikka, P.K.E.; Adhikari, H.; Maeda, E.E. Climatic impacts of bushland to cropland conversion in Eastern Africa. *Sci. Total Environ.* **2020**, *717*, 137255. [CrossRef]
93. Aalto, I. Assessing the Cooling Impact of Tree Canopies in an Intensively Modified Tropical Landscape. Master's Thesis, University of Helsinki, Helsinki, Finland, 2020.
94. Liedloff, A.C.; Cook, G.D. Modelling the effects of rainfall variability and fire on tree populations in an Australian tropical savanna with the Flames simulation model. *Ecol. Modell.* **2007**, *201*, 269–282. [CrossRef]
95. Kenya Wildlife Service. *Tsavo Conservation Area Management Plan 2008–2018*; Kenya Wildlife Service: Nairobi, Kenya, 2008.
96. Stoldt, M.; Göttert, T.; Mann, C.; Zeller, U. Transfrontier Conservation Areas and Human-Wildlife Conflict: The Case of the Namibian Component of the Kavango-Zambezi (KAZA) TFCA. *Sci. Rep.* **2020**, *10*, 1–16. [CrossRef]
97. Durant, S.M.; Becker, M.S.; Creel, S.; Bashir, S.; Dickman, A.J.; Beudels-Jamar, R.C.; Lichtenfeld, L.; Hilborn, R.; Wall, J.; Wittemyer, G.; et al. Developing fencing policies for dryland ecosystems. *J. Appl. Ecol.* **2015**, *52*, 544–551. [CrossRef]
98. Dupuis-Desormeaux, M.; Davidson, Z.; Pratt, L.; Mwololo, M.; MacDonald, S.E. Testing the effects of perimeter fencing and elephant exclosures on lion predation patterns in a Kenyan wildlife conservancy. *PeerJ* **2016**, *2016*, 1–19. [CrossRef]

99. Hayward, M.W.; Kerley, G.I.H. Fencing for conservation: Restriction of evolutionary potential or a riposte to threatening processes? *Biol. Conserv.* **2009**, *142*, 1–13. [CrossRef]
100. Ringma, J.L.; Wintle, B.; Fuller, R.A.; Fisher, D.; Bode, M. Minimizing species extinctions through strategic planning for conservation fencing. *Conserv. Biol.* **2017**, *31*, 1029–1038. [CrossRef]
101. Soto-Shoender, J.R.; McCleery, R.A.; Monadjem, A.; Gwinn, D.C. The importance of grass cover for mammalian diversity and habitat associations in a bush encroached savanna. *Biol. Conserv.* **2018**, *221*, 127–136. [CrossRef]
102. Staver, A.C.; Bond, W.J. Is there a "browse trap"? Dynamics of herbivore impacts on trees and grasses in an African savanna. *J. Ecol.* **2014**, *102*, 595–602. [CrossRef]
103. Sankaran, M.; Augustine, D.J.; Ratnam, J. Native ungulates of diverse body sizes collectively regulate long-term woody plant demography and structure of a semi-arid savanna. *J. Ecol.* **2013**, *101*, 1389–1399. [CrossRef]
104. Miller, G.R. *The Effects of Mammalian Herbivores on Natural Regeneration of Upland, Native Woodland*; Scottish Natural Heritage: Perth, Scotland, 2015.
105. Leuthold, W. Changes in tree populations of Tsavo East National Park, Kenya. *Afr. J. Ecol.* **1977**, *15*, 61–69. [CrossRef]

land

MDPI

Article

Mapping of the Land Cover Changes in High Mountains of Western Carpathians between 1990–2018: Case Study of the Low Tatras National Park (Slovakia)

Michaela Žoncová [1], Pavel Hronček [2] and Bohuslava Gregorová [1,*]

[1] Department of Geography and Geology, Faculty of Natural Sciences, Matej Bel University in Banská Bystrica, Tajovského 40, 97401 Banská Bystrica, Slovakia; michaela.zoncova@umb.sk

[2] Department of Geo and Mining Tourism, Institute of Earth Resources, Faculty of Mining, Ecology, Process Control and Geotechnologies, Technical University of Kosice, Němcovej 32, 04001 Košice, Slovakia; pavel.hroncek@tuke.sk

* Correspondence: bohuslava.gregorova@umb.sk; Tel.: +421-907-559-818

Received: 13 November 2020; Accepted: 30 November 2020; Published: 1 December 2020

check for updates

Abstract: At present, the protection of nature and landscape in the high mountains of the Western Carpathians, protected as national parks, is becoming increasingly at the forefront of society's interests in connection with the development of their economic use and the development of mass tourism. Our research was focused on analyzing the extent and character of land cover changes in the Low Tatras National Park in Slovakia over the last 30 years (1990–2018) using CORINE land cover (CLC) data. The period captures almost the entire existence of the Slovak Republic. Therefore, it was possible to evaluate the landscape changes in the protected area and to identify barriers and possibilities of its long-term sustainable development. Based on computer modeling, the main areas of the land cover changes were identified, and on the basis of historical-geographical and field research, land cover flows were determined and justified in the studied landscape of the national park. Changes were monitored using three methods: by comparing CLC maps over the years, by analyzing land cover flows, and by comparing landscape metrics obtained through the PatchAnalyst. Land cover changes occurred on up to 20% of the national park area in the given period. The most significant change was observed in the CLC class coniferous forests, with almost a 12% decrease. Conversely, there was an increase of more than 11% in the CLC class transitional woodland-shrub.

Keywords: CORINE land cover; mapping of changes; GIS tools; land cover flows; protected areas; Low Tatras National Park

1. Introduction

Land cover changes (whether natural or occurring by anthropogenically affected development) are a continuous process worldwide [1–5], especially in developing countries of Asia and South America, but also in the ("post-socialist") countries of Central and Eastern Europe, including Slovakia. Socio-political reforms occurred after 1989, and subsequent transformations that began after 2004, when Slovakia joined the European Union, can be considered as the leading causes of land cover changes in this region. Issues of environmental protection also came to the forefront, and in 2002 the Act No. 543/2002 Coll. on Nature and Landscape Protection was adopted. This act strengthened the protection of natural balance, the protection of the diversity of natural conditions and life forms, natural values and creates the preconditions for the sustainable use of natural resources and ecosystem services, taking into account the economic, social and cultural needs as well as regional and local circumstances. The act defines five zones of protection, the fifth being the highest.

Large-scale protected areas, which include national parks and protected landscape areas, are declared in Slovakia within the scope of nature conservation. A national park is defined as an area of over 1000 ha, predominantly with ecosystems substantially unchanged by human activity or with a unique and natural landscape structure constituting supra-regional biocentres and the most significant natural heritage, in which nature protection and conservation are superior to other activities. Territories of national parks fall under the third zone of protection.

One prominent body of research on the mountain environment in Slovakia focuses on geodynamic and geomorphological processes [6,7]. The second prominent body of research focuses on changes in the landscape structure, but this research captures especially changes at the turn of the 18th, 19th and 20th centuries. The main processes in this period were mainly the development of urbanization and agriculture [8] or the transformation of scattered settlements into recreational areas [9]. Changes in the landscape structure are partly a reflection of social changes that occurred after 1949 and later after 1989 in Slovakia [10]. All of these processes resulted in changes in ecological stability, both in the mountain environment and in the lowlands [11,12]. In recent years, the disproportionate spatial expansion of recreational infrastructure [13–15] has had a significant impact on the landscape, and it is, therefore, essential to set the limit of unbearability and to develop tourism inside the protected areas only up to a point, where no disruption of elemental links between ecosystems can be guaranteed [16]. This phenomenon often becomes irreversible when humans disproportionately affect the natural environment, affecting water [17], soil [18], flora, fauna, and the overall biodiversity of the landscape [19,20] in the protected areas. Changing climate, frequent temperature fluctuations, fires [21], windstorms, torrential rains and storms, droughts, and other natural phenomena also affect changes in nature. The most important anthropogenic influences are grazing, intensification and extensification of agriculture, mining of raw materials, areal growth of rural settlements, the development of recreational infra, and suprastructure.

The monitoring of landscape changes in Europe is mostly conducted using the CORINE land cover (CLC) database [7,10,22–24], which is considered to be the most complex database of spatial-temporal data. The functioning of the CLC database is financed by the Member States of the European Union, and it is managed by the European Environmental Agency (EEA) and is one of the products of the Copernicus Land Monitoring Service (CLMS). The extent and quality of information within the CLC database are different across European countries. The CLC database of Slovakia is one of the most accurate and complete. The quality and detail result from the use of a minimum scale (1:50,000) as well as from the adaptation of its legend to the specific local conditions [25].

The main aim of our research is to evaluate land cover changes in the Low Tatras National Park between 1990 and 2018. The selection of time periods was primarily based on data availability. These data also capture significant historical context since social-political changes in 1989, accession to the European Union in 2004 to nowadays. We have specified three sub-objectives within our research:

1. to find out the character of landscape structure based on a comparative analysis of land cover maps from 1990 and 2018 using the CORINE land cover (CLC) data;
2. to assess land cover flows based on CORINE land cover change layers in period 1990–2000, 2000–2006, 2006–2012, 2012–2018;
3. to describe landscape structure changes based upon landscape metrics calculations.

The following primary research questions should be answered:

- What type of landscape changes prevailed in the Low Tatras National Park during the research period?
- What was the intensity of landscape changes in the observed period in the Low Tatras National Park?
- How do the land cover changes influence the ecological attributes of a national park?
- What were the causes and processes (drivers) of landscape changes in the Low Tatras National Park within the monitored period?

The research of land cover changes is necessary to identify the negative anthropogenically conditioned and created processes and phenomena in the protected landscape, their prediction, analysis, prevention, and elimination concerning the active management of the landscape.

2. Materials and Methods

2.1. *Territory of Interest*

The Low Tatras National Park is located in the central part of the Western Carpathians (Figure 1). It spreads in the top part of the Low Tatras Mountains, which is the second-highest in the Carpathian Arc. The asymmetric vault of the central mountain ridge, located in the center of Slovakia, is significantly extended in the west-east direction. The highest point of the mountain range is the summit of Ďumbier (2043 m). From the orographic point of view, the territory of the national park consists mainly of the beforementioned Low Tatras, while some parts of Veľká Fatra Mountains, Staré Hory Mountains, Zvolen Basin, Horehronské Podolie Basin, Podtatranská Basin, Kozie chrbty Mountains and Spiš-Gemer Karst extend partially to this area.

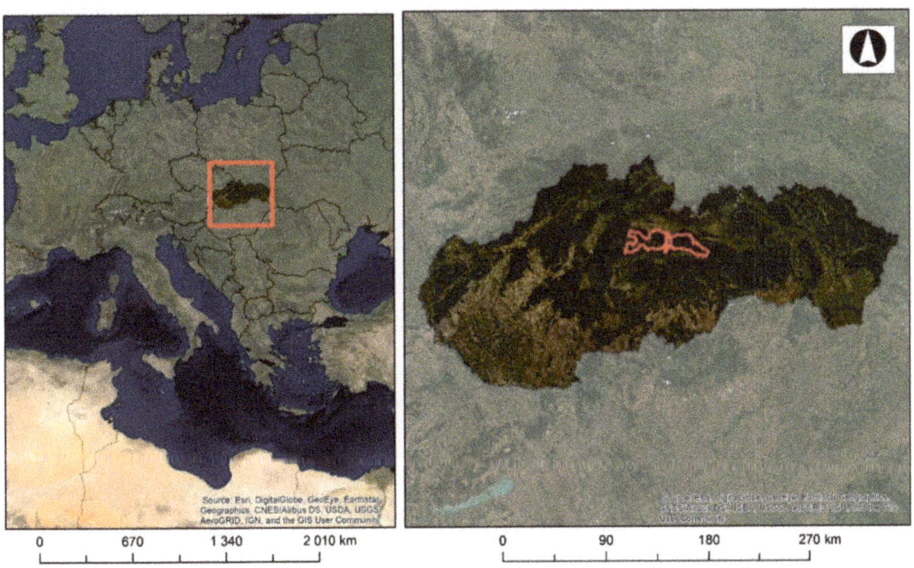

Figure 1. The location and demarcation of Low Tatras National Park in Slovakia.

The main ridge of the Low Tatras is built by the crystalline core (Tatrikum). It consists mainly of rocks formed in the Paleozoic when large rock complexes metamorphosed at the sea bottom during the younger geological times. During the long-term geological-tectonic development, these rocks were lifted and folded into a complex vaulted structure. The core was encapsulated initially by sandstone, shale, dolomite, and limestone strata, which were gradually removed by denudation, but mainly by the Quaternary glacial activity. The process has been particularly evident on the northern slopes of the mountain range, which is why the main range has a distinctly asymmetric profile (the northern slopes are much steeper and shorter than the southern ones) [26,27].

Glacial relief is characterized by steep rock walls, glacial cirques, moraines, and glacial lakes. There were 16 glaciers in the territory of the Low Tatras National Park. The longest one was in Dúbravská Valley and reached a length of more than 6000 m. The largest glacier lake ("pleso") is Vrbické Lake with an original area of 0.73 ha, which is situated in Demänovská Valley and was created on the frontal moraine after the glacier retreat [28].

Up to 5 climatic-geographical types (the highland climate-very cold, cold, cool and moderately warm, and the basin climate-cold) can be distinguished in the Low Tatras due to high vertical articulation (from 500 to 2043 m a.s.l.). Mountain locations above 1500 m a.s.l. have a very cold mountain climate with average temperatures of −7 °C to −8 °C in January and approximately −9 °C on the main ridge [29,30].

The highest annual rainfall values are reached in the area between the peaks Prašivá and Dereše, which is caused by the global western streaming (1400–1700 mm on average, maximum annual totals are 1900–2300 mm, and the minimum annual rainfall is about 1000 mm) [31].

The streaming has a strongly variable direction due to local relief shapes. In the basins oriented in the west to east direction, the wind generally flows in the same direction, while the north-south streaming prevails in the mountain ridge part. However, winds from the northwest and southwest are also frequent. The least occurring are south-eastern and northeastern winds. The average wind speed increases with the increasing altitude reaching an average of 9.6 m/s at Chopok (2023 m) and only 1.2 m/s at Jasná (1200 m) [29].

Large areas of the National Park territory had been covered mostly by beech and beech-fir primeval forests until the 15th century, except for the peak areas of the highest part of Chopok and Ďumbier massifs. The vast extent of the mountain range and the articulation of the relief enabled primeval forests to retain their natural character for quite a long time. The gradual settlement, development of mining and metallurgy, and since the 13th century, the related logging and pasturing since the 14th century, have significantly accelerated the deforestation process [32]. A substantial part of today's national park was mostly a clear-cutting in the 16th century [33]. The current composition of forest vegetation was significantly affected by artificial restoration since the 19th century, when the emphasis on spruce monocultures began.

Heavily anthropogenically affected beech and fir-beech forests nowadays cover the western edges of the national park mountain range, lined with oak stands in contact with basins. A wide belt of monocultural spruce stands extends above this level, which completely prevails in the eastern part (Kráľovohoľské Tatras) and creates the timberline at an altitude of 1500–1600 m a.s.l. Forest pine grows on the rocky northern slopes. A dwarf mountain pine belt gradually changes into human-made mountain grasslands occurring from an altitude of about 1400 m a.s.l. [30,34].

The Low Tatras National Park was declared (as the third national park in Slovakia) by Decree of the Government of the SSR no. 119/1978 Coll [35]. Subsequently, the Ministry of Culture of the SSR issued Decree No. 120/1978 [36], establishing its status on 17 October 1978. It was confirmed by Act No. 287/1994 Coll. on Nature and Landscape Protection [37], as well as Act no. 543/2002 Coll [29,38,39]. The area of the largest national park in Slovakia was limited to 205,085 ha, including a protection zone (81,095 ha was the area of the national park itself, and 123,990 ha was the area of the protection zone). The border length was 340 km and up to 575 km together, including the boundaries of the protection zone [39]. There were eight small-scale specially protected areas at the time of its declaration, while currently there are 48 of them. There are also protected areas of NATURA 2000:2 sites—birds directive and 10 sites—habitats directive. Almost the whole area of the national park is under specific protection (except the part of the Demänová valley and Bystrá valley) (Figure 2).

National park boundaries were changed mainly due to property-law relations 19 years later, in 1997. The area of the national park was 72,842 ha, and the area of the protection zone was 110,162 ha. Approximately 11,000 ha of the most valuable parts of its territory are strictly protected (included within the 4th and the 5th zone of protection-A and B zone).

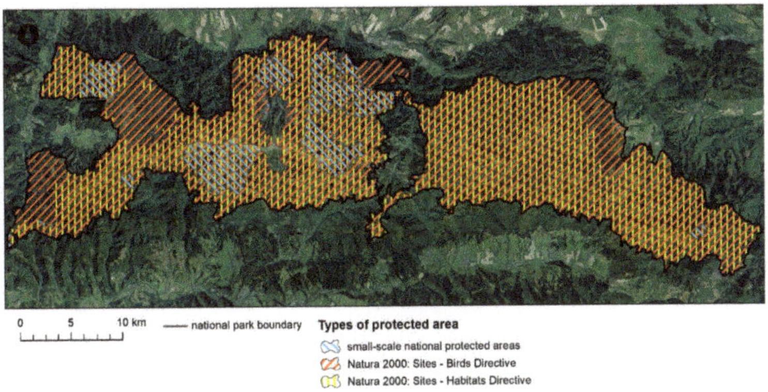

Figure 2. Protected areas of Low Tatras National Park.

Despite the lengthy and complicated process, the declaration of the national park had been a successful culmination of many years of effort of a large number of professional and voluntary nature conservationists, various experts, as well as simple supporters of this unique mountain range in the middle of Slovakia.

2.2. *Data*

For a long time, aerial and satellite imagery has been used to detect and classify landscape transformations over time, as it is useful to capture the impacts of many processes causing natural (e.g., fires, wind disasters) and anthropogenic (e.g., deforestation, urbanization, agriculture) changes [40]. The CORINE land cover (CLC) data for 1990 and 2018, created by visual interpretation of high-resolution satellite images, were applied in our research. CLC data uses a minimum mapping unit (MMU) of 25 hectares (ha) for areal phenomena and a minimum width of 100 m for linear phenomena. The CLC change layers, which highlight changes in land cover with an MMU of 5 ha, have been used to monitor land cover changes. Different MMUs mean that the change layer has a higher resolution than the status layer (Table 1) [41].

Table 1. The specifications of the CORINE land cover data in the reference years 1990 and 2018.

Specification	CLC 1990	CLC 2018
Geometric accuracy, satellite data	≤50 m	≤10 m
Min. mapping unit/width	25 ha/100 m	25 ha/100 m
Geometric accuracy, CLC	100 m	Better than 100 m
Thematic accuracy, CLC	≥85%	≥85%

Using this data, we were able to capture and analyze changes occurring in almost 30 years. We have identified 13 land cover classes within the studied area (Table 2). We have used RGB color codes defined by EEA in the maps of land cover.

Table 2. CORINE land cover classes in the studied area.

Level 1	Level 2	Level 3
1 Artificial surfaces	11 Urban fabric	112 Discontinuous urban fabric
	14 Artificial, non-agricultural vegetated areas	142 Sport and leisure facilities
2 Agricultural areas	21 Arable land	211 Non-irrigated arable land
	23 Pastures	231 Pastures
	24 Heterogeneous agricultural areas	243 Land principally occupied by agriculture, with significant areas of natural vegetation
3 Forest and semi-natural areas	31 Forests	311 Broad-leaved forest 312 Coniferous forest 313 Mixed forest
	32 Scrub and/or herbaceous vegetation associations	321 Natural grasslands 322 Moors and heathland 324 Transitional woodland-shrub
	33 Open spaces with little or no vegetation	333 Sparsely vegetated areas
5 Water bodies	51 Inland waters	512 Water bodies

2.3. Methods

We have used the ArcMap 10.5 software in our research, in which the land cover maps from 1990 and 2018 and maps of land cover changes using available CLC change layers had been made. We have focused mainly on the percentual evaluation of changes in the individual elements of the land cover and analyzed them statistically and spatially. We have made a cross-table to express qualitative relationships between the two variables [42,43]. Using this method, we have found out which land cover classes had been changed and to which classes they had been modified at the same time. Thus, we were able to identify the core processes that took place in the landscape. At the same time, we can identify the period in which the most changes took place and which left the most significant consequences on the landscape.

Landscape changes were categorized into so-called "land cover flows (LCFs)", i.e., classes that reflect processes taking place in the observed area. The definition of these changes was studied by many scientists, who have defined different amounts and types of land cover flows [44–47].

The most extensive and detailed categorization was introduced by Haines-Young and Weber in 2006 [48], defining nine types of changes:

- LCF1 Urban land management—internal transformation of urban areas
- LCF2 Urban residential sprawl—land uptake by residential buildings altogether with associated services and urban infrastructure (classified in CLC111 and 112) from non-urban land (extension over sea may happen)
- LCF3 Sprawl of economic sites and infrastructures—sprawl of economic sites and infrastructures: Land uptake by new economic sites and infrastructures (including sport and leisure facilities) from non-urban land (extension over sea may happen)
- LCF4 Agriculture internal conversions—conversion between farming types. Rotation between annual crops is not monitored by CLC
- LCF5 Conversion from forested and natural land to agriculture—extension of agriculture land use
- LCF6 Withdrawal of farming—farmland abandonment and other conversions from agriculture activity in favor of forests or natural land
- LCF7 Forests creation and management—creation of forests and management of the forest territory by felling and replanting. Due to the CLC cycle of 10 years, only one part of the shrubs is tall

enough to be identified as trees. In order to take stock of all recent plantations, conversions of semi-natural land to CLC324 are conventionally recorded as afforestation (although some natural colonization may take place)

- LCF8 Water bodies creation and management—creation of dams and reservoirs and possible consequences of the management of the water resource on the water surface area
- LCF9 Changes of land cover due to natural and multiple causes—changes in land cover resulting from natural phenomena with or without any human influence.

This categorization was chosen for its detail and complexity for our research. Dominant processes in the landscape of the Low Tatras National Park can be identified based on the percentual data of individual changes. Land cover flows summarize and interpret all possible one-to-one changes between the CORINE land cover classes. The changes are grouped into so-called flows and are classified according to major land-use processes. We have focused on the main class of land cover flows (e.g., LCF7), which consist of several subclasses of land cover flows (LCF71, LCF72, LCF73 and LCF74) (Figure 3).

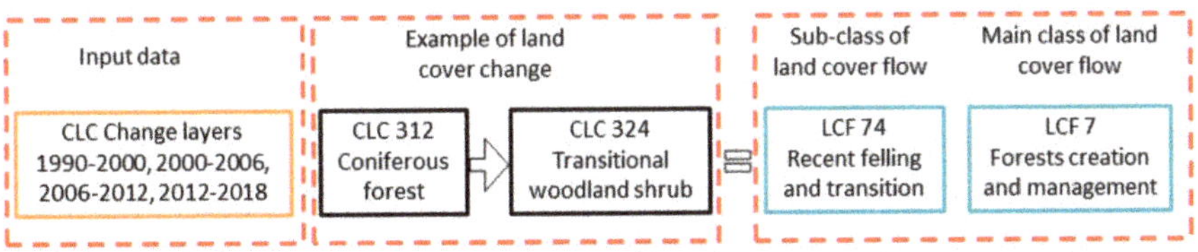

Figure 3. Methodology of the land cover flows.

The last part of our research consisted of assessing landscape structure changes based on landscape metrics calculations using the PatchAnalyst tool. Patch Analyst is an ArcGIS extension that facilitates spatial analysis of landscape patches. It is used for spatial pattern analysis, often in support of habitat modeling, biodiversity conservation, and forest management. The software offers analyses of several types of landscape-ecological metrics, which are often used in scientific research, primarily to assess landscape fragmentation in Slovakia [49–51] and abroad [52–54]. In addition to national or regional spatial data [55], CORINE land cover data, which are also applicable to regional research, are suitable and frequent input data for the calculation of landscape metrics. However, they are not suitable for research of a smaller area due to their lower accuracy. Krajewski [56] distinguishes three approaches to the study of landscape changes: identification of spatiotemporal changes [57], identification of driving forces of changes [2], and identification of landscape changes based on landscape metrics [52,58]. In our research, we combine all three approaches and make a comprehensive analysis of changes in land cover in the Low Tatras National Park. Following Kumar et al. [52], Obeidat et al. [59], Singh et al. [53], we have not used all of the indices from the PatchAnalyst tool because some of them are closely related and some are redundant. The selection was made following the intention of the study based on the knowledge of previous research. An important principle is to select unrelated indices. The following six indices were selected:

- Number of patches (NumP)—a simple indicator that indicates the total increase, respectively, a decrease in the number of patches in all categories in the observed area.
- Mean patch size (MPS)—average patch size. This indicator shows the disintegration of the spatial structure of the landscape.
- Total edge (TE)—an indicator that represents the sum of perimeters of all patches.
- Area weighted mean shape index (AWMSI)—an index that reflects the shape complexity of patches. The index is equal to 1 if the patches have a circular or square shape. The index value increases if the shape is irregular. It differs from the "mean shape index" metric by assigning different weights to individual patches (the larger the area, the higher the weight).

- Shannon's diversity index (SDI)—an index that determines landscape diversity calculated as the proportions of the land cover classes across the total area. SDI increases by the number of patches in the landscape feature categories. The higher the index value, the higher the landscape heterogeneity, i.e., the landscape is more abundant in the number of categories of landscape features and the number of patches [60]. The index will be equal to 0 when there is only one patch in the landscape and increases as the number of patch types or the proportional distribution of patch types increases.
- Shannon's evenness index (SEI)—an index that determines the distribution of patches and their abundance. A proportional reduction in the number of patches and categories also causes a reduction in the overall balance. The landscape metrics balance within the observed landscape is better when the value of this index converges to one.

It is important to evaluate changes in the landscape, especially in national parks, from the ecological point of view, too. The interpretation of the quantified data is important to determine ecological signification and the current state of the landscape, e.g., [52,53,56,58,59].

3. Results

3.1. Land Cover of Low Tatras National Park in 1990 and 2018

There were 13 land cover classes in 1990 and one less in 2018 in the observed area. The land cover of the Low Tatras National Park in 1990 and 2018 is shown in Figure 4.

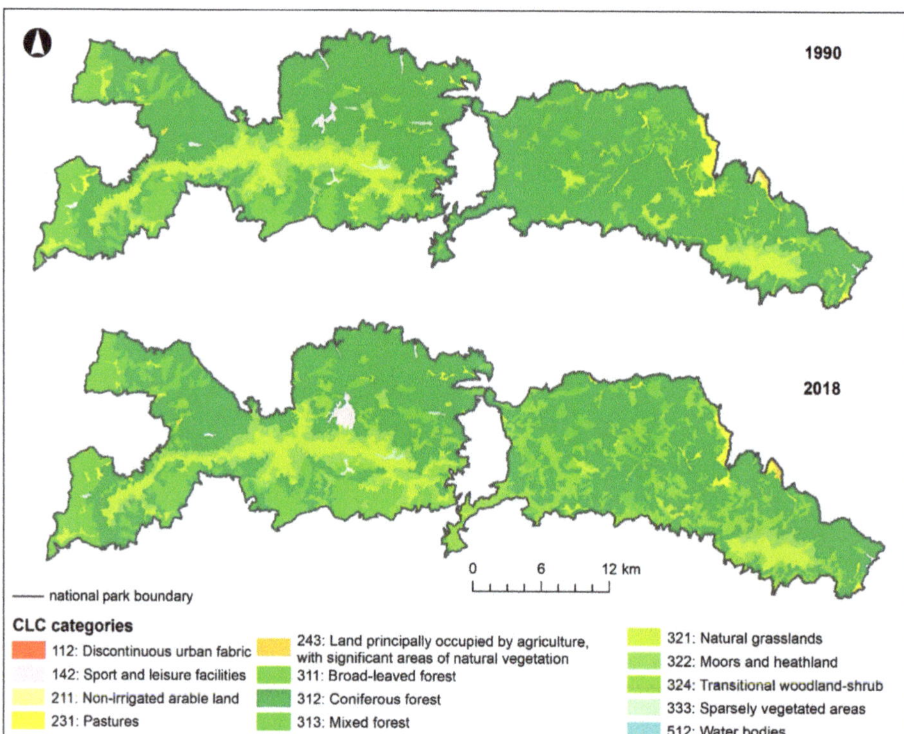

Figure 4. Land cover of Low Tatras National Park in 1990 and 2018.

Almost two-thirds of the National Park area was covered with coniferous forests (CLC 312) in 1990, which consisted of various closed formations of conifers-spruce, fir, pine, and larch. A sample of the class was formed by islands of trees of the abovementioned species, alternating individually or in groups, represented by several types of conifers. They were complemented by fragments of incidental deciduous trees, grasslands of forest meadows, shrubs, transitional woodland-shrubs, forest roads, or parts of recreational facilities, scattered settlements, croplands [61] and abandoned mining sites. Approximately 11% of the National Park area was covered by natural grasslands (CLC 321), consisting mainly of alpine meadows. These areas were not primarily suited for agricultural function, and their natural development was not inhibited by human influence. Meadows were sometimes supplemented by dwarf mountain pine growths, rocks, or groups of trees and shrubs [60]. Nearly 9% of the area was occupied by mixed forests (CLC 313), consisting mainly of spruce, pine, larch, beech, oak, maple, birch, and other tree species. The sample of mixed forests consisted of groups of alternating solitary individuals or islands of coniferous and deciduous trees [61]. Approximately 6% of the area was covered by dwarf mountain pine (CLC 322) sporadically interrupted by enclaves of rock relief forms and alpine meadows. Its occurrence was related to the top parts of the Low Tatras mountain ridge [61].

Almost 5% of the area was covered with transitional woodland-shrub. Young forest trees (deciduous and coniferous) planted after loggings or various calamities were mainly represented in this class, together with forest nurseries, naturally developed forest formations (shrubs and herbaceous vegetation with scattered trees), or shrub formations on abandoned meadows, pastures and forest cuttings for high-voltage power lines. A representative sample of the class consisted of alternating coppice belts and forest remnants (the areal representation of coppice within the respective patches reached 60% or more) [61]. Other land cover classes had not reached values higher than 3% (Table 3).

Table 3. Land cover classes of Low Tatras National Park in 1990 and 2018.

CLC	1990		2018	
	Area in ha	%	**Area in ha**	%
112	6.39	0.01	0.46	0.00
142	421.92	0.56	576.05	0.76
211	47.48	0.06	-	0.00
231	1641.91	2.16	1004.76	1.32
243	203.73	0.27	228.45	0.30
311	1458.41	1.92	1286.17	1.69
312	48,913.12	64.35	39,958.69	52.57
313	6814.73	8.96	9114.03	12.00
321	8392.56	11.04	6632.16	8.72
322	4330.54	5.70	5062.95	6.66
324	3597.34	4.73	12,021.51	15.81
333	185.37	0.24	131.21	0.17
512	3.05	0.00	0.11	0.00
Total	76,016.55	100.00	76,016.55	100.00

Legend: 112 Discontinuous urban fabric, 142 Sport and leisure facilities, 211 Non-irrigated arable land, 231 Pastures, 243 Land principally occupied by agriculture, with significant areas of natural vegetation, 311 Broad-leaved forest, 312 Coniferous forest, 313 Mixed forest, 321 Natural grasslands, 322 Moors and heathland, 324 Transitional woodland-shrub, 333 Sparsely vegetated areas, 512 Water bodies.

3.2. *Land Cover Flows in Low Tatras National Park between 1990 and 2018*

Five main processes were identified in the studied area between 1990 and 2018. LCF7 had the largest share in terms of the proportion of the total area changed (Table 4, Figure 5). This process was clearly dominant in all monitored periods and is mainly represented by two processes: wind calamities and grazing.

Table 4. Land cover flows (LCF) in the Low Tatras National Park between 1990 and 2018.

LCF		1990–2000	2000–2006	2006–2012	2012–2018	1990–2018	CLC Classes Changes
LCF3	ha	0.00	0.00	13.80	193.32	207.12	312–142
	%	0.00%	0.00%	0.19%	6.50%	1.38%	
LCF4	ha	0.00	54.01	0.00	0.00	54.01	211–231
	%	0.00%	1.80%	0.00%	0.00%	0.36%	
LCF5	ha	9.07	0.00	0.00	0.00	9.07	243–211
	%	0.48%	0.00%	0.00%	0.00%	0.06%	
LCF6	ha	133.98	66.00	0.00	0.00	199.98	231–324
	%	7.14%	2.20%	0.00%	0.00%	1.33%	
LCF7	ha	1732.54	2901.05	7170.29	2778.76	14,582.64	311–324, 312–324, 313–324, 313–311, 321–324, 324–312, 324–313, 324–311
	%	92.37%	96.03%	99.81%	93.50%	96.88%	
change of total area (%)		2.47%	3.97%	9.45%	3.91%	19.80%	-

An increase of CLC class 324 (transitional woodland-shrub) on one hand, and a significant decline of CLC 312 (coniferous forests) on the other, was observed due to a frequent occurrence of wind calamities in the recent years, which are a result of widespread climate change, not only on a global but also on a local scale in the last three decades. Mountain ranges of the Carpathian Arch are no exception. Recurring extreme climatic situations, which occur in the observed area, are becoming increasingly frequent [62], and the forest stands are destructively affected mostly by windstorms [63].

Several massive windstorms have swept through the mountain ridge of the Low Tatras belonging to the Low Tatras National Park over the past 25 years, causing vast windfalls in the spruce monocultures forest growths, especially in the eastern part (Kráľovohoľská part) of the mountain range (National Park). Since the beginning of the studied period, windstorms with an impact on the spread of spruce monocultures had been recorded on 8 July 1996 (wind calamity Ivan), 27–28 October 2002 (wind calamity Sabina), 16–17 November 2002 (wind calamity Klaudia), 19 November 2004 (wind calamity Alžbeta), 18–19 January 2007 (wind calamity Kyrill), 23–24 August 2007 (wind calamity Filip), 17–19 May 2010 (wind calamity Gizela) [63–65]. The last more extensive one occurred on 14–15 May 2014 (windstorm Žofia) [66].

It is logical to conclude that the consequences of windstorms manifested by large-scale windfalls will be reduced significantly in the future years, since the critical relief sites overgrown with monocultural spruce forests, which were exposed to impact air currents, have been replaced mainly by transitional woodland-shrub.

Relatively extensive changes in vegetation have also been recorded in the zone of the (anthropogenically created) timberline in the subalpine level, in addition to the extensive area changes caused by windstorms scattered throughout the National Park (mountain range). The dwarf mountain pine belt (CLC 312) experienced an area increase of almost 1%. The phenomenon of windstorms was also marginally present here, causing a slight retreat of monocultural coniferous forests which were replaced by transitional woodland-shrub (CLC 324) on the timberline, especially in the area of Veľký Gápeľ and Malý Gápeľ, as well as in the area of the northern and southern slopes of Lajštroch.

Land cover flows in the anthropogenically lowered (current) timberline had been affected by grazing of sheep and cattle in the past decades (before the beginning of the studied period). Currently, climate change is seen as a significant phenomenon of dwarf mountain pine expansion and acceleration of succession in the subalpine level of mountain meadows, as well as the shifting of the timberline to its original altitude.

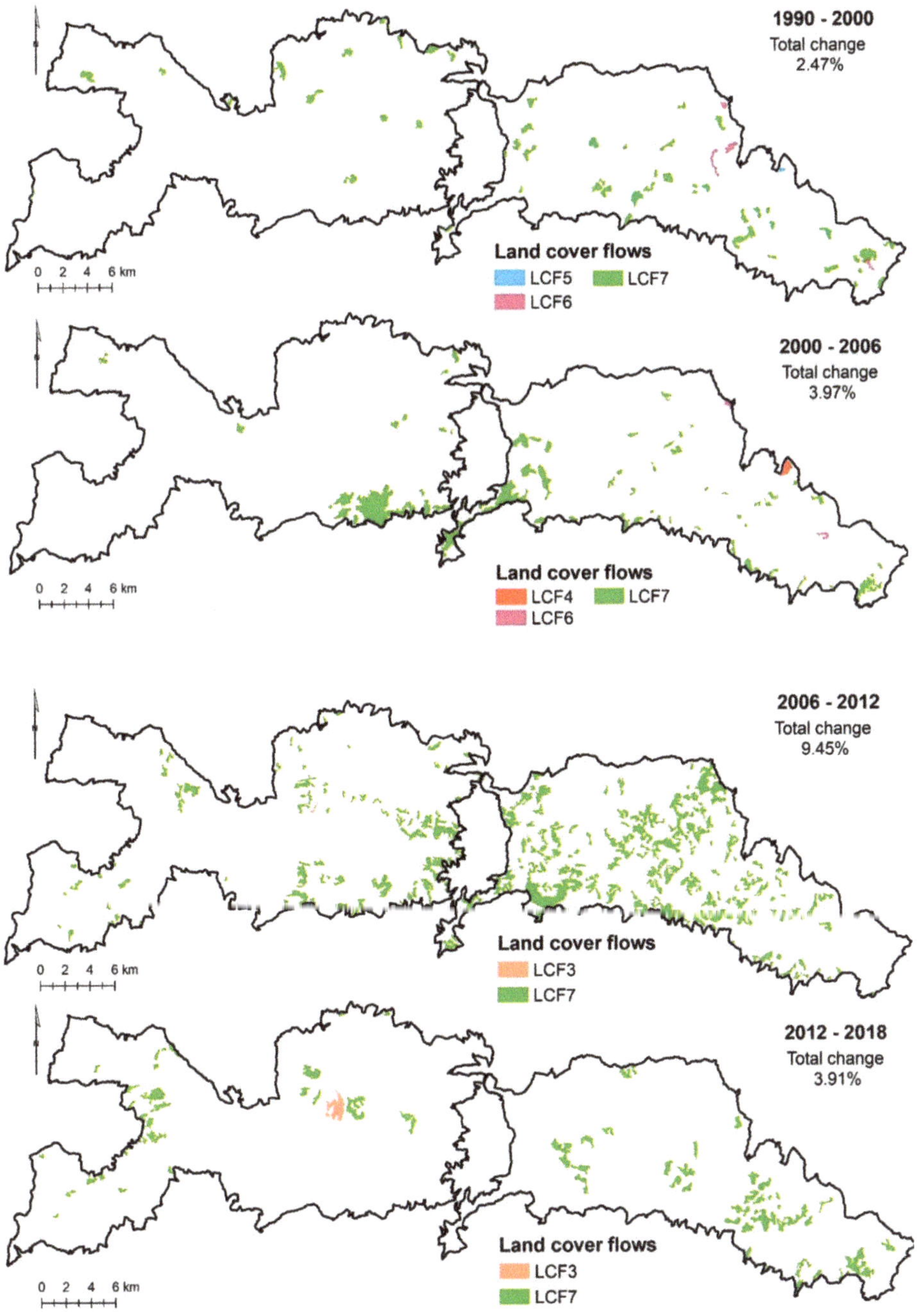

Figure 5. Land cover flows in Low Tatras National Park between 1990, 2000, 2006, 2012 and 2018.

Relatively significant changes in the area extent of individual types of land cover have occurred on the timberline in the high mountain ranges of the Western Carpathians, including the Low Tatras, in recent decades. According to R. Midriak's research, the tree and dwarf mountain pine vegetation at the timberline expanded by up to 6% at the expense of the subalpine mountain meadows level in the decade from 1990 to 2000 [67]. These changes occurred due to the end of sheep and cattle grazing above the timberline. Attenuation of grazing began during the 60 s and 70 s of the 20th century, especially in the Kráľovohoľská (eastern) part of the Low Tatras. Decrease of grazing is the reason why the succession at the timberline and the shift of the anthropogenic timberline to higher altitudes are more pronounced in the eastern than in the western (Ďumbierska) part of the mountain range [68], where grazing finally disappeared only after 1989.

Up to the 14th century, the original-natural timberline, which existed before anthropogenic interventions of shepherds, ascended to an altitude of 1600 m a.s.l., even up to 1700 m a.s.l. in some parts of the mountain range [69–71]. Artificial timberline in the Low Tatras oscillates nowadays between altitudes of 1300–1520 m a.s.l. [67,70]. Climate change also plays a role in the land cover development and thus significantly shifts the timberline to higher altitudes in addition to the end of grazing and the natural course of succession. These processes create much more favorable ecological conditions for the growth of dwarf mountain pine or spruce trees on the timberline. For this reason, we expect an increase in dwarf mountain pine formations, as well as the expansion of transitional woodland-shrub with the potential of their conversion to coniferous (spruce) forest formations to higher altitudes in the coming years or decades [72,73]. The average temperature in mountain areas of the Western Carpathians should increase by at least 2 °C over 30 to 50 years, according to climatological models [74]. Longer periods of droughts should also occur. Under such conditions, the montane zones (800–1200 m) will be unsuitable for the natural occurrence of spruce, which will shift to higher altitudes.

The increase of the mixed forest (CLC class 313 mixed forest) during the observed period is linked to montane and foothill areas, which were used as agricultural areas with 80% permanent grassland in the second half of the 20th century [75].

Significant changes in the long-term use of pastures in this landscape area occurred after 1989. The transformation of the political-economic situation and the ownership relations has led to a rapid reduction in the numbers of cattle grazing to almost zero and a gradual reduction of sheep herds, which are currently grazed here only rarely. This trend has caused the beginning of a rapid succession downwards into the landscape parts, from which the mixed or deciduous forests have been pushed out as a result of extensive agriculture in the recent centuries [76].

These rapid changes in the long-term use of permanent grassland-agricultural land after 1989 triggered the abandonment of foothills and mountain pastures, which caused a rapid spontaneous overgrowth of the landscape (succession). The result can be seen as a 3% increase in the CLC class 313 mixed forest after less than thirty years. All of these factors are understood as driving forces within the DPSIR model developed by the European Environment Agency [77]. They represent the triggering mechanisms of landscape processes, which induce area extent changes of each CLC category.

3.3. Assessment of Land Cover Changes Based on Landscape Metrics

Landscape metrics is the most commonly used tool to compare the evolution of land cover changes over time. The values of the selected landscape indices calculated at the landscape level are shown in Table 5.

A small increase in the number of patches (NumP) was observed, indicating only a slight increase in the heterogeneity of the landscape. The MPS index values point to a higher degree of landscape fragmentation in 2018. This finding is also reflected in the value of the TE index, which has shown a visible increase. The AWMSI index indicates a decrease in the heterogeneity of patch shapes, which may be caused by regular rectangular or oval shapes of patches created in the process of salvage collection. The Shannon Diversity Index was higher in 2018 than in 1990. However, only a minimal increase can be seen, which points to a higher balance in the proportions of the landscape features. The Shannon Equilibrium Index, which is complementary to the SDI, increased by almost 0.1 over the studied period.

Thus, we can speak of a slight trend in an increasing uniformity of patch expansion in the landscape mosaic classes.

Table 5. Landscape metrics changes between 1990 and 2018 at the landscape level.

Landscape Metric	1990	2018	Change	Manifestation in Landscape
NumP	275	281	+6	a small increase in the number of patches
MPS	276.536	270.521	−6.015	a more fragmented landscape
TE	2,998,750	3,609,590	+610,840	an increase in the total perimeter of patches
AWMSI	9.65107	6.16634	−3.48473	a decrease in the heterogeneity of patch shapes and a higher regularity of the shapes of larger patches
SDI	1.27467	1.46898	+0.19431	an increase in landscape heterogeneity
SEI	0.496958	0.591159	+0.094201	an increase in landscape balance

4. Discussion

The main research questions asked in the introduction of this study can be answered based on analyses made during the research.

As already mentioned and presented in Table 4, five significant land cover flows have been identified in the Low Tatras National Park between 1990 and 2018. The most significant land cover flow (96.88%) was the LCF 7 Forest creation and management, in which up to eight types of processes took place (Table 4). Area changes of all classes are spatially located throughout the national park, but to a greater extent in the Kráľovohoľská part (Figure 5). The main reason for their predominant location in the eastern part of the National Park lies in its geographical location. Land cover in the form of monocultural forests is the result of intensive forestry during the second half of the 20th century. Spruce monocultures on the exposed mountain slopes cannot withstand the wind as the original anthropogenically removed forests [32,62,67,73].

The massive windstorms in the prevailing north-western streaming naturally oscillated on the northern slopes of the mountain range, after running off the Tatra's massive vault, to impact the large-scale artificial monocultural spruce formations located at inappropriate relief sites with catastrophic consequences. This process was dominant in all four monitored periods, but mostly in 2006–2012, which is the result of frequent wind disasters.

LCF 6 Withdrawal of farming (1.33%) was shown in the CLC class 231 pastures (meadows and pastures), which was transformed into CLC 324 (transitional woodland-shrub). The end of the almost 500 years of farm animals grazing was very rapid at the end of the 20th century. The grazing ban was related to legal regulation in connection with the protection of the landscape within the national park. Farm animals have disappeared from alpine pastures, which have been subject to intense succession since then [67,68,76,78–80].

This process is closely related and complementary to the previous land cover flow. It is reflected in a greater extent again in the Kráľovohoľská part of the national park, significantly in the entire valley of Ipoltica, in the broader hinterland of Liptovská Teplička, in the vicinity of the Čierny Váh water reservoir and the transformed areas to the east extend up to the main ridge near Kráľova Hoľa. This process occurred only in the period 1990 to 2006. In addition to wind disasters, the reason is also the human factor (logging).

LCF 5 Conversion from forested and natural land to agriculture is the inverse process to the previous one, although it is of lower intensity (0.06%). CLC class 243 land principally occupied by agriculture, with significant areas of natural vegetation, has been transformed into CLC class 211 non-irrigated arable land. This process was identified only in the period 1990–2000 in the vicinity of the village Liptovská Teplička as a conversion from agriculture–nature mosaics to continuous agriculture [81,82]. The cause of this transformation must be sought within links of ownership in the second half of the 20th century. After the transformation of the original strip fields into large agricultural areas used as meadows and pastures, these were abandoned after 1989. The original owners have gradually begun to use them again during the period studied. Original meadows and pastures in higher altitudes succumbed to succession and have been transformed into forests, while the original fields of arable land in the lower altitudes have been transformed and are currently used as grasslands–meadows, and pastures [83].

A surprising result of our research was the finding that the development of touristic centers with high demands on recreational infra and suprastructure does not manifest itself as a significant land cover flow. LCF 3 Sprawl of economic sites and infrastructures has transformed the CLC class 312 coniferous forests into the CLC class 142 sport and leisure facilities, but this impact is not significant compared to changes caused by wind calamities or agricultural land use. The most significant changes have been identified on the southern and northern slopes of Chopok [84–86], where the ski resort Jasná was built. Tatry mountain resorts, Inc. operates 23 lifts and 39 ski slopes in the largest resort in Slovakia. Other location of transformations can be found in the territory of Demänovská Valley (Demänovská Ice Cave).

LCF 4 Agriculture internal conversions (0.36%) was recorded only in the period 2000–2006 in the vicinity of the village Liptovská Teplička. Arable land was transformed into meadows and pastures in this locality. Based on the landscape-ecological metrics results, the studied area showed an increase in the land cover heterogeneity, although the shape of patches was more regular. These results also showed that the landscape of the national park has significantly lost its forest potential at the expense of less valuable forest formations over the studied period. Analyses have shown that the timberline shifted to higher altitudes, and there is a trend of a continual succession of alpine grasslands. Although most of the landscape metrics results were positive, it is not possible to draw more profound conclusions from them. According to Ružičková et al. [78], the resulting values of the landscape diversity index do not describe the ecological stability and quality of the assessed area and do not take into account the internal differentiation of landscape structure features. For this reason, we do not refer to our calculation results as absolute, and we consider the different (and changing) quality and structure of landscape features over time as well. We can expect extensive linear and areal interventions into the current land cover based on the expected future changes that will occur due to the planned construction of new transportation projects and the technical infrastructure connected to tourism (ski cableways in Demänovská Dolina resort and sports hall in Donovaly resort). Combined with the anthropogenically predisposed development of the timberline and the progress of succession, the CLC class 142 Sport and leisure facilities will increase at the expense of classes 312 Coniferous forests, 324 Transitional woodland-shrub, and 333 Sparsely vegetated areas in the critical construction localities. The construction of express road R1 section Slovenská Ľupča-Liptovská Osada will affect CLC classes at lower altitudes [87]. A decrease in the areal spread of the classes 231 Pastures, 311 Broad-leaved forest, and 313 Mixed forests is predicted.

Landscape metrics were used in this research because they have been providing a backbone for spatial pattern analysis in landscape ecology for more than three decades. It is very important to select the correct approach, or combination of approaches, for investigating the issue [88]. On the basis of landscape metrics results, we could contend that area of the National Park is more heterogenous, uniform and balanced. These conclusions are very one-sided; therefore, it is required asses the landscape changes according to several approaches. Changes in the landscape of the national park caused by radical interventions (natural and anthropogenic) were reflected in an increase of diversity, but they may also have an impact on ecological stability. These values should be interpreted sensitively because it does not take the internal differentiation of land cover classes into account. Assessment of land cover changes is especially relevant for protected areas where long-term ecosystem stability is a critical aspect of protecting and maintaining high levels of biodiversity and ecosystem functions [89].

Based on the results of our research, we can formulate basic recommendations for the management of the National Park concerning the negative processes caused by unwanted changes in the land cover resulting from the results and conclusions of the study. The priorities should be:

- To stop the deterioration of habitat status, in particular, for the habitats of European and national importance, maintain their current state, and then take steps towards a measurable improvement. Therefore, the National Park Administration should give priority to the detailed mapping of the habitat status, complete an overall map of the National Park and provide operational data for decision-making by state administration authorities, in a particular state and private forest managers;
- To map in detail the natural forests and primeval forests relics of the National Park in the shortest possible time to ensure that their area extent is maintained and to gradually increase their extent of areas with a potential for natural forest development;
- Prevent further fragmentation of forests and encourage their regeneration while ensuring compensatory mechanisms to cover the loss of forest management and favoring alternative uses of high nature value forests;
- Implement measures to preserve and improve habitats of European importance, particularly in Natura 2000 sites and habitats of national importance within the National Park;
- Improve the effectiveness of communication between the environmental and agricultural departments;
- Define or revise the nature and landscape conservation objectives in the National Park in more detail.

5. Conclusions

Based on the analyses of changes in land cover transformation over the observed period, we can conclude that wind calamities were the main transformation factor of national park landcover changes between 1990 and 2018. Their destructive power stems from improper forest management in the second half of the 20th century in combination with anthropogenic climate change. The ending of livestock grazing on foothills, but also montane pastures, was also an essential factor. The end of grazing triggered succession towards lower, as well as higher altitudes, to the original forest habitats. At the end of the 1980s, this disrupted the landscape balance of the National Park, maintained by humans since the Middle Ages. Changes in agricultural management at the foothills of the National Park were another essential impulse in the transformation of land cover. Last but not least, the development of tourism and the growth of recreational infrastructure have been among the most intensive transforming factors of land cover changes in recent decades.

Analyses of land cover changes over the last 30 years in the Low Tatras National Park have clearly pointed to the inaccuracy of forest management and planting of monocultural forests, especially on the northern slopes, which are most exposed to extreme wind situations. The combination of these two factors has the most negative effect on the alpine country. When managing the forests of a national park, emphasis must be placed on the species composition of forests, which should be as close as possible to the composition of the original forests. After wind calamities, monocultural spruces should be replaced by beech and beech-fir forests.

A plan for making the zones of the national park in terms of its economic use should also be drawn up. Currently, the Ministry of the Environment of the Slovak Republic is working intensively on it. The proper delimitation of individual zones of use, especially in relation to tourism, could significantly prevent the expansion of recreational areas at the expense of the surrounding countryside, where the main example is the Demänovská valley on the northern side of the mountain range.

Our conclusions clearly show that analyses of land cover and land cover flows can contribute to the proper planning of land use in national parks and thus to its stabilization and sustainability.

Author Contributions: Conceptualization, M.Ž.; methodology, M.Ž., B.G. and P.H.; software, M.Ž.; validation, M.Ž., B.G. and P.H.; formal analysis, M.Ž. ; investigation, B.G. and P.H.; resources, M.Ž., B.G. and P.H.; data curation, M.Ž.; writing—Original draft preparation, M.Ž., B.G. and P.H.; writing—Review and editing, M.Ž., B.G. and P.H.; visualization, M.Ž.; supervision, M.Ž.; project administration, B.G.; funding acquisition, B.G. All authors have read and agreed to the published version of the manuscript.

Funding: This research and the APC were funded by Slovak Research and Development Agency, grant number APVV-18-0185 "Land-use changes of Slovak cultural landscape and prediction of its further development" and by VEGA, grant number 1/0236/18 "Environmental aspects of mining localities settings in Slovakia in Middle Ages and the beginning of Modern history".

Conflicts of Interest: The authors declare no conflict of interest. The funders had no role in the design of the study; in the collection, analyses, or interpretation of data; in the writing of the manuscript, or in the decision to publish the results.

References

1. Fiorini, L.; Zullo, F.; Marucci, A.; Romano, B. Land take and landscape loss: Efect of uncontrolled urbanization in Southern Italy. *J. Urban Manag.* **2019**, *8*, 42–56. [CrossRef]
2. Tesfamariam, Z.; Nyssen, J.; Ghebreyohannes, T.; Tafere, K.; Zenebe, A.; Poesen, J.; Deckers, S.; Van Eetvelde, V. Transhumance as a driving force of landscape change in the marginal grabens of northern Ethiopia. *Singap. J. Trop. Geogr.* **2019**, *40*, 476–495. [CrossRef]
3. Mendoza-Ponce, A.; Corona-Nunez, R.O.; Galicia, L.; Kraxner, F. Identifying hotspots of land use cover change under socioeconomic and climate change scenarios in Mexico. *AMBIO* **2019**, *48*, 336–349. [CrossRef] [PubMed]
4. Kedra, M.; Szczepanek, R. Land cover transitions and changing climate conditions in the Polish Carpathians: Assessment and management implications. *Land Degrad. Dev.* **2019**, *30*, 1040–1051. [CrossRef]
5. Duan, C.; Shi, P.L.; Song, M.H.; Zhang, X.Z.; Zong, N.; Zhou, C.P. Land Use and Land Cover Change in the Kailash Sacred Landscape of China. *Sustainability* **2019**, *11*, 1788. [CrossRef]
6. Hreško, J.; Bugár, G.; Boltižiar, M.; Kohút, F. The dynamics of recent geomorphic processes in the alpine zone of the Tatra Mountains. *Geogr. Pol.* **2008**, *81*, 53.
7. Hreško, J.; Boltižiar, M.; Bugár, G. Spatial structures of geomorphic processes in high-mountain landscape of the Belianske Tatry Mts. *Ekológia Bratislava* **2003**, *22*, 341.
8. Lieskovský, J.; Kaim, D.; Balász, P.; Boltižiar, M.; Chmiel, M.; Grabska, E.; Király, G.; Konkoly-Gyuró, E.; Kozak, J.; Antalová, K.; et al. Historical land use dataset of the Carpathain region (1819–1980). *J. Maps* **2018**, *14*, 644. [CrossRef]
9. Petrovič, F.; Muchová, Z. The potential of the landscape with dispersed settlement (Case study Čadca town). In *Book Series: Public Recreation and Landscape Protection—With Sense Hand in Hand . . .* , 1st ed.; Fialová, J., Ed.; Mendel University: Brno, Czech Republic, 2013; pp. 199–204.
10. Boltižiar, M.; Olah, B.; Gallay, I.; Gallayová, Z. Transformation of the Slovak cultural landscape and its recent trends. In Proceedings of the 17th International Symposium on Landscape Ecology: Landscape and Landscape Ecology, Nitra, Slovakia, 27–29 May 2015; Halada, L'., Bača, A., Boltižiar, M., Eds.; Institute of Landscape Ecology: Bratislava, Slovakia, 2016; pp. 57–67.
11. Ivanová, M.; Michaeli, E.; Boltižiar, M.; Fazekašová, D. The analysis of changes ecological stability of landscape in the contrasting region of the mountain range and a lowland. In *Book Series: International Multidisciplinary Scientific GeoConference-SGEM*, 1st ed.; STEF92 Technology Ltd.: Sofia, Bulgaria, 2013; pp. 925–938.
12. Muchová, Z.; Tarniková, M. Land cover change and its influence on the assessment of the ecological stability. *Appl. Ecol. Environ. Res.* **2018**, *16*, 2169. [CrossRef]
13. Sweetman, B.M.; Cissell, J.R.; Rhine, S.; Steinberg, M.K. Land Cover Changes on Ambergris Caye, Belize: A Case Study of Unregulated Tourism Development. *Prof. Geogr.* **2019**, *71*, 123–134. [CrossRef]
14. Stankov, U.; Klaučo, M.; Dragicevic, V.; Vujicic, M.D.; Solarevic, M. Assessing land-use changes in tourism area on the example of Cajetina municipality (Serbia). *Geogr. Pannonica* **2016**, *20*, 105–113. [CrossRef]
15. Raczkowska, Z. Human impact in the Tatra mountains. *Cuad. de Investig. Geogr.* **2019**, *45*, 219–244. [CrossRef]
16. Krtička, L.; Tomčíková, I.; Rakytová, I. Development versus conservation: Evaluation of landscape structure changes in Demanovska Valley, Slovakia. *J. Mt. Sci.* **2018**, *15*, 1153–1170. [CrossRef]
17. Li, S.; Yang, H.; Lacayo, M.; Liu, J.; Lei, G. Impacts of Land-Use and Land-Cover Changes on Water Yield: A Case Study in Jing-Jin-Ji, China. *Sustainability* **2018**, *10*, 960. [CrossRef]
18. Ozsahin, E.; Duru, U.; Eroglu, I. Land Use and Land Cover Changes (LULCC), a Key to Understand Soil Erosion Intensities in the Maritsa Basin. *Water* **2018**, *10*, 335. [CrossRef]

19. Sharma, R.; Nehren, U.; Rahman, S.A.; Meyer, M.; Rimal, B.; Aria Seta, G.; Baral, H. Modeling Land Use and Land Cover Changes and Their Effects on Biodiversity in Central Kalimantan, Indonesia. *Land* **2018**, *7*, 57. [CrossRef]
20. Sánchez-Reyes, U.J.; Niño-Maldonado, S.; Barrientos-Lozano, L.; Treviño-Carreón, J. Assessment of Land Use-Cover Changes and Successional Stages of Vegetation in the Natural Protected Area Altas Cumbres, Northeastern Mexico, Using Landsat Satellite Imagery. *Remote Sens.* **2017**, *9*, 712. [CrossRef]
21. Millones, M.; Rogan, J.; Li, B.T.; Parmentier, B.; Harris, R.C.; Griffith, D.A. Fire Data as Proxy for Anthropogenic Landscape Change in the Yucatán. *Land* **2017**, *6*, 61. [CrossRef]
22. Muchová, Z.; Konc, Ľ.; Petrovič, F. Land plots valuation in land consolidation in Slovakia: A need for a new approach. *Int. J. Strateg. Prop. Manag.* **2018**, *22*, 372. [CrossRef]
23. Havlíček, M.; Skokanová, H.; Dostál, I.; Vymazalová, M.; Pavelková, R.; Petrovič, F. The consequences of establishing military training areas for land use development-A case study of Libava, Czech Republic. *Land Use Policy* **2018**, *73*, 84. [CrossRef]
24. Pechanec, V.; Machar, I.; Pohanka, T.; Opršal, Z.; Petrovič, F.; Švajda, J.; Šálek, Ľ.; Chobot, K.; Filippovová, J.; Cudlín, P. Effectiveness of Natura 2000 system for habitat types protection: A case study from the Czech Republic. *Nat. Conversat. Bulg.* **2018**, *24*, 21. [CrossRef]
25. Gómez, J.N.; Loures, L.; Castanho, R.A.; Cabezas, J.; Fernández-Pozo, L.; Lousada, S.; Escórcio, P. Assessing Land-Use Changes in European Territories: A Retrospective Study from 1990 to 2012. In *Land Use—Assessing the Past, Envisioning the Future*, 1st ed.; Loures, L.C., Ed.; InTech: London, UK, 2018; pp. 243–581. [CrossRef]
26. Biely, A.; Bezák, V.; Bujnovský, A.; Vozárová, A.; Klinec, A.; Miko, O.; Halouzka, R.; Vozár, J.; Beňuška, P.; Hanzel, V.; et al. *Explanations to the Geological Map of the Low Tatras 1: 50,000*; GÚDŠ: Bratislava, Slovakia, 1992. (In Slovak)
27. Biely, A.; Bezák, V. *Explanations to the Geological Map of the Low Tatras 1: 50,000*; Geologická služba Slovenskej republiky; Vydavateľstvo Dionýza Štúra: Bratislava, Slovakia, 1997. (In Slovak)
28. Jasík, M.; Kaliská, Z.; Kaliský, M.; Galfyová, Z.; Hubková, M.; Dzúriková, Ľ.; Hančin, M.; Kuna, F. *Low Tatras National Park, Yearbook 2004*, 1st ed.; Správa Národného parku Nízke Tatry: Liptovský Hrádok, Slovakia, 2004; 61p. (In Slovak)
29. Turis, P.; Jasík, M. *Low Tatras, Natural Values, History and Current State of Protection of the Area*, 1st ed.; Správa národného parku Nízke Tatry: Banská Bystrica, Slovakia, 2007; 116p. (In Slovak)
30. *Atlas of Landscape of Slovakia*; Ministerstvo životného prostredia SR: Bratislava, Slovakia, 2002; CD ROM. (In Slovak)
31. Šoták, Š.; Borsányi, P. Climate monitoring of SHMÚ in the Low Tatras. In *Príroda Nízkych Tatier 1, Zborník Referátov a Posterov z Konferencie Usporiadanej pri Príležitosti 25. Výročia Vyhlásenia Národného parku Nízke Tatry*; Jasík, K., Kochjarová, S., Eds.; Štátna ochrana prírody Slovenskej republiky, Správa Národného parku Nízke Tatry: Banská Bystrica, Slovakia, 2004; pp. 275–282. (In Slovak)
32. Hronček, P. Environmental history of the forest landscape in the Štiavnička valley in the Low Tatras. *Quaest. Rerum Nat.* **2015**, *2*, 91–115. (In Slovak)
33. Madlen, J. Constitutio Maximiliana. In *Sborník prác Lesníckeho a Drevárskeho Múzea v Antole*; Osveta: Martin, Slovakia, 1962; pp. 9–45.
34. Lukniš, M.; Plesník, P. *Lowlands, Basins and Mountains of Slovakia*, 1st ed.; Osveta: Bratislava, Slovakia, 1961; 134p. (In Slovak)
35. SSR Government Regulation No. 119/1978 Coll. Available online: https://www.slov-lex.sk/pravne-predpisy/SK/ZZ/1978/119/19781019.html (accessed on 13 January 2020).
36. Decree No. 120/1978. Available online: https://www.slov-lex.sk/pravne-predpisy/SK/ZZ/1978/120/vyhlasene_znenie.html (accessed on 13 January 2020).
37. Act of the National Council of the Slovak Republic, No. 287/1994 Coll. on Nature and Landscape Protection. Available online: https://www.noveaspi.sk/products/lawText/1/42354/1/2/zakon-c-287-1994-zz-o-ochrane-prirody-a-krajiny (accessed on 13 January 2020).
38. Act of the National Council of the Slovak Republic No. 543/2002 Coll. On Nature and Landscape Protection. Available online: https://www.zakonypreludi.sk/zz/2002-543 (accessed on 13 January 2020).

39. Burkovský, J. History of the Low Tatras National Park. In *Príroda Nízkych Tatier 1, Zborník Referátov a Posterov z Konferencie Usporiadanej pri Príležitosti 25. Výročia Vyhlásenia Národného Parku Nízke Tatry*; Jasík, K., Kochjarová, S., Eds.; Štátna Ochrana Prírody Slovenskej Republiky, Správa Národného parku Nízke Tatry: Banská Bystrica, Slovakia, 2004; pp. 3–10. (In Slovak)
40. Verbesselt, J.; Hyndman, R.; Newnham, G.; Culvenor, D. Detecting trend and seasonal changes in satellite image time series. *Remote Sens. Environ.* **2010**, *114*, 106–115. [CrossRef]
41. Büttner, G.; Kosztra, B.; Soukup, T.; Sousa, A.; Langanke, T. *CLC2018 Technical Guidelines*; EEA: Copenhagen, Denmark, 2017; Available online: https://land.copernicus.eu/user-corner/technical-library/clc2018technicalguidelines_final.pdf (accessed on 20 October 2020).
42. Rodríguez-Rodríguez, D.; Martínez Vega, J.; Echavarría, P. A twenty year GIS-based assessment of environmental sustainability of land use changes in and around protected areas of a fast developing country: Spain. *Int. J. Appl. Earth Obs. Geoinf.* **2019**, *74*, 169–179. [CrossRef]
43. Pontius, R.G.; Shusas, E.; McEachern, M. Detecting important categorical land changes while accounting for persistence. *Agric. Ecosyst. Environ.* **2004**, *101*, 251–268. [CrossRef]
44. Stott, A.; Haines-Young, E. Linking land cover, intensity of use and botanical diversity in an accounting framework in the UK. In *Environmental Accounting in Theory and Practice*, 1st ed.; Uno, K., Bartelmus, P., Eds.; Kluwer: Dordrecht, The Netherlands, 1998; pp. 245–260.
45. Feranec, J.; Jaffrain, G.; Soukup, J.; Hazeu, G.W. Determining changes and flows in European landscapes 1990–2000 using CORINE land cover data. *Appl. Geogr.* **2010**, *10*, 285–301. [CrossRef]
46. Perdigao, V.; Christensen, S. *The LACOAST Atlas: Land Cover Changes in European Coastal Zones*; Joint Research Centre: Milan, Italy, 2000.
47. Köhler, R.; Olschofsky, K.; Gerard, F. *Land Cover Change in Europe from the 1950'ies to 2000*; Institute for Worldforestry, University of Hamburg: Hamburg, Germany, 2004.
48. Haines-Young, R.; Weber, J.L. Land Account for Europe 1990–2000. Towards Integrated Land and Ecosystem Accounting. EEA Report No 11/2006. 2006. Available online: http://www.eea.europa.eu/publications/eea_report_2006_11 (accessed on 14 October 2019).
49. Boltižiar, M. Analysis of development of heterogenity and chosen qualities of the spots of structure of alpine land (on example of chosen model land of High Tatra). *Folia Geogr.* **2010**, *16*, 105–124.
50. Gajdoš, A.; Klaučo, M.; Škodová, M. Ecological significance of the forest ecotone vegetation in the Starohorské Hills. *Geogr. J.* **2012**, *64*, 253–266.
51. Olahová, J.; Vojtek, M.; Boltižiar, M. Application of geoinaromation technologies for the assesment of landscape structure using landscape-ecological indexes (case study of the Handlová landslide). *Tájökológiai* **2013**, *11*, 351–366.
52. Kumar, M.; Denis, D.M.; Singh, S.K.; Szabó, S.; Suryavanshi, S. Landscape metrics for assessment of land cover change and fragmentation of a heterogeneous watershed. *Remote Sens. Appl. Soc. Environ.* **2018**, *10*, 224–233. [CrossRef]
53. Singh, S.K.; Basommi, B.P.; Mustak, S.K.; Srivastava, P.K.; Szabo, S. Modelling of land use land cover change using earth observation data-sets of Tons River Basin, Madhya Pradesh, India. *Geocarto Int.* **2017**, *33*, 1202–1222. [CrossRef]
54. Li, Z.; Li, X.; Ma, W.Y.A.; Wang, J. Land-use change analysis in Yulin prefecture, northwestern China using remote sensing and GIS. *Int. J. Remote Sens.* **2004**, *25*, 5691–5703. [CrossRef]
55. Araya, Y.H.; Cabral, P. Analysis and Modeling of Urban Land Cover Change in Setúbal and Sesimbra, Portugal. *Remote Sens.* **2010**, *2*, 1549. [CrossRef]
56. Krajewski, P. Monitoring of Landscape Transformations within Landscape Parks in Poland in the 21st Century. *Sustainability* **2019**, *11*, 2410. [CrossRef]
57. Feranec, J.; Kopecka, M.; Vatseva, R.; Stoimenov, A.; Oťahel, J.; Beták, J.; Husár, K. Landscape change analysis and assessment (case studies in Slovakia and Bulgaria). *Cent. Eur. J. Geosci.* **2009**, *1*, 106. [CrossRef]
58. European Commission (EC). From Land Cover to Landscape Diversity in the European Union. Available online: https://enviro-edu.sk/pluginfile.php/137/mod_resource/content/6/Ucebnica_Ochrana_a_vyuzivanie_krajiny/Text/Eiden_et_al_2000.pdf (accessed on 2 April 2020).
59. Obeidat, M.; Awawdeh, M.; Lababneh, A. Assessment of land use/land cover change and its environmental impacts using remote sensing and GIS techniques, Yarmouk River Basin, north Jordan. *Arab. J. Geosci.* **2019**, *12*, 685. [CrossRef]

60. Klaučo, M.; Weis, K.; Gregorová, B.; Anstead, L. *Geographic Information Systems 3*; Vydavateľstvo Univerzity Mateja Bela Belianum: Banská Bystrica, Slovakia, 2014. (In Slovak)
61. Feranec, K.; Oťahel, J. *Land cover of Slovakia*; Veda: Bratislava, Slovakia, 2001. (In Slovak)
62. Pecho, J. Climate change: A global problem with local impacts. In *Klimatická Zmena—Výzva pre Lokálny Rozvoj na Slovensku*, 1st ed.; Šteiner, A., Hegyi, L., Eds.; Karpatský Rozvojový Inštitút: Spišská Nová Ves, Slovakia, 2012; pp. 9–26. (In Slovak)
63. Kunca, A.; Zúbrik, M.; Vakula, J.; Galko, J. *Occurrence of Harmful Factors in the Forests of Slovakia in the Years 1960–2014, in 2015 and a Forecast of Their Development*, 1st ed.; Národné Lesnícke Centrum: Zvolen, Slovakia, 2016; 139p. (In Slovak)
64. Kunca, A.; Galko, J.; Zúbrik, M. Significant calamities in the forests of Slovakia over the last 50 years. In *Zborník Referátov z 23. Ročníka Medzinárodnej Konferencie, Ktorá sa Konala 23. a 24. Apríla 2014 v Novom Smokovci*; Kunca, A., Ed.; Národné Lesnícke Centrum: Zvolen, Slovakia, 2014; pp. 25–31. (In Slovak)
65. Konôpka, J.; Kunca, A. Accidental logging in the years 1960–2014. In *Výskyt Škodlivých Činiteľov v Lesoch Slovenska v Rokoch 1960–2014, v Roku 2015 a Prognóza ich Vývoja*, 1st ed.; Kunca, A., Zúbrik, M., Vakula, J., Galko, J., Eds.; Národné Lesnícke Centrum: Zvolen, Slovakia, 2016; pp. 19–25. (In Slovak)
66. Kunca, A.; Vakula, J.; Galko, J.; Zúbrik, M.; Leontovyč, R.; Nikolov, C.; Rel, S.; Konôpka, B.; Longauerová, B.; Maľová, M. How to proceed in the processing of the wind calamity of Žofia? *Les Letokruhy* **2014**, *11*, 6–9. (In Slovak)
67. Midriak, R. *Eroded and Run-Down Soils of Slovakia*; Ústav Vedy a Výskumu UMB: Banská Bystrica, Slovakia, 2010. (In Slovak)
68. Hronček, P.; Štrba, L.; Gregorová, B. Heritage of the medieval human activity in the present landscape of the national park Low Tatras (Slovakia). In *Book Series: Public Recreation and Landscape Protection—With Sense Hand in Hand ...* , 1st ed.; Fialová, J., Ed.; Mendel University: Brno, Czech Republic, 2019; pp. 172–176.
69. Plesník, P. Upper forest border in Slovakia. In *Sborník prác Lesníckeho a Drevárskeho Múzea*; Osveta: Martin, Slovakia, 1961; pp. 84–98. (In Slovak)
70. Zatkalík, F. The upper border of the forest in the Prašivá group in the Low Tatras. *Geogr. Časopis* **1973**, *255*, 148–164. (In Slovak)
71. Midriak, R. *Morphogenesis of the Surface of High Mountains*; Veda: Bratislava, Slovakia, 1983. (In Slovak)
72. Schlaepfer, R. (Ed.) *Long-Term Implications of Climate Change and Air Pollution on Forest Ecosystems*, 1st ed.; IUFRO: Vienna, Austria, 1993; 132p.
73. Midriak, R.; Zaušková, Ľ. The country of Slovakia and its use in relation to climate change. In *Klimatická Změna—Zmierňujúce a Adaptačné Opatrenia v Sektore Pôdohospodárstva*; Národné Poľnohospodárske Centrum a Potravinárske Centrum: Nitra, Slovakia, 2014; pp. 64–72. (In Slovak)
74. Mello, M.; Lapin, M.; Kapolková, H.; Pecho, J.; Križicová, A. Climate trends in the Slovak part of the Carpathians. In *The Carpathians—Integrating Nature and Society towards Sustainability*, 1st ed.; Kozak, J., Katarzyna, O., Bytnerowicz, A., Wyżga, B., Eds.; Springer: Berlin/Heidelberg, Germany, 2013; pp. 131–150.
75. Krajčovič, V. The structure of the agricultural land fund and the evaluation of permanent grasslands in the foothills and mountain areas. In *Spustnuté Pôdy a Pustnutie Krajiny Slovenska*, 1st ed.; Midriak, R., Ed.; UMB: Banská Bystrica, Slovakia, 2011; pp. 230–241. (In Slovak)
76. Midriak, R.; Zaušková, Ľ. Non-forest and desolate soils in Slovakia, their afforestation and desolation. In *Obnova Lesního Prostředí při Zalesnění Nelesních a Devastovaných Stanovišť, Kostelec nad Černými lesy*; Česká zemědelská Univerzita, Lesy České republiky: Praha, Česká Republika, 2008; pp. 39–45. (In Slovak)
77. Gabrielsen, P.; Bosch, P. Environmental Indicators: Typology and Use in Reporting. EEA Internal Working Paper. 2003. Available online: http://costabalearsostenible.es/PDFs/AMYKey%20References_Indicators/EEA%202003.pdf (accessed on 4 April 2020).
78. Ružičková, J.; Moravčíková, Z.; Lehotská, B. *Landscape Protection and Use*; Univerzita Komenského v Bratislave: Bratislava, Slovakia, 2009. (In Slovak)
79. Midriak, R. Anthropogenic influences on vegetation and soil in the high mountains of the Western Carpathians (Part I). In *Zborník Lesníckeho, Drevárskeho a Poľovníckeho Múzea v Antole, Nr. VII.*; Osveta: Martin, Slovakia, 1973; pp. 148–185. (In Slovak)
80. Midriak, R. Anthropogenic influences on vegetation and soil in the high mountains of the Western Carpathians (Part II - Regional analysis). In *Zborník Lesníckeho, Drevárskeho a Poľovníckeho Múzea v Antole. Nr. IX.*; Osveta: Martin, Slovakia, 1977; pp. 141–182. (In Slovak)

81. Ružičková, H.; Dobrovodská, M.; Valachovič, M. Landscape-ecological evaluation of vegetation in relation to the forms of anthropogenic relief in the cadastre of the Liptovská Teplička Village, the Nízke Tatry Mts. *Ekol. Bratisl.* **1999**, *18*, 381–400.
82. Dobrovodská, M. *Landscape Ecological Research of the Historical Agricultural Landscape in the Villages of Liptovská Teplička, Osturna and Mala Franková*; VEDA: Bratislava, Slovakia, 2014; 149p. (In Slovak)
83. Midriak, R. *Lowered soils and the desolation of the country of Slovakia. Fakulta Prírodných vied UMB B.*; Bystrica: Banská Bystrica, Slovakia, 2011; 401p. (In Slovak)
84. Gregorová, B. Analysis of selected factors influencing the origin and development of tourism in the Low Tatras. *Geogr. Cassoviensis* **2010**, *4*, 53–56. (In Slovak)
85. Gregorová, B. The use of archival sources and geoinformation technologies in the research of the history of environmental protection on the example of the Bystra valley in the Low Tatras. *Acta Reg.* **2018**, *3*, 75–83. (In Slovak)
86. Čech, V.; Gregorová, B. Tourism in the Low Tatras National Park — Basic prerequisites. In *Міждисциплінарні ІнтеграційніПроцеси уСистеміГеографічної, Туризмологічної та Екологічної Науки*; Міністерство Освіти і Науки України: Тернопіль, Ukraine, 2020; pp. 264–275. (In Slovak)
87. Károlyová, I.; Sabo, P.; Rossová, L. The impact of tourism on the natural environment of the Ďumbier Tatras and on hiking trails and their immediate surroundings. In *Krajina, História a Tradície Čipkárskych obcí Horehronia*, 1st ed.; Hronček, P., Ed.; Ústav Vedy a Výskumu: Banská Bystrica, Slovakia, 2007; pp. 161–178. (In Slovak)
88. Frazier, A.E.; Kedron, P. Landscape Metrics: Past Progress and Future Directions. *Curr. Landsc. Ecol. Rep.* **2017**, *2*, 63–72. [CrossRef]
89. Guerra, C.A.; Rosa, I.M.D.; Pereira, H.M. Change versus stability: Are protected areas particularly pressured by global land cover change? *Landsc. Ecol.* **2019**, *34*, 2779–2790. [CrossRef]

Publisher's Note: MDPI stays neutral with regard to jurisdictional claims in published maps and institutional affiliations.

MDPI
St. Alban-Anlage 66
4052 Basel
Switzerland
Tel. +41 61 683 77 34
Fax +41 61 302 89 18
www.mdpi.com

Land Editorial Office
E-mail: land@mdpi.com
www.mdpi.com/journal/land

www.ingramcontent.com/pod-product-compliance
Lightning Source LLC
Chambersburg PA
CBHW061837210926
51850CB00006B/912